T0302197

Process Engineering and Plant Design

Process Engineering and Plant Design

The Complete Industrial Picture

Siddhartha Mukherjee

CRC Press
Taylor & Francis Group
Boca Raton London New York

CRC Press is an imprint of the
Taylor & Francis Group, an **informa** business

First edition published 2022
by CRC Press
6000 Broken Sound Parkway NW, Suite 300, Boca Raton, FL 33487-2742

and by CRC Press
2 Park Square, Milton Park, Abingdon, Oxon, OX14 4RN

© 2022 Taylor & Francis Group, LLC

CRC Press is an imprint of Taylor & Francis Group, LLC

Library of Congress Cataloging-in-Publication Data
Names: Mukherjee, Siddhartha (Researcher in fluid mechanics), author.
Title: Industrial process engineering and plant design / Siddhartha
    Mukherjee.
Description: First edition. | Boca Raton, FL : CRC Press, [2022] | Includes
    bibliographical references and index.
Identifiers: LCCN 2021020834 (print) | LCCN 2021020835 (ebook) | ISBN
    9780367248413 (hbk) | ISBN 9781032119915 (pbk) | ISBN 9780429284656
    (ebk)
Subjects: LCSH: Chemical plants--Design and construction. | Chemical
    processes.
Classification: LCC TP155.5 .M836 2022 (print) | LCC TP155.5 (ebook) |
    DDC 660--dc23
LC record available at https://lccn.loc.gov/2021020834
LC ebook record available at https://lccn.loc.gov/2021020835

ISBN: 978-0-367-24841-3 (hbk)
ISBN: 978-1-032-11991-5 (pbk)
ISBN: 978-0-429-28465-6 (ebk)

DOI: 10.1201/9780429284656

Typeset in Times
by SPi Technologies India Pvt Ltd (Straive)

In fond memory of my parents, Bholanath and Sudhira, and my parents-in-law, Sitanshu and Sovana Banerjee, for their blessings and guidance and for giving me the strength to face the world.

To my wife, Indrani, for inspiring me to write the book and standing by me until completion of the manuscript.

Special thanks to my son, Aditya, for his silent support and belief in my efforts.

To all my colleagues and students for igniting my passion to share my knowledge and experience.

Contents

Preface

Going back 31 years, when I had just earned my degree in chemical engineering, I had knowledge of the fundamentals, but it was theoretical. I did not have any feel for the industry. Today, after more than three decades of core engineering experience, I realize that there is a distinct gap between what is covered as part of university curricula and what is required of an engineer to become part of a team that executes design of large chemical process plants.

Unlike certain disciplines where a recent graduate can immediately hit the ground running after starting a job, the situation in the chemical process industry is different. After carrying out process calculations and designing the equipment, there are several steps before which all equipment are despatched to site and the erection of the plant and machinery is complete. Therefore, unless a recent graduate engineer goes through this complete cycle, he or she would be missing the bigger picture while retaining knowledge of the fundamentals. The book therefore provides a practical perspective to a graduating chemical engineer and tries to answer remaining questions raised while at the university which are somehow not quite answered to his or her satisfaction.

At the outset, let me also briefly discuss the two terms: chemical engineering and process engineering. "Chemical engineering" is the term that is used in university curricula. Let us delve a little deeper into this terminology. If one looks at the operations that take place in a chemical plant, they mainly include pumping, compression, conveying, storage, heat exchange, adsorption, fractionation and chemical reaction. Of course, there are reactors as well, but most of the operations are physical in nature. Therefore, across the chemical process industry, this field is referred to as "process engineering." In this book, also, the term "process engineering" would be used rather than "chemical engineering."

It has been assumed that the reader has prior knowledge of the fundamentals of fluid mechanics, heat transfer, mass transfer, reaction engineering, process equipment design, etc. Therefore, I have avoided providing detailed theoretical explanations of various chemical processes, since these are already covered in established textbooks.

Chapter 1 directly sets the tone. It describes how a process plant is executed from concept-to-completion, starting from arriving at an investment proposal, preparing a design basis, selecting a process licensor, and awarding the project. Thereafter, stages of basic engineering, detail engineering, procurement, construction and commissioning as they appear in various stages of a project execution are touched upon. The role of a project management consultant in the execution of a project is also covered.

In the subsequent chapters, the book focusses on the various aspects of process engineering such as hydraulics, viz., pump and compressor head calculations, types of drivers and selection of motor rating. The concepts of process flow diagrams, piping and instrumentation diagrams, process data sheets, equipment lists, line lists, etc.

are described. The book provides the reader with an insight on equipment layout and plot plan. Fundamental mechanical aspects such as pipe classes are discussed. The section on columns is covered in detail. An illustration is provided on how from a process simulation, one generates a process data sheet to be sent for detailed engineering. A lot of information is provided on column internals. In addition, the book covers topics on heat exchangers, storage tanks, instrumentation and controls, all from a practical point of view.

The book also covers the guidelines and standards followed in the process industry and how engineering documents are generated using these standards. There are some chapters that are unique in this book with regard to Piping and Piping Components, Plot Plan and Equipment Layout, Relief System Design, Hazardous Area Classification, Revamp Engineering, Interaction Detail Engineering Disciplines, Pre-commissioning and Commissioning, Execution of Large Projects and Cost Engineering. Several illustrative and practical examples are provided in the book.

I sincerely believe if one has the passion for the subject, he or she would definitely enjoy reading this book. I wish my readers the very best and hope they feel proud and happy to have chosen this field of engineering.

Acknowledgments

At the outset, I am thankful to my employer Air Liquide Engineering and Construction for supporting my efforts to publish this book. I am thankful to Tarun Kumar Sharma, Alexander Schriefl, Thomas Wurzel, Peter Trabold, Dorothea Buttgereit and Nathalia Anjaparitze for their cooperation.

Over the years, there have been many people who have shared their ideas and experiences with me. Of special mention are Apurba Lal Das and Anjan Ray, with whom I have had many discussions. Their ideas and words of encouragement were very valuable to me. In addition, there are several colleagues, friends and well-wishers who have been a constant source of inspiration. Their encouragement and support will always be cherished by me. The list is long and my sincere apologies for not being able to cover their names here.

From the academic world, I am grateful to Souvik Bhattacharyya, Parameshwar De, Sudip Das, Arvind Sharma, Suresh Gupta, Hare Krishna Mohanta, Shalini Gupta, Uttam Mandal, Abhishek Sharma, Subhajit Majumder and Dhiraj Garg. Some of them invited me to their universities to deliver guest lectures which gave me the opportunity to interact with the students.

I thankfully acknowledge Heat Transfer Research Inc. and Koch-Glitsh for their heat exchanger and tower calculation software, respectively, which I utilized in some of the examples. I also convey my thanks to the various publishing houses and companies who have allowed me to use material from their published literature.

I am grateful to Gagandeep Singh and his team, Mouli Sharma, Lakshay Gaba and Aditi Mittal, who were my first points of contact and with whom I was in constant touch with all my queries until the submission of my manuscript. My sincere thanks to the Production team of Robin Lloyd-Starkes, Thivya Vasudevan, Samar Haddad, and Lisa Wilford for their highly efficient work in seeing through the final product.

It is apparent that writing a book calls for the involvement of several individuals who work as a close-knit team in order to ensure the closure what may be termed as "years of effort." As they say, "it is the team that wins and not the individual." So, my heartfelt thanks to the entire team for all the cooperation and support in this endeavor.

About the Author

Dr. Siddhartha Mukherjee has a B.Tech. and a Ph.D. in Chemical Engineering from the Indian Institute of Technology (IIT), Kharagpur. He has worked in key positions in two reputed engineering companies in India: Air Liquide Engineering & Construction, New Delhi (formerly known as Lurgi India Company Pvt Ltd), and Development Consultants Ltd., Kolkata.

He has extensive experience of more than 30 years in the design, engineering, pre-commissioning and commissioning of refineries and petrochemical plants and in general process engineering while working on several projects during his career. In addition, he has knowledge in gasification technologies including the complete gas-ification block for the synthesis of downstream products. He also worked on several international assignments in the area of oil, gas and chemicals.

Dr. Mukherjee is a Life Fellow of the Institute of Engineers (India) and the Indian Institute of Chemical Engineers. He has published several articles in national and international journals and has delivered several invited lectures at various universities and institutions. He is listed in the *Marquis Who's Who* in Science and Engineering (2006-2007). He has also been an Air Liquide Group International Expert.

List of Abbreviations

ANSI	American National Standards Institute
APH	Air preheater
API	American Petroleum Institute
ARC	Automatic recirculation
ASME	American Society of Mechanical Engineers
BEP	Basic engineering package
BFD	Block flow diagram
DCS	Distributed Control System
DN	Nominal diameter
DP	Differential pressure
EPC	Engineering, procurement and construction
ESD	Emergency Shutdown System
FD	Forced draft
FEED	Front-end engineering and design
FOW	Full of Water
HAZOP	Hazard and Operability
HDS	Hydrodesulphurization
HDM	Hydrodenitrification
HETP	Height equivalent to a theoretical plate
HIPS	High Integrity Protective System
HLL	High liquid level
HP	High pressure
ID	Inside diameter
ITB	Invitation to bid
L-EPC	Licensing, engineering, procurement and construction
LLL	Low liquid level
LP	Low pressure
LSTK	Lumpsum turnkey
MAWP	Maximum allowable working pressure
MC	Mechanical completion
MDS	Mechanical data sheet
MSD	Material selection diagram
MTO	Material takeoff
NFPA	National Fire Protection Association
NGL	Natural gas liquid
NLL	Normal liquid level
NPSH	Net positive suction head
NPSH$_a$	Net positive suction head (actual)
NPSH$_r$	Net positive suction head (required)
OP	Operating pressure

OT	Operating temperature
PDS	Process data sheet
PFD	Process flow diagram
PGTR	Performance guarantee test run
P&ID	Piping and instrumentation diagram
PMC	Project management consultant
PMS	Piping material specifications
PRV	Pressure relief valve
TDS	Thermal data sheet
TEMA	Tubular Exchanger Manufacturers Association

1 Executing a Process Plant: In a Nutshell

1.1 INTRODUCTION

Since the middle of the twentieth century, the design and construction of chemical plants have become more and more specialized. The activities start with the potential triggering of the idea of a new facility. Complex market conditions and interconnections with the economy demand a critical analysis of the project regarding feasibility, economic relevance and environmental impact. This is normally carried out with the aid of a feasibility study.

Based on the results of the feasibility study, once the decision has been made to proceed with the project, the owner/client prepares an elaborate definition of the plant also called the "process design basis" (explained later). The process design basis is used to invite bids from process licensors. The selection of the process licensor marks the end of the conceptual phase of the project.

The selected process licensor is asked to prepare the basic engineering package (BEP). The BEP serves as the basis for detail engineering and construction. The project execution phase starts with the process engineers carrying out the front-end engineering. The engineering team then transforms these into detailed drawings and specifications for all the components of the plant. The procurement group orders the equipment and all other components from specialist manufacturers. Construction companies are then contracted to build and install the plant and components. The final activity involves the commissioning and start-up of the plant. This involves several parties, particularly the process engineers from the process licensing company which has provided the technology. After a successful test run, the plant is handed over to the owner/client [1].

1.2 FEASIBILITY STUDY

At the beginning of the project, there is very little or only preliminary knowledge planning. No company would like to incur costs unless it knows whether the project is feasible. A small group is formed consisting of engineers from various disciplines to carry out a pre-feasibility study. Usually, this group later becomes responsible for implementation of the project in case the company makes a decision to proceed.

DOI: 10.1201/9780429284656-1

1.2.1 PRE-FEASIBILITY STUDY

Before a project moves into the detailed feasibility study stage, companies sometimes conduct a so-called "pre-feasibility study." Such a study is carried out if the economics of the investment appears doubtful and the investors wish to be sure before incurring time and cost in carrying out a detailed feasibility report. Further, such a study is par-ticularly useful when there is more than one route for a particular product, and the company wants to know which one is the best, both technically and financially [2].

A comprehensive pre-feasibility study should include design and description of the plant and its operation for each of the routes, as well as cost estimates, project risks, safety issues and other important information.

1.2.2 DETAILED FEASIBILITY STUDY

If the selected scenario is considered feasible, it is recommended to carry out a detailed feasibility study to get a deeper analysis of the selected project scenario.

In a detailed feasibility study, the technical, commercial, financial, economic and environmental pre-requisites for the proposed project are critically examined on the basis of the findings of the pre-feasibility study.

A detailed feasibility study typically consists of the following:

- Preparation of basis of study
- Identification of the most appropriate location for the facility
- Description of technology including process description, block flow diagram, raw materials and utilities, catalysts and chemicals consumption
- Identification of utilities and offsites facilities
- Assessment of plot area requirements
- Preparation of project organization chart, project description, project schedule and execution methodology
- Estimation of major quantities of material for civil, mechanical and electrical instrumentation items
- Assessment of human resources, managerial staff, labor cost and other over-head costs
- Estimation of fixed and variable costs for the proposed facility
- Conduction of financial analysis
- Recommendations for Implementation

1.3 APPOINTMENT OF THE PROJECT MANAGEMENT CONSULTANT

At this stage, the owner needs to decide whether he has enough manpower and exper-tise to handle all the complexities involved in the execution of the project including experienced design engineers, procurement personnel and construction experts. In most cases, the owner appoints an outside agency to act as the project management consultant, widely called PMC. The PMC works as an extended arm of the owner in the execution of the project.

1.4   PREPARING THE PROCESS SPECIFICATIONS

Once the desired operating hours per annum have been finalized (say 8000 hours per annum), the design capacity of the plant is defined. The next step is to compile the specifications for the feedstock, end products, utilities, catalysts and chemicals. These specifications include relevant physical and chemical properties as well as battery limit conditions of all streams entering and leaving the proposed complex. Appendix 1 illustrates typical process specifications.

1.5   LICENSED TECHNOLOGIES AND SELECTION OF THE PROCESS LICENSOR

"Process licensor" is an individual or a company that has exclusive rights of a particular proprietary technology. It allows another party, called the "licensee," to use the technology in its process plant. In return for the process license, the licensee pays the licensor a "license fee" that is normally a percentage based on the plant capacity.

The selling of a process license alone, however, does not carry any meaning. It is normally sold along with a BEP which contains all the details about the technology in the form of drawings, descriptions and data sheets. For this, an additional fee is charged by the licensor called the "basic engineering fee." The BEP is explained in detail in Section 1.7.

1.6   TYPES OF EXECUTION STRATEGIES

A chemical plant can be executed in several ways. Accordingly, different types of contracts are signed with the client depending on the type of execution strategy. For small and simple plants, while clients normally get the basic engineering (discussed later) done by process licensors, they would rather get the detail engineering executed by local contractors who are sufficiently competent to execute such plants. In this way, they save on the detail engineering cost. For such small plants, clients normally have sufficient manpower to get the procurement and construction executed by their own resources. In another option, the client gets the basic and detail engineering carried out by the same party. In this way, while they incur a higher detail engineering cost, they minimize the number of interfaces.

For larger plants, while the clients get the basic engineering executed by the process licensor, they club the detail engineering (E), procurement (P) and construction (C) in a single contract. Such a contract is called an "EPC contract." Such contracts are nowadays more popular. In large complexes such as refineries or petrochemical complexes, such type of contracts are quite common. The whole complex in clubbed into sections depending mainly on the process flow or proximity to each other. A particular section may have three units provided by three different process licensors. However, the client may appoint one EPC contractor to execute the entire section.

In another type of contract, the entire range of activities are awarded to one EPC contract, including the process license and basic engineering. In this case, instead of the client selecting the process licensor, it is the EPC bidder who collaborates with a process licensor and offers the entire package. Such a contract is called an "L-EPC contract," with 'L' standing for process licensing.

1.7   BASIC ENGINEERING

Now that the PMC has been appointed, the process licensor selected and the process design basis prepared, it is time to commence the basic engineering activities. In some cases, there can be two or more units in the project whereby the product from one unit feeds the next one. If there are different process licensors for the various units, the situation becomes even more complicated. In such cases, the execution of the basic engineering from the different process licensors would have to be executed in a staggered and partially overlapping manner. The PMC needs to play a major role in the coordination of such an activity in order not to waste time in the overall project cycle.

1.7.1   PROCESS SIMULATION

Process design calculations for processes involving multiple units and the interconnected processes with material and energy balance may become very complex. Nowadays, computer programs are available that model unit operations mathematically and allow them to be interconnected [1]. A plant can thus be simulated as a network of unit operations with material and energy balance. Often in such programs, the licensors build up their own proprietary thermodynamic packages that have been developed after years of laboratory work and operating experience.

1.7.2   PROCESS FLOW DIAGRAM

A process flow diagram describes the unit from the process engineer's point of view. All the major lines and equipment are shown. Each line has a stream number. All the stream numbers are linked to a separate document called the "heat and mass balance." In the heat and mass balance, each stream is described with respect to its composition, temperature, pressure and relevant important properties. The major control loops are also illustrated.

1.7.3   PIPING AND INSTRUMENTATION DIAGRAM

Piping and instrumentation is an extension of the process flow diagram and provides detailed information for down-the-line engineering to follow. For example, unlike a process flow diagram, the piping and instrumentation diagram (P&ID) shows all the lines in the unit. For each line, the line number, line size, pipe class (refer to Chapter 2 for details) and the insulation requirements are provided. Slopes required from the process point of view are shown. Connection of the lines to the respective equipment is shown, and hence all nozzles of every equipment are shown. Control valves are shown as well. Wherever the line size changes, corresponding expanders and reducers are also shown. Various vents, drains, sampling points, etc. are also shown (refer to Chapter 3 for details).

1.7.4   Equipment Process Data Sheets

Such documents provide all process data with respect to equipment. Equipment dimensions and capacities are specified. Data from the process flow diagram and material balance are used for the sizing of static equipment, rotating equipment, piping, etc. For example, column diameters are calculated using established procedures, the number of trays is derived from the process simulation and tray spacing is selected from process guidelines. Vessel diameters and heights are established from the residence times and guidelines for length-to-diameter ratio. Tanks are sized based on the number of days of storage and height restrictions.

Data for such static equipment are reproduced in the form of a process data sheet that typically has a table providing the overall dimensions, operating and design temperatures and pressures, material of construction, corrosion allowance and insulation thickness. The data sheet also has a table showing all the nozzles with their sizes and designations. There is also a sketch showing all the dimensions. For columns, in addition to the above, there is also a section that provides the vapor liquid traffic that is required for the vendor to confirm the column diameter and other performance parameters (refer to Chapter 9 for details on such data sheets).

For rotating equipment, e.g., pumps, the process data include the type of pump, nominal, maximum and minimum flow rates, inlet and outlet pressures, material of construction and physical properties of the medium [1]. The flowrates are taken from the material balance, and the inlet and outlet pressures (and hence the pump head) are arrived at by carrying out pump head calculations (refer to Chapter 5 for details).

1.7.5   Equipment Layout

The equipment layout is the arrangement of various equipment in a process unit. It shows all equipment, main piperack, buildings, major structures, roads, access ways and other items of importance to the process unit. Equipment must be grouped within common process areas to suit independent operation and shutdown. Unless required for common operation, or safety, the most economical layout is that in which equipment are located in process sequence to minimize interconnecting piping and structural steel.

Equipment should normally be located at minimum height from grade to facilitate operational and maintenance requirements. However, for process reasons, certain equipment may need to be elevated, and they must be supported in technological structures (refer to Chapter 3 for details).

1.7.6   Piping Material Specifications

The basic engineering package (BEP) has a section on "Piping Material Specifications" (PMS). A PMS is a document which contains the definition of the various pipes, fittings and all related components that are used under various pressure, temperature conditions and for various services. A typical definition contains the material specification, type, rating and dimensional data for each service (viz., hydrocarbons, cooling water, steam, condensate, flare, etc.). Chapter 4 discusses PMS in greater detail.

1.7.7 OTHER DOCUMENTS

There are a number of other documents that form part of the BEP. They include:

- Process Data Sheets of Instruments
- Cause and Effect Diagram
- Hazardous Area Classification Drawings
- Supervisory Operating Manual

For detailed descriptions, the reader may refer to Chapters 10 and 11.

1.8 AWARD OF PROJECT TO ENGINEERING CONTRACTOR

The initial step in the award of the project to an engineering contractor is the preparation of an "Invitation to Bid" document (typically called the ITB). The ITB is normally prepared by the PMC in consultation with the client and typically contains the following:

- General description of the project with site layout plan
- Contractors' scope of work
- Licensors' Basic Engineering Package (BEP)
- Engineering specifications for various disciplines
- Standard drawings
- Applicable codes and standards
- Project schedule
- Guarantees
- Terms of payment
- Commercial terms and conditions
- Deadline for bid submission

For large process plants, the ITB could run a few thousand pages. In case of projects involving several process technologies, there could be more than one ITB, each dealing with two or three process plants for example. Such a grouping of process plants is called "EPC packages." A contractor can put in a bid for one or more EPC packages.

In the next 4 to 6 months, what follows is the so-called "proposal engineering" carried out by the bidders for the various EPC packages. The proposal engineering entails complete cost estimation of civil and structural components, piping components, static equipment, rotating equipment, and electrical and instrumentation components. In case of critical equipment like compressors, high pressure reactors, etc., the bidder may even have to get the price from competent vendors. In addition, the bidder includes his engineering cost for carrying out detail engineering, procurement, construction and commissioning activities. After compiling his total price, the bidder submits his bid to the client.

Clients have the following important criteria for selecting a bidder who will execute the project [1]:

- Does the bidder have requisite experience in the tasks he will have to perform?
- Does the bidder have experience in the country and location where the plant is to be built?

- Is the bidder conversant with the language of the place where the plant is to be built?
- What is the status of the bidder with regard to manpower strength?
- Can the bidder ensure compliance with local standards that may be unfamiliar to him?
- What references does the bidder have with respect to execution of similar plants?
- Is the bidder's financial strength adequate for the requirements of such a contract?

1.9 EXECUTION OF THE PROJECT

As mentioned above, for large units, clients normally place orders as EPC contracts. An EPC contract needs a workforce from across all engineering and procurement disciplines. A large complex could involve several EPC contractors working together. Before a project starts, the project team should be in place. There could be two categories of project teams. One category is where the PMC takes the leading role with the client/owner supporting him. Figure 1.1 illustrates a project team led by the PMC. The other category is where the client's/owner's resources are sufficiently diverse to carry out the service of the PMC himself, i.e., there is no separate PMC. Figure 1.2 illustrates a project team led by the owner.

1.9.1 PROJECT KICK-OFF MEETING

The first meeting in any particular project is started with the so-called "kick-off meeting." This could be held in the client's office or the contractor's office. The meeting helps all stakeholders have a better understanding of the project's objectives, assumptions, constraints and challenges. In other words, it is a stage to get the engineering contractor and the client on the same page and agree on the steps to achieve effective execution of the project. A successful kick-off meeting can set the tone for the rest of the project and helps build a relationship between the client's and contractor's teams and helps them understand how the project will proceed.

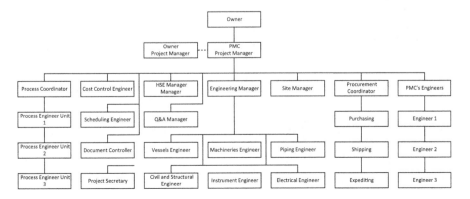

FIGURE 1.1   Organization of a project team led by the PMC.

FIGURE 1.2 Organization of a project team led by the owner.

The participants from the engineering contractor's and the client's teams include the project manager and one or more members from each of the disciplines. For large projects, some people from the senior management level could also be present in the inaugural session. Thereafter, the teams from various disciplines break into different groups and continue discussion pertaining to their respective disciplines as detailed in the meeting agenda. The following is a brief typical agenda for a project kick-off meeting:

Process Engineering

- Discussion on the Process Specifications (Appendix 1) and Basic Design Data (Appendix 2)
- Discussion and confirmation of battery limits as well as tie-in points of raw materials, utilities, products, bypass products and effluents
- Early delivery of critical process data sheets
- Detail engineering documents to be reviewed by the process licensor

Detail Engineering

- Various engineering specifications
- Codes and standards
- Plot plan and layout
- Battery limits
- Deliverables
- Approval of documents and turnaround time
- Approved vendors/suppliers of equipment

Project Management

- Discussion and confirmation of document delivery schedule
- Discussion on overall project time schedule
- Meeting schedule, including location and method
- Method of progress measurement
- Payment milestones

- Establishing protocol for communication (names of contact persons)
- Confirmation of contact addresses

Procurement

- Approved vendors' list
- Review of vendor drawings by the client
- Turnaround period for vendor document review
- Approval procedure for vendor documents

Miscellaneous

- Construction
- Commissioning
- Commercial issues

1.9.2 PROJECT SCHEDULE

Figure 1.3 illustrates a typical project schedule for basic and detail engineering of a small process plant.

Figure 1.3 shows basic engineering as part of the schedule. It may be noted that an EPC contract may or may not include the basic engineering part. In some cases, the EPC contractor is asked to tie up with a process licensor and keep the basic engineering in his scope. In other cases, the client gets the basic engineering done through a process licensor after carrying out his licensor selection process.

Figure 1.4 illustrates a typical project schedule for procurement, construction and commissioning of a small process plant.

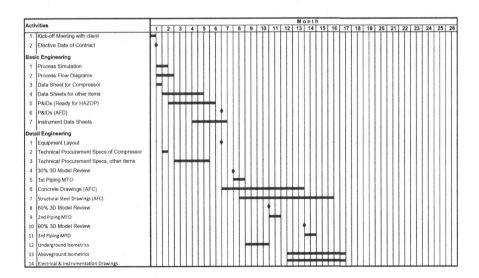

FIGURE 1.3 Project schedule for basic and detail engineering.

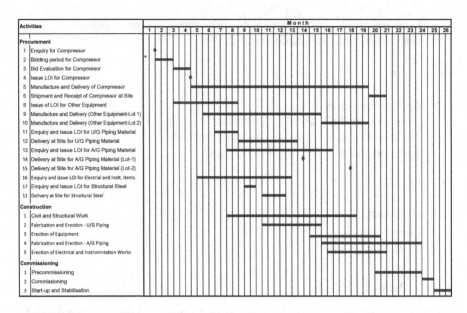

FIGURE 1.4   Project schedule for procurement, construction and commissioning.

1.9.3   DETAIL ENGINEERING

As the name suggests, "detail engineering" entails carrying out various engineering calculations, preparation of detailed specifications and drawings. In summary, the following activities are carried out [1]:

Process Engineering: Most of the process engineering work is already done in the basic engineering stage. However, the process engineer still needs to generate all information not contained in the BEP, particularly the various blowdown drums, flare knockout drums, interconnecting P&IDs, various missing line sizes, tankages, etc. In addition, the process engineer constantly needs to provide support to various engineering disciplines.

Piping Engineering: The piping engineer generates the detailed equipment layout and updates the P&IDs based on information from the BEP. Chapters 2 and 3 provide further details about these activities. In addition, he generates the piping material specifications (refer to Chapter 4 for details). All lines are identified by line numbers and are contained in the piping line list generated by the piping engineer. This activity is carried out together with the P&IDs. Isometric drawings that show both the geometry of the piping run and its location in the plant are done. In parallel, a piping model is also generated and includes pipe racks, all equipment, buildings, structures, platforms and ladders and staircases.

When a P&ID has reached a certain level of completeness and the equipment layout has been prepared, the material takeoff (MTO) is generated. The MTO provides details of all pipe lengths, valve, elbows, flanges and fittings of various sizes and various pipe classes. In addition, it provides details of special piping elements as required. The objective of this MTO estimate is to prepare bids and place orders for piping components. Sufficient material can thus be made available at the site when piping installation begins.

Static and Machinery Engineering: The static and machinery engineers carry out similar activities with respect to static equipment, viz., exchangers, columns vessels and rotating equipment such as pumps compressors, agitators, mixers, etc. The static engineer prepares mechanical data sheets of columns, vessels, heat exchangers, etc. To generate such documents, he carries out shell thickness calculations, flange rating, nozzle rating, wind and seismic load calculations and nozzle load calculations. Likewise, the machinery engineer prepares specification of pumps, compressors and other rotating equipment.

In addition, the following specifications are generated as part of the enquiry document:

- Welding specs
- Insulation specs
- Painting specs
- Standard drawings

Civil and Structural Engineering: Work on civil and structural at a project site starts fairly early in the project, typically as soon as the engineering work is 25–30% complete [1]. Therefore, engineering work in this discipline should also start once the equipment layout is frozen and there is sufficient progress in static and machinery engineering. Typical activities include:

- Analysis of structure and foundation
- Design of structure and foundation
- Preparation of piling layout (if required)
- Preparation of foundation formwork and reinforcement drawings
- Preparation of structural steel drawings

Electrical and Instrumentation and Electrical Engineering: The instrumentation engineer carries out the selection of instruments based on information in the BEP and prepares data sheets of all instruments. In addition, he carious out engineering of cables, cable trays, junction boxes and hook-up drawings. He also prepares the instrument list and logic diagrams.

The electrical engineer prepares the electrical consumer list, the single line diagram, cable routing, lighting and earthing layout.

1.9.4 PROCUREMENT SERVICES

Procurement activities in a process plant consist of the following three main steps [1]:

- Order plant and machinery items and bulks
- Expedite fabrication of the plant components (i.e., supervising the fabrication sequence according to the agreed schedule)
- Ship plant components to the construction site

In addition, there are a number of important preparatory steps as follows:

- Prepare commercial documents
- Prepare strategy for sourcing (this includes determining local and overseas suppliers)
- Determine number of suppliers
- Prepare logistics strategy, which includes determining the number of lots of shipments, optimizing the number of shipments and identifying the critical path.
- Float enquiries
- Review vendor documents by engineering disciplines
- Carry out commercial negotiations
- Place orders

1.9.5   CONSTRUCTION

Construction Planning: Planning the installation of plant and machinery starts at a relatively early stage of the project. The schedule for engineering and procurement has a strong dependence upon the sequence of installation. The overall project execution is therefore developed backward from the date of mechanical completion. Selection of proper crane and lifting equipment is an important part of construction planning [1].

Team Mobilization and Site Setup: Manpower selection is an important part of team mobilization. Personnel with requisite experience and specialized knowledge of site safety need to be carefully selected as part of the construction team. The site setup should comprise facilities such as office, stores, storage areas, roads, fencing, piping pre-fabrication yard, communication facilities, toilets, first-aid facilities, security, etc. The construction manager and his team supervise and coordinate construction activities. The construction manager reports to the project manager and to the client.

Excavation and Foundation: Construction work at a site starts with the land survey followed by preparation of the terrain. The plot is surveyed, graded and terraced if necessary. Old structures are demolished and access is laid out. Coordinates are marked. If the soil is soft, piles are driven. The sequencing of laying foundation depends upon the order in which equipment are to be installed as per the schedule [1].

Erection of Equipment and Machinery: Once the site facilities have been set up and the land excavation and foundations have been done, erections of heavy equipment, viz., columns and reactors, are executed. In parallel, steel structures are erected on which smaller equipment like reflux drum, heat exchangers, etc. are installed. Pipe racks are also erected at this time (refer to Chapters 3 and 4). Tall columns are sometimes delivered in two or more pieces and are welded at the site. Large tanks are field fabricated, and this activity is also carried out. Vendors of package items are called at the site to supervise the erection of items like compressors, refrigeration units, membrane systems, etc.

Piping Erection: Once the equipment and machinery have been installed, the piping components are erected. The activities include pre-fabrication of pipe spool and their erection, erection of pipe supports, erection of valves and fittings, and testing which could be a hydrotest or pneumatic test.

Electrical and Instrumentation: After piping erection, installation of electrical and instrumentation items is carried out. Instrumentation activities typically include installation of field instruments, analyzers, valves, cable trays and cables, instrument air distribution systems, cabinets, consoles and junction boxes. Electrical activities typically include installing cables and cable trays, fixing lighting fixtures and wiring, fixing panels and plants and, in the substation, laying trenches (in case cables run underground), etc.

1.10   PRE-COMMISSIONING AND COMMISSIONING

Commissioning of a plant comprises all activities after mechanical completion up to the certification of guarantees embodied in the contract between the EPC contractor and the owner. There are two main activities which fall under "commissioning":

Pre-commissioning: Pre-commissioning is a post-construction activity that brings the plant to a state of readiness for commissioning. This includes all activities ranging from removal of rust, loading of lubricants, installation of mechanical seals, removal of temporary bracings, alignment of rotating equipment, installation of column/vessel internals, loading of catalysts and chemicals, flushing and chemical/mechanical cleaning, drying out, pressure/vacuum hold tests, up to inertization. In addition, pre-commissioning includes box-up of columns/vessels, test run of rotating equipment, dry-out of fired heaters, and calibration and loop/functional check of instrumentation. Commissioning of utilities is also carried out at this stage. Once these activities are completed, the plant is ready for start-up.

Start-up: Once the pre-commissioning activities are completed, the plant is ready for feed-in. The activities for a hydrocarbon processing plant typically include the following:

- Take hydrocarbon into the system
- Generate hydrocarbon circulation in the system by
  - Charging steam to the reboiler
  - Charging cooling water to the column overhead cooler
  - Commissioning the system on total reflux
- Take feed into the system
- Stabilize the plant
- Gradually ramp up the load to the plant design capacity

1.11 PERFORMANCE GUARANTEE TEST RUN

Once the plant operation has stabilized and it is generating product of the desired specification, the contractor invites the client for carrying out the performance guarantee test run (PGTR). During the PGTR, the plant is run at the designed capacity for a certain duration (normally 48–72 hours) depending upon the procedure agreed in the contract. During this period, all parameters described in the procedure are monitored. At the end of the test run, if it is found that all parameters are as per the specifications, the PGTR is deemed to be complete. If, however, some of the parameters are not met, the PGTR is repeated until all parameters are as per the specifications.

1.12 PLANT ACCEPTANCE AND HANDOVER

Following the successful completion of the PGTR, the Plant Acceptance Certificate is given to the contractor. Thereafter, final payment to the contractor is made. This marks the conclusion of the contract.

REFERENCES

1. Mosberger, E., Chemical Plant Design and Construction, *Ullmann's Encyclopedia of Industrial Chemistry*, Volume B4, Wiley-VCH Verlag GmbH, 1992.
2. What are Prefeasibility and Feasibility Studies? Investing News Network, (investingnews.com), January 19, 2017.

2 Flow Sheets, Equipment Lists and Line Lists

2.1 INTRODUCTION

After an education on the fundamentals of chemical engineering, a few things that a process engineer needs to learn include exposure to the following fundamental process documents:

- Process flow diagram
- Piping and instrumentation diagram
- Material selection diagram
- Equipment list
- Line list

The concept of a heat and material balance is normally known to a new process engineer. However, he would most likely lack knowledge of the remaining documents listed above, especially the last four in most cases. This chapter discusses with some degree of detail the above five fundamental documents.

2.2 BLOCK FLOW DIAGRAM

A block flow diagram (BFD) is normally not used for a particular unit but is used to represent a chemical complex where each process unit is represented as a block. It outlines the complete process complex in the form of blocks in proper sequence, joined by interconnecting lines showing the direction of flow of feed, intermediate products and final product along with by-products, if any. It does not have any engineering value but is used in high-level meetings and presentations to outline the overall process, or it is used as material in technology brochures to present an overview. Figure 2.1 illustrates an overall BFD of a chemical complex.

BFDs can also be used to illustrate an individual process unit of a chemical complex, for example a sulfur recovery unit (Figure 2.2) showing the three sections of a typical acid gas removal process. Apart from high-level presentations, such diagrams

DOI: 10.1201/9780429284656-2

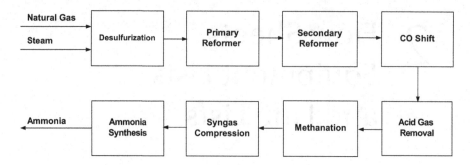

FIGURE 2.1   Block flow diagram of a chemical complex.

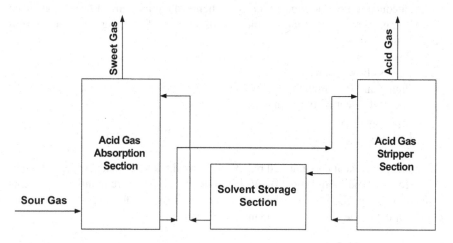

FIGURE 2.2   Block diagram of an acid gas removal unit.

do not add much value to further understanding and are not normally used in routine process engineering activities.

2.3   FLOWSHEET

A flowsheet is the diagrammatic description of a process. It pictorially and graphically identifies the chemical process steps in proper sequence. It is done in such a manner that an engineer can understand the basic process flow. The flowsheet in also not considered as an engineering document. It is mainly used in presentations, in company technical brochures and as basic training material to convey a basic understanding of the process. A flowsheet is, however, one level higher than the block flow diagram in terms of complexity and normally consists of vessels, columns, heat exchangers and reactors, pumps, compressors, etc. Instrumentation of the plant is not covered in such documents. Figure 2.3 illustrates a typical flowsheet of a stripper column.

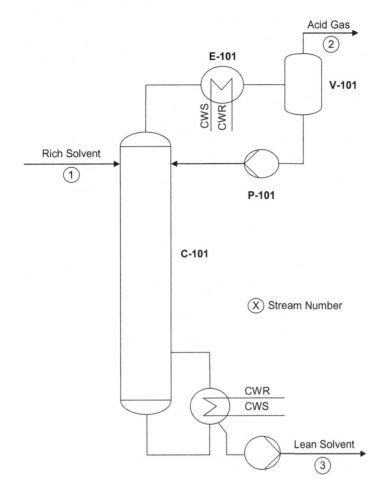

FIGURE 2.3   Process flowsheet of a stripper column.

2.4   PROCESS FLOW DIAGRAM

A process flow diagram (PFD) is the key document in describing a process. Such diagram is read in conjunction with the process description and in addition contains stream details which are linked to the heat and material balance of the process. The material balance calculations are carried out using commercial software also called "flow-sheeting programs." The sequence of equipment in a PFD follows the flow of material as it will occur in the actual process plant [1]. The following are major constituents of a PFD:

- Item numbers of all equipment
- Designations of all equipment
- Flow route and process flow direction
- Major process control loops
- Designations of feed, product and co-product produced

- Stream numbers of all major process lines. The stream numbers are linked to the heat and mass balance.
- Temperature, pressure and flow of certain specific lines are also shown in some cases

Figure 2.4 illustrates a typical PFD of a stripper column where a solvent loaded with sour gases coming from an absorber is stripped in a stripper column to yield sour gases from the top of the column and lean solvent from the bottom.

The material balance that is linked to the PFD illustrated in Figure 2.4 can be shown as a table at the bottom of each PFD. However, for material balance tables consisting of several components, the space within the PFD is normally not sufficient. In such cases, the material balance is included as a separate document. Table 2.1 illustrates a typical material balance of the stripper column illustrated in Figure 2.4.

2.5   PIPING AND INSTRUMENTATION DIAGRAM

While the PFD is a document that is of interest only to the process engineer, the piping and instrumentation diagram (P&ID), in addition to showing the process flow, combines piping and instrumentation details as well. Unlike the PFD, the P&ID shows each and every process line, instrument and valve. In fact, the P&ID is the basic document used while the plant is under the engineering phase, through construction, pre-commissioning and all the way to start-up and operation.

2.5.1   BASIC ENGINEERING P&ID

While there are no strict guidelines, P&IDs generally come in two variants: (1) the basic engineering level document and (2) the detailed engineering level document. Basic engineering level P&IDs are issued by the process licensor. The following information is included in typical basic engineering level P&IDs:

- Item numbers of all equipment and details matching with the respective data sheets
- Designations of all equipment
- Flow route and process flow direction
- All instrumentation and controls on local panel, central control room, etc.
- All control valves with condition on instrument air/power failure
- All shutdown interlocks
- All piping components, valves and fittings
- Pipe sizes
- Insulation and tracing requirements
- Elevations of equipment related to process requirements
- Special piping requirements regarding piping configuration, viz., slopes, minimum distances, etc.
- Safety systems, viz., pressure relief valves, rupture discs, etc. indicating the set pressures

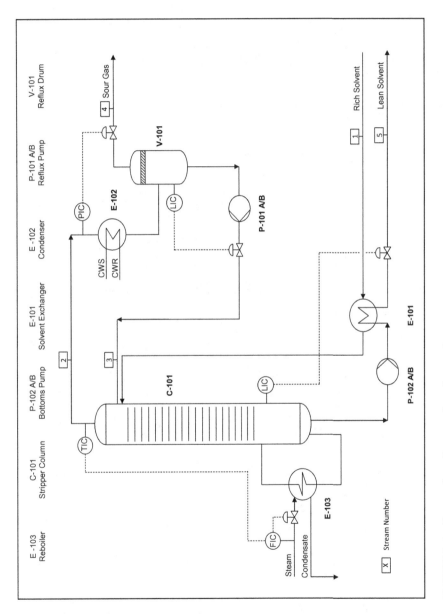

FIGURE 2.4 Process flow diagram of a stripper column.

TABLE 2.1
Mass Balance of a Stripper Column

Components	Rich Solvent 1				Sour Gas 4				Lean Solvent 5			
Stream Number	kg/hr	kgmol/hr	mass %	mole %	kg/hr	kgmol/hr	mass %	mole %	kg/hr	kgmol/hr	mass %	mole %
Methane	12.56148	0.7850925	0.0026838	0.0053382	12.56148	0.7850925	0.0369992	0.0973135	0	0	0	0
Ethane	4.924296	0.126264	0.0010521	0.0008585	4.924296	0.126264	0.0145042	0.0156506	0	0	0	0
Propane	3.15354	0.0716714	0.0006738	0.0004873	3.15354	0.0716714	0.0092886	0.0088838	0	0	0	0
i-Butane	0.4409484	0.0076026	9.421E-05	5.169E-05	0.4409484	0.0076026	0.0012988	0.0009423	0	0	0	0
n-Butane	0.9047496	0.0155991	0.0001933	0.0001061	0.9047496	0.0155991	0.0026649	0.0019335	0	0	0	0
i-Pentane	0.0571798	0.0007942	1.222E-05	5.4E-06	0.0571798	0.0007942	0.0001684	9.844E-05	0	0	0	0
n-Pentane	0.0683671	0.0009495	1.461E-05	6.456E-06	0.0683671	0.0009495	0.0002014	0.0001177	0	0	0	0
Hexane	0.0163662	0.0001903	3.497E-06	1.294E-06	0.0163662	0.0001903	4.821E-05	2.359E-05	0	0	0	0
CO_2	32768.52	744.73909	7.0012271	5.0638177	32424.6	736.92273	95.504905	91.342789	344.0208	7.8186545	0.0792512	0.056248
H_2S	577.2552	16.978094	0.1233347	0.1154417	563.0292	16.559682	1.658372	2.0526	14.226	0.4184118	0.0032772	0.0030101
Water	216914.4	12050.8	46.345303	81.938836	940.9608	52.2756	2.7715491	6.479647	215973.6	11998.533	49.753275	86.318442
Solvent	217757.36	1893.5423	46.525408	12.87505	6.426E-06	5.588E-08	1.893E-08	6.926E-09	217757.36	1893.5423	50.164196	13.6223
Total	468039.67	14707.068	100	100	33950.7	806.77	100	100	434089.21	13900.313	100	100

Properties	Rich Solvent	Sour Gas	Lean Solvent
	1	4	5
Temperature, °C	72.5	49.0	95.4
Pressure, kg/cm²g	3.60	0.85	0.88
Mole fraction vapor, %	0	100	0
Mole fraction liquid, %	100	0	100
Mass flow rate, g/hr	468039.7	33950.7	434089.2
Molar flow rate, kmol/hr	14707.07	806.77	13900.31
Density, kg/m³	1105.5	2.78	999.8
Viscosity, cP	2.75	0.016	1.14

Figure 2.5 illustrates a typical basic engineering level P&ID of a purification column.

There are certain other features of P&IDs which need to be discussed in more detail as follows:

Equipment Numbering: Equipment numbers normally consist of a combination of alphanumeric characters such as XX-Y-ZZ, where

XX is the unit number in the plant complex
Y is the equipment notation
ZZ is the serial number

For example, 03-V-02 denotes vessel number 02, in unit 03 of a complex. Table 2.2 provides a list of equipment notations typically used in chemical/petrochemical industries.

Line Numbering: Process and utility lines are also numbered is a similar fashion. Lines are shown by an identification number such as AA"-BB-CCC-DDD-EE. The first field, i.e., AA, conveys the line size. The next field, BB, consists of one or two letters that indicate the service. (Refer to Table 2.3 for details). The third field, CCC, is the pipe sequence number. The fourth field, DDD, is the pipe class, i.e., C10P. The last segment, EE, indicates the insulation type, viz., WH for heat, WP for personnel protection, etc.

Thus, a line designated as 12"-PL-C10P-006-WH is a 12" line carrying gaseous hydrocarbons through a pipe of specification C10P and is insulated for heat conservation.

Instrument Numbering: Instrumentation is more complicated. The numbering appears inside balloons and consists of two elements. The element at the top portion of the balloon consists of between two and four letters which define the function of the instrument, viz., flow indication, pressure control, etc. (refer to Table 2.4 for details). The element at the bottom portion of the balloon indicates the sequence number of the instrument [2].

The first letter inside the balloon indicates the parameter that the instrument measures. For example, a balloon with PIC indicates pressure measuring instrument. The second letter represents the function. For example, PI indicates a pressure indicate, and PS indicates a pressure switch. The third letter indicates the output, viz., PIC represents pressure indicating controller. A balloon with PIC with L and H outside the balloon indicates a pressure indicating controller with alarm at high and low pressures.

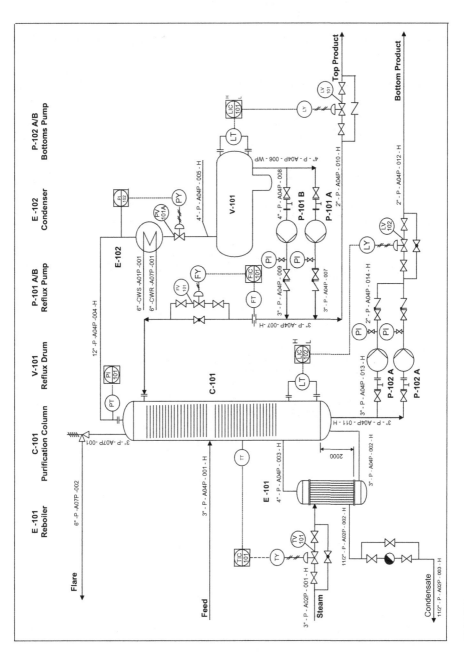

FIGURE 2.5 Basic engineering level P&ID.

TABLE 2.2
Equipment Notations in Chemical/ Petrochemical Industries

Sl. No.	Equipment	Notation
1	Agitators	A
2	Air Coolers	AC
3	Columns	C
4	Heat Exchangers	E
5	Fired Heaters	H
6	Compressors	K
7	Pumps	P
8	Reactors	R
9	Tanks	T
10	Vessels	V
11	Filters	Z
12	Miscellaneous	X

TABLE 2.3
Fluid Service Designations

Sl. No.	Service	Designation
1	Process	P
2	Cooling Water Supply	CWS
3	Cooling Water Return	CWR
4	Low Pressure Steam	LS
5	Low Pressure Condensate	LC
6	Medium Pressure Steam	MS
7	Medium Pressure Condensate	MC
8	High Pressure Steam	HS
9	High Pressure Condensate	HC
10	Plant Air	PA
11	Instrument Air	IA
12	Nitrogen	N
13	Demineralized Water	DM
14	Drinking Water	DW
15	Fire Water	FW
16	Waste Water	WW
21	Flare	FL

Instrument Loops: A control loop is the fundamental building block in industrial control systems. It consists of the process sensor, the controller function, and the final control element (required for closed loops). In an open loop, the system only measures and indicates the measured parameter. There is no facility for an output signal since this is an open-ended system. In contrast, in a closed loop system, apart from proving an indication of the measured parameter, the process is accurately controlled to a desired set point without human intervention. In this

TABLE 2.4
Instrument Nomenclature

	Indicator	Controller	Switch	Alarm High	Alarm Low	Alarm High High	Alarm Low Low
Pressure							
PI	o						
PIC (H and L outside balloon)	o	o		o	o		
PIC (HH and LL outside balloon)	o	o				o	o
PS (H and L outside balloon)			o	o	o		
Temperature							
TI	o						
TIC (H and L outside balloon)	o	o		o	o		
TIC (HH and LL outside balloon)	o	o				o	o
TS (H and L outside balloon)			o	o	o		
Level							
LI	o						
LIC (H and L outside balloon)	o	o		o	o		
LIC (HH and LL outside balloon)	o	o				o	o
LS (H and L outside balloon)			o	o	o		
Flow							
FI	o						
FIC (H and L outside balloon)	o	o		o	o		
FS (H and L outside balloon)			o	o	o		

system, the controller monitors the process variable and compares it with the desired value, called "set point." The difference in these two values called the "error signal" generates a control action that restores the controlled variable back to its desired value or the set point. Figure 2.6 illustrates some typical instrument loops.

Symbols: Symbols are used in P&IDs that identify equipment, valves, instruments and control loops. International Standard symbols for instruments, controllers and valves are given by the Instrumentation Systems and Automation Society design code ISA-5.1-1984 (R1992). Other well-known standards are BS 1646 (UK) and DIN 19227 and DIN 2429 (Germany). Many companies also use their own symbols, and while symbols used by one company may be different

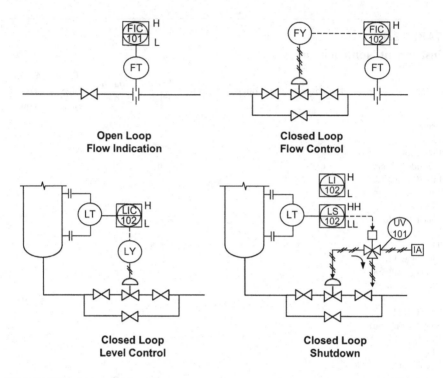

FIGURE 2.6 Instrument loops.

from another, the differences are not significant and there is an element of commonality in the various symbols used. Figures 2.7 illustrates symbols for equipment. Figures 2.8(a) and 2.8(b) illustrate symbols for valves, instrumentation and piping items.

2.5.2 DETAILED ENGINEERING P&ID

Detailed engineering P&IDs are generated from activities carried out during detailed engineering. The changes/additions are due to the following reasons:

- Inclusion of missing information in the basic level P&IDs
- Inclusion of utility lines, viz., cooling water, nitrogen, drains, closed blowdown, relief outlet, etc.
- Changes arising out of compliance with project codes and standards
- Changes arising out of the results of the HAZOP study (refer to Chapter 13 for details)
- Changes due to inclusion of vendor information from pumps, control valves, relief valves, etc. (viz., pump minimum flow lines, reducers at inlet and outlet lines, etc.)
- Changes arising out of inclusion of information from package units, viz., refrigeration, compressor, ejector system, etc.

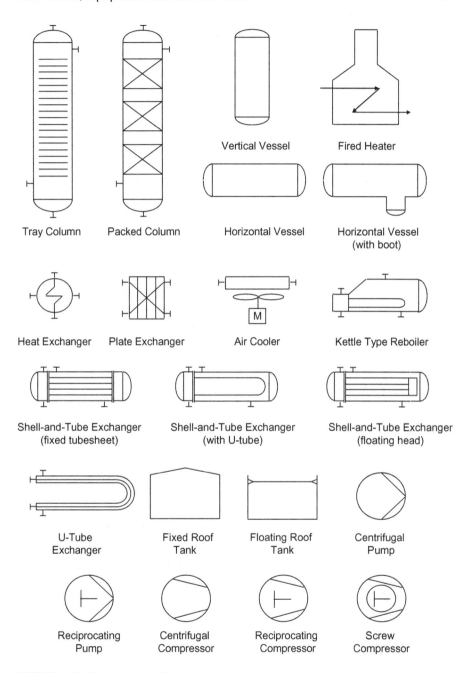

FIGURE 2.7 Equipment symbols.

The following additional information is included in typical detailed engineering level P&IDs:

- Stand-by pumps which may not be shown in basic level P&IDs
- Stand-by pressure relief valves as required

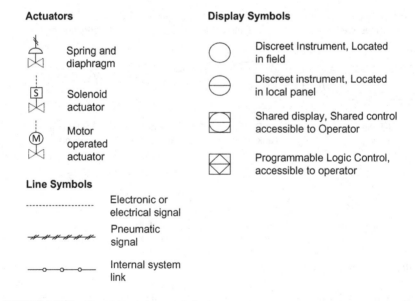

Valves

- Gate Valve
- Globe Valve
- Ball Valve
- Angle Valve
- Angle Globe Valve
- Angle Ball Valve
- 3 Way Valve General
- 3 Way Globe Valve
- 3 Way Ball Valve
- Butterfly Valve
- Check Valve General
- Swing Check Valve

Piping

- Gate Valve
- Globe Valve
- Ball Valve
- Angle Valve
- Angle Globe Valve
- Angle Ball Valve
- 3 Way Valve General
- 3 Way Globe Valve
- 3 Way Ball Valve
- Butterfly Valve
- Check Valve General
- Swing Check Valve

FIGURE 2.8(a)   Piping symbols.

Actuators

- Spring and diaphragm
- Solenoid actuator
- Motor operated actuator

Line Symbols

- Electronic or electrical signal
- Pneumatic signal
- Internal system link

Display Symbols

- Discreet Instrument, Located in field
- Discreet instrument, Located in local panel
- Shared display, Shared control accessible to Operator
- Programmable Logic Control, accessible to operator

FIGURE 2.8(b)   Instrumentation symbols.

Source: Adapted from ISA Standards, Recommended Practices and Technical Reports, Copyright © 1997, Instrument Society of America.

-   Isolation valves (in pressure gauges, level instruments, relief valves, etc.)
-   Reducers at inlet and outlet lines of pumps, control valves and pressure relief valves based on vendor information
-   Integration of utility and other lines, viz., cooling water, relief valve outlet, control valve drains, pump drains, etc.
-   Vendor package interfaces

Figure 2.9 illustrates a typical detailed engineering level P&ID of the same purification column, the basic engineering version of which was illustrated in Figure 2.5.

2.6   UTILITY P&ID

The utility distribution diagram, or the utility P&ID, is a diagrammatic representation of the distribution of utilities to various consumers in a process plant. In principle, the components in a utility P&ID are similar to those for a main plant P&ID, viz., item numbers of all equipment, (wherever applicable), flow route and process flow direction, all instrumentation and controls (to the extent applicable), all control valves, all piping components, valves and fittings, pipe sizes, insulation and tracing requirements, etc. In general, the utility P&IDs are less complicated.

Unless otherwise stated in the contract, utility P&IDs are not normally prepared by the process licensor. They form part of normal detail engineering deliverables and are executed by the company carrying the detail engineering.

Utility P&IDs are typically prepared to cover the following utility hook-ups:

-   Cooling water
-   Steam and condensate
-   Plant air and instrument air
-   Fuel oil and fuel gas
-   Nitrogen
-   Flare system

Figure 2.10 illustrates a typical utility P&ID for a cooling water system.

2.7   MATERIAL SELECTION DIAGRAM

In the hydrocarbon world, a material selection diagram (MSD) is a PFD in which the various sections of the plant are marked showing the specified materials of construction. In addition, corrosion allowances and sour service requirements are also mentioned. This includes lines as well as equipment. For established processes, this information along with information on design temperature and pressures is enough for the piping engineer to select pipe classes.

For new processes, a more detailed MSD is required. This includes information on valve types including trims as well. Figure 2.11 illustrates a typical MSD.

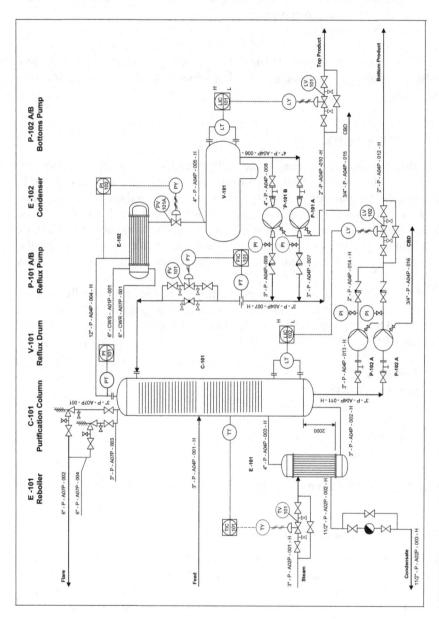

FIGURE 2.9   Detail engineering level P&ID.

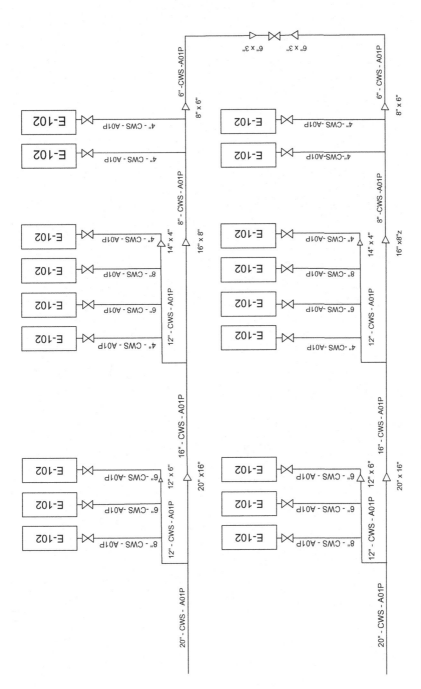

FIGURE 2.10 Utility P&ID for a cooling water system.

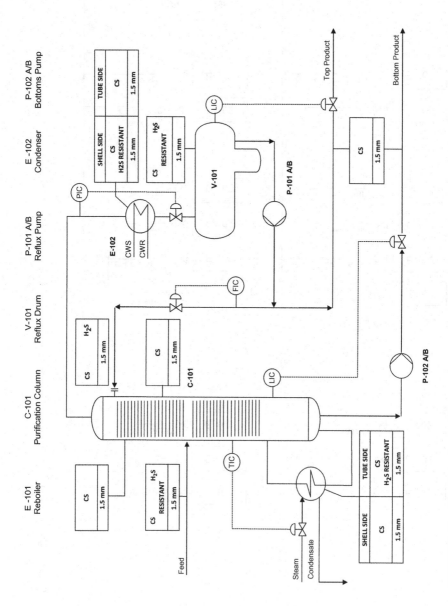

FIGURE 2.11   Material selection diagram.

2.8 LINE LIST

Line list is a document predominantly generated by the process engineer (with some inputs from the piping engineer). The line list as a name implies that it is a list of all lines in a process plant. Each line is listed with its size, pipe class, sequence number, operating and design conditions, insulation requirements, testing medium, chemical cleaning requirements, etc. Table 2.5 illustrates a typical line list developed on the basis of the P&ID illustrated in Figure 2.9.

A line list is very useful during the construction phase of a project. The insulation quantities are arrived at and ordered based on information provided in the line list. During the installation of various pipe spools, the line list is used as a guiding document. It serves as a guide to the test pressures (hydraulic or pneumatic) for various loops. Once the piping is installed, the document is referred to in order to determine those lines in the plant that require chemical cleaning (viz., compressor suction lines). The document is also referred to in order to identify which lines require stress calculations (depending upon the temperature and pressures). The document is also used by the project managers to estimate the percentage of progress during the construction phase of the project.

During pre-commissioning of the plant, the commissioning engineers frequently mark the various lines that have been cleaned and flushed in the line list.

Once the plant has been commissioned and handed over, the line list is preserved as a valuable document and is again referred to during future expansions programs, debottlenecking studies and modifications.

2.9 EQUIPMENT LIST

An equipment list, as the name implies, is a list of equipment (category-wise) along with broad specifications of each of the equipment. The categories consist of vessels and columns, pumps and compressors, reactors, heat exchangers, air coolers, package items, etc.

The equipment list, as such, is not of much use to the core disciplines, but it is regularly used by the project management to track progress as far as release of deliverables. Particularly, in case of EPC or L-EPC type of jobs, such a list is very useful because all activities with respect to the procurement of the various equipment can be tracked using this list; for example, release of process data sheets, release of mechanical data sheets, release of enquiry specifications, receipt of vendor offers, review of vendor offers, negotiations with vendors, ordering of equipment, expediting, receipt at site and, finally, installation. Table 2.6 illustrates a typical equipment list.

TABLE 2.5

Line List

Line Size, inch	Fluid Code	Pipe class	Sequence number	Route From	Route To	Operating Temperature °C	Operating Pressure kg/cm²g	Design Temperature °C	Design Pressure kg/cm²g	Insulation Heat	Insulation Cold	Insulation Personnel Protection	Insulation Thickness, mm	Testing Hydraulic	Testing Pneumatic	Testing Pressure kg/cm²g
3"	P	A04P	001	–BL	C-101	174	1.2	204	3.2	H				Y		4.2
3"	P	A04P	002	C-101	E-101	189	1.7	219	3.7	H				Y		4.8
4"	P	A04P	003	E-101	C-101	191	1.5	221	3.5	H				Y		4.6
12"	P	A04P	004	C-101	E-102	163	1.0	193	3.0	H					P	3.3
4"	P	A04P	005	E-102	V-101	163	1.0	193	3.0	H				Y		3.8
4"	P	A04P	006	V-101	P-101A	163	1.3	193	3.3	H				Y		4.3
3"	P	A04P	007	P-101A	C-101	163	5.0	193	7.0	H				Y		9.1
4"	P	A04P	008	V-101	P-101B	163	1.3	193	3.3	H				Y		4.3
3"	P	A04P	009	P-101B	3"-P-A014P-007	163	5.0	193	7.0	H				Y		9.1
3"	P	A04P	010	3"-P-A014P-007	BL	163	5.0	193	7.0	H				Y		9.1
3"	P	A04P	011	C-101	P-102A	189	1.7	219	3.7	H				Y		4.8
2"	P	A04P	012	P-101A	BL	189	5.0	219	7.0	H				Y		9.1
3"	P	A04P	013	3"-P-A014P-011	P-102B	189	1.7	219	3.7	H				Y		4.8
3"	P	A04P	014	P-102B	3"-P-A014P-013	189	5.0	219	7.0	H				Y		9.1
3/4"	P	A04P	015	P-101A/B	BL	163	atm	193	2.0	H				Y		2.6
3/4"	P	A04P	016	P-102A/B	BL	189	atm	219	2.0	H				Y		2.6

(Continued)

Line Size inch	Fluid Code	Pipe class	Sequence number	Route		Operating		Design		Insulation				Testing		
				From	To	Temperature °C	Pressure kg/cm²g	Temperature °C	Pressure kg/cm²g	Heat	Cold	Personnel Protection	Thickness, mm	Hydraulic	Pneumatic	Pressure kg/cm²g
3"	LPS	A02P	001	BL	E-101	200	15.0	230	17.0	H				Y		22.1
11/2"	LPC	A02P	002	E-101	BL	200	15.0	230	17.0	H				Y		22.1
6"	CWS	A01P	001	Header	E-102	33	6	63	8					Y		10.4
6"	CWR	A01P	001	E-102	Header	43	5.5	73	7.5					Y		9.8
3"	P	A07P	001	C-101	PSV-101A	163	1.0	193	3.0	H				Y		3.9
6"	P	A07P	002	PSV-101A	Flare Header	55	0.3	85	2.3			HP			P	2.5
3"	P	A07P	003	C-101	PSV-101B	163	1.0	193	3.0	H				Y		3.9
6"	P	A07P	004	PSV-101B	Flare Header	55	0.3	85	2.3			HP			P	2.5

TABLE 2.6
Equipment List

Vessels Columns and Tanks

Item No.	Designation	Reqd.	Dimensions		Design Conditions		Basic Material	Remarks
			Diameter mm	TL-TL mm	Press kg/cm²g	Temp °C		
C-101	Prefractionation Column	1	1500	40000	5.5	110	CS	60 valve trays
C-102	Final Distillation Column	1	2000	35000	4.5	90	CS	50 valve trays
V-101	Prefractionation Column Reflux Drum	1	1500	3000	5.5	110	CS	
V-102	Final Distillation Column Reflux Drum	1	1650	3500	4.5	90	CS	

Heat Exchangers

Item No.	Designation	Reqd.	Specifications		Design Conditions				Basic Material		Remarks
			Heat Duty MMCal/hr	Surface Area m²	°C Shell	Tubes	kg/cm²g Shell	Tubes	Shell	Tubes	
E-101	Prefractionation Column Condenser	1	0.58	178	110	70	5.5	5.5	CS	CS	TEMA Type: BEM
E-102	Prefractionation Column Reboiler	1	0.78	190	160	130	6.5	5.5	CS	CS	TEMA Type: AES
E-103	Final Distillation Column Condenser	1	0.64	183	90	70	4.5	5.5	CS	CS	TEMA Type: BEM

(Continued)

TABLE 2.6. (Continued).
Equipment List

Pumps

Item No.	Designation	Reqd.	Type	Flow Rate m³/hr	Suction Conditions Pressure kg/cm²g	Suction Conditions Temp. °C	Differential head m of liquid	Basic Material Casing	Basic Material Impeller	Remarks
P-101	Prefractionation Column Reflux Pump	1+1	Centrifugal	120	5.7	110	28	CI	CS	
P-102	Prefractionation Column Bottoms Pump	1+1	Centrifugal	152	5.9	130	32	CI	CS	
P-103	Final Distillation Column Reflux Pump	1+1	Centrifugal	105	4.7	90	31	CI	CS	
P-104	Final Distillation Column Bottoms Pump	1+1	Centrifugal	135	4.9	110	36	CI	CS	

REFERENCES

1. Towler, G. and Sinnott, R., *Chemical Engineering Design - Principles, Practice and Economics of Plant and Process Design*, Butterworth-Heinemann, 2008.
2. Walker, V., "Designing a Process Flowsheet," *Chemical Engineering Progress*, May 2009.

3 Plot Plan and Equipment Layout

3.1 INTRODUCTION

Plot plan and equipment layout are two important aspects that a process engineer should have knowledge of. First, let us have a quick definition of these two terms since there is sometimes confusion between the two. The "overall plot plan" is a scale drawing that gives an overview of an entire plant complex and is also called a "site master plan." In contrast, the "detailed plot plan" provides the top view of various equipment in a process unit. The "equipment layout" (sometimes called "equipment arrangement drawings") shows both the top and side views of the various equipment. It also covers pipe racks, staircases, ladders, etc.

Although neither of the documents is generated by the process engineer, the basic rules, guidelines and good practices should be known to him. In the equipment layout, while the plan arrangement is not a major concern and can normally be generated by the piping engineer, the elevation calls for inputs by the process engineer, since this is largely decided by gravity flow of the process fluid.

There are two occasions when the piping engineer consults the process engineer during the generation of the equipment layout. First when, for certain decisions, the view of the process engineer is sought regarding the placement of one or more equipment. The other occasion is when after the generation of the first issue of the layout drawing, it is sent to the process engineer for his comments. It is therefore mandatory that the latter has the understanding and the competence to meet this expectation.

3.2 OVERALL PLOT PLAN

A plot plan is a scale drawing that gives an overview of the entire plant complex. It covers all process units, pipe racks, roads, buildings, tank farms, control rooms, employee entrance, etc. It also provides the true north and plant north, port address, and sometimes the prevailing winds. An overall plot plan of a chemical complex covers the following facilities:

- Raw material storage facilities
- Various process units
- Control room
- Product storage/loading/unloading facilities
- Blowdown and flare system
- Utilities generation and distribution facilities
- Wastewater treatment facilities
- Administrative and service buildings, laboratory, workshop, canteen, fire station, first aid area, parking lot, etc.

DOI: 10.1201/9780429284656-3

3.3   OVERALL PLOT PLAN DEVELOPMENT

While developing a plot plan of a complex, it is recommended that the various facilities follow the general route of raw material storage, to process unit and product storage. The product of one unit may be a raw material to another unit.

Process units and diked areas for storage tanks should be identified as separate blocks. The units should be located on high ground to avoid flooding during rain. Roads should be planned encircling each block in a systematic manner to serve all process areas requiring access for operation, maintenance and firefighting. Primary traffic roads should be clear of hazard classified areas. Two road approaches from major roads (or highways) should be provided, one for movement of personnel and the other for movement of products or materials [1, 2].

Hazardous areas should be arranged with the required safe spacing to safety facilities and other plant units, facilities with a high density of personnel, plant sections and process equipment. Typical hazardous areas are systems which process, store or handle the following substances:

- Flammable liquefied gases
- Explosive substances
- Fire-propagating substances
- Flammable substances
- Toxic and corrosive substances

In addition, the following guidelines should be followed [1]:

- Facilities which generate utilities like cooling water, plant air, instrument air, plant water, etc. should preferably be located next to the process blocks.
- Cooling towers should be located downstream of the process units such that the mist formed does not corrode process equipment or obstruct vision. Further, cooling towers should be located away from fired heaters, flare stacks or other heat-generating equipment. Cooling towers are to be located where the prevailing wind is directed towards the small side. This allows both long sides to intake equal amounts of circulating fresh air. If the prevailing wind is directed towards the long side, one-half of the tower will intake fresh air, while the other half is starved [5].
- Area should be considered for construction activities.
- Flare stacks should be located upwind of the process complex.
- Effluent treatment facilities should be located at least one block away from the main process area, downwind of process units to avoid odor problems.
- Control buildings should be outside hazardous areas and upwind of process plants. The nearest process equipment should be at least 15 meters away.
- Laboratories, workshops and storage facilities should preferably be located at the shortest possible distance from process units.
- Location of administrative buildings should be such that visitors do not have to enter the fenced process area.
- Future expansion should not be overlooked. Space provision for known and unforeseen needs shall be provided.

Figure 3.1 Typical overall plot plan.

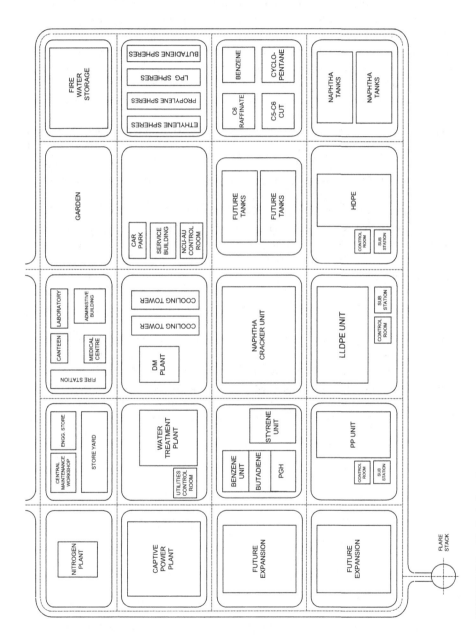

FIGURE 3.1 Typical plot plan.

3.4   EQUIPMENT LAYOUT

The equipment layout is the arrangement of various equipment in a process unit. It shows all equipment, main pipe racks, buildings, major structures, roads, access ways and other items of importance to the process unit. Equipment must be grouped within common process areas to suit independent operation and shutdown. Unless required for common operation or safety, the most economical layout is that in which the equipment are located in process sequence to minimize interconnecting piping and structural steel. The factors that should be kept in mind while setting the equipment layout include the following [3]:

Economics: The layout should be compact to the extent possible with provision for access and safety requirements. Location of equipment should be in sequential order of process flow so that piping runs are minimum. Equipment should preferably be located at grade level to facilitate operational and maintenance requirements. However, for process reasons, certain equipment may need to be elevated. These must be supported in technological structures. Structural steel work should be optimized in a way that it caters to more than multiple equipment. Figure 3.2 shows how a distillation column is illustrated as a flowsheet and how it looks in reality.

Safety: Hazardous equipment should be grouped and located separately from other areas of the plant. Safe routes should be planned for movement of forklift trucks. Clear routes must be provided for access by firefighting equipment.

Construction: Long delivery items which are likely to arrive late during construction should be given due consideration. Insulation should be reviewed during the layout of the plant, since insulation decreases the available space for access. The plot area should be so selected that adequate access is available to lift large equipment to its proper location. Addition of equipment at a later date (e.g., during expansion or

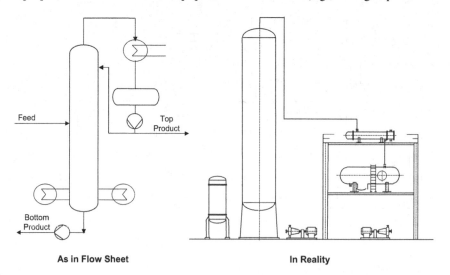

As in Flow Sheet   **In Reality**

FIGURE 3.2   Technological structure carrying overhead condenser and reflux drum.

revamping) requires space. A recommended good practice is to provide about 30% spare space for such future expansion.

3.5   EQUIPMENT SETTING GUIDELINES

Equipment should be spaced based on certain specifications like access widths, elevation clearances for operator and maintenance access and safety spacing requirements. Table 3.1 illustrates typical plant layout specifications. The distances specified in the table are the recommended distances that the industry should adhere to. These, however, could be modified as required to suit space constraints and relevant local engineering standards. In addition, special recommendations from the process licensor also need to be followed. The following guidelines describe how to locate equipment in an equipment layout drawing [1, 4, 5]:

Pipe Rack: The pipe rack is the main artery of a chemical process plant and connects all equipment with lines that cannot run through adjacent areas. The pipe rack is normally made of structural steel either in a single tier or a multi-tier, depending on the quantum of piping. During the initial stages of the design, the piping engineer, using the process flow diagram and preliminary plot plan, establishes the process lines and utility lines. The width of the pipe rack is arrived at based on the approximate sizing of these lines and information on insulation requirements. In addition, the electrical and instrument engineers provide their respective cable tray requirements. Typically, a 20% margin is added for future requirements [4]. The unit pipe rack is normally located at the center, thereby splitting the unit into two or more areas of equipment [1]. This is called a "straight-through arrangement." However, to meet the specific requirements of the project, the final pipe rack layout could be T-, L- or U-shaped (Figure 3.3).

Beneath the pipe rack, a minimum clearance of 4.0 m should normally be maintained. For pipe racks which support air coolers, a width of 9 m is desirable [4]. Pumps are accordingly arranged in two rows on either side of the pipe rack [1].

Distillation Columns: Distillation columns should normally be located on a common centerline about 3.5–4.0 m from the pipe rack columns [5]. The columns should be located close to the road side for unobstructed erection and ease of maintenance. Tall columns requiring frequent operating attention at upper levels should preferably be located close to each other with a common connecting platform [1].

Heat Exchangers: Back head of heat exchangers should be located on a line 2.5 m from the pipe rack column. This is called the "equipment line." Shell and tube heat exchangers having a removable tube bundle must have a tube pulling or rod cleaning area at the channel end. This should be the tube length plus a distance of 1.5 m from the tube sheet [5].

Air Coolers: Air fin coolers should be installed above the pipe rack. An air cooler can overhang equally on either side of the pipe rack. Thus, an air cooler with a bundle length of 12 m can be adequately supported on a pipe rack having a width of 10.5 m [4]. Pumps

TABLE 3.1

Distances between Equipment in a Hydrocarbon Service

Sl. No.	From/To	Fired Heaters	Reactors	Distillation Columns	Vessels/ Drums	Compressors	Hydrocarbon Day Tanks	Hydrocarbon Pumps	Heat Exchangers	Air Fin Coolers
1	Fired Heaters	x	15	15	15	15	15	15	15	15
2	Reactors	15	x	4.5	4.5	9	15	7	M	4.5
3	Distillation Columns	15	4.5	x	3	7	15	5	M	3
4	Vessels/Drums	15	4.5	3	x	M	15	M	M	3
5	Compressors	15	9	7	M	x	15	7.5	7	7
6	Hydrocarbon Day Tanks	15	15	15	15	15	x	15	15	15
7	Hydrocarbon Pumps	15	7	5	M	7.5	15	x	M	M
8	Heat Exchangers	15	M	M	M	7	15	M	x	M
9	Air Fin Coolers	15	4.5	3	3	7	15	M	M	x

Note: M: Minimum required for operator or for maintenance access.
Source: Adapted from OISD [1] and Bausbacher and Hunt [4].

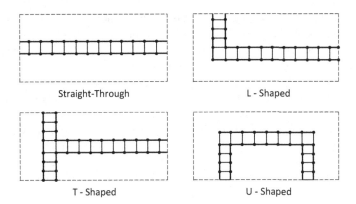

Straight-Through   L - Shaped

T - Shaped   U - Shaped

FIGURE 3.3   Pipe rack configurations.

Source: Adapted from Bausbacher and Hunt [4].

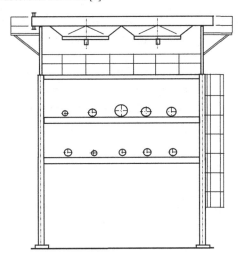

FIGURE 3.4   Air cooler on a pipe rack.

handling hydrocarbons and materials above 230°C should not be installed underneath the air fin coolers [1]. Figure 3.4 illustrates an air cooler supported on the pipe rack.

Vessels: Vessels having large liquid hold-up should be installed preferably at grade unless the process design dictates an elevated location. Vertical vessels should be located such that the shell of the vessel with the largest diameter lies approximately 0.5 m away from the equipment line or 3 m from the columns of the pipe rack. For horizontal vessels, the heads should fall on the equipment line [5].

Fired Heaters: Fired heaters should be located at least 15 m from equipment containing hydrocarbons. Further, they should be located upwind at one corner of the unit. The fired heaters must have access by road for machinery that are needed for tube repair or replacement. Tubes mounted vertically are pulled up from the top with a crane. Horizontal box type heaters must have tube removal space allocated behind the heater, equal to the tube length plus 3.5 m. A light duty road can be utilized as part of this maintenance area [5]. Forced draft fans should be located away from process equipment from where they could suck hydrocarbon vapors.

Reboilers: Reboilers should be located next to the respective columns. Kettle type reboiler is determined by the column liquid. Vertical thermosiphon reboilers are located at elevations with regard to the column and are guided by thermosiphon calculations calculated during the reboiler design stage. Small thermosiphon reboilers may be mounted on the column. Large reboilers may require a separate supporting structure [5]. Forced circulation reboilers, however, need to be located depending upon the location of the circulation pump and the associate piping.

Compressors: Compressors should be located downwind of the heaters so that leaked gases do not get directed towards the heaters. Compressor sheds should have a roofing but should be open from the sides so that heavier vapors/gases do not accumulate on the floor of the compressor house. The compressor house should preferably be located near battery limits for ease of maintenance and operation [1].

Pumps: Figure 3.5 shows possible locations of pumps. One option is to locate the pumps under the main overhead pipe rack. This is the normal location in many process units where there is minimum possibility of hydrocarbon leaks into electric

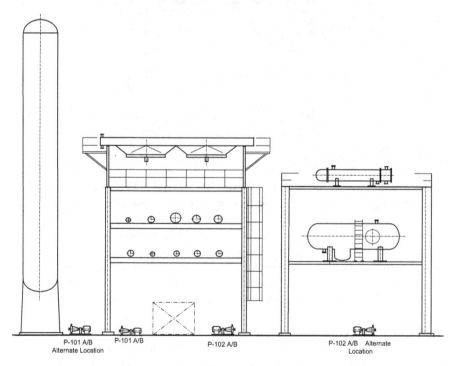

FIGURE 3.5 Location of pumps.

motors (P-101 A/B for Column Bottom Pumps and P-102 A/B for Reflux Pumps). Hydrocarbon bearing air coolers located on pipe racks are, however, a concern to many clients. In such cases, the pumps are located outside the confines of the pipe rack. For Reflux Pumps (P-102 A/B) these could be directly under the suction equipment supported on the structure above. Reflux drums and shell-and-tube heat exchangers are less likely to cause hydrocarbon spills and therefore may have pumps located directly beneath them.

Storage Tanks: Storage tanks should be grouped according to product classification; however, only day tanks shall be provided within battery limits of any process unit.

Miscellaneous: The following additional guidelines should be followed [1, 5]:

- Clearance from main roads should be 5.5 m and from secondary roads 4.5 m.
- Minimum horizontal clearance between equipment and/or piping should be 0.75 m. However, clearance between exchanger flanges can be 0.5 m.
- Flare knockout drum should be located at the battery limit of the unit and upwind of process units.
- Blowdown facilities/buried drum should be located at one corner of the plant, farthest from any fired equipment and on the leeward side of the plant. Vent from the blowdown facility shall be at least 6 m clear from the highest equipment located within a radius of 15 m from the vent stack.
- Main power receiving station should be close to the fence line with minimum overhead power transmission lines passing through the installation.

3.6   PREPARING EQUIPMENT LAYOUT

Figure 3.6 shows a flowsheet of a typical fractionation unit consisting of a series of columns. Such configurations are common in natural gas liquid (NGL) fractionation units or fractionation units downstream of steam crackers in a petrochemical complex. The reader is, however, informed that this is only a generic configuration meant to demonstrate how an equipment layout is prepared and not to compare it with the flowsheet of any established unit in operation.

Figures 3.7(a) and 3.7(b) show the plan and elevation drawings of the equipment layout of such a unit. The procedure is described following these steps:

Step 1: Locate column C-101 at one end of the plot. Locate the reboilers E-101 adjacent to the column. Locate the column bottoms pump P-101 A/B close to the column, at grade level, below the pipe rack. Locate the overhead condenser E-102 above the reflux drum V-101 in the adjacent structure, keeping in mind the net positive suction head (NPSH) requirements of the reflux pumps P-102 A/B. Locate the pumps P-102 A/B close to the reflux drum V-101, at grade level below the pipe rack.

Step 2: Locate column C-102 next to C-101. Locate the reboilers E-103 adjacent to the column. Locate the column bottoms pump P-103 A/B close to the column, at grade level, below the pipe rack. The structure which houses E-102

and V-101 is also shared by E-104 and V-102. This structure is located opposite to the pipe rack. Keeping in mind the NPSH requirements, locate the
reflux pumps P-104 A/B close to the reflux drum V-102, at grade level,
below the pipe rack.

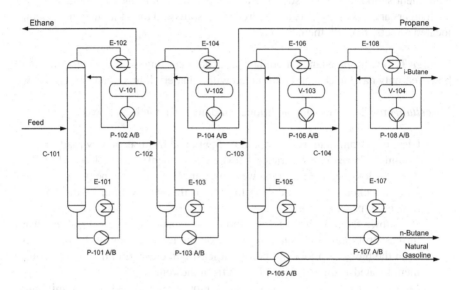

FIGURE 3.6 Process flowsheet of a fractionation unit.

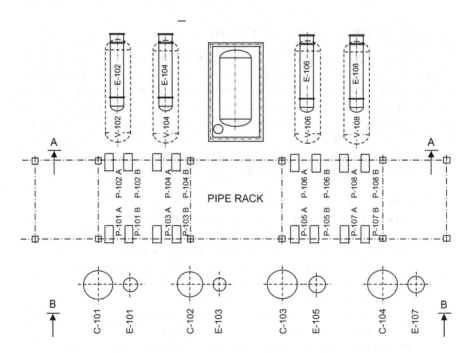

FIGURE 3.7(a) Equipment layout – plan.

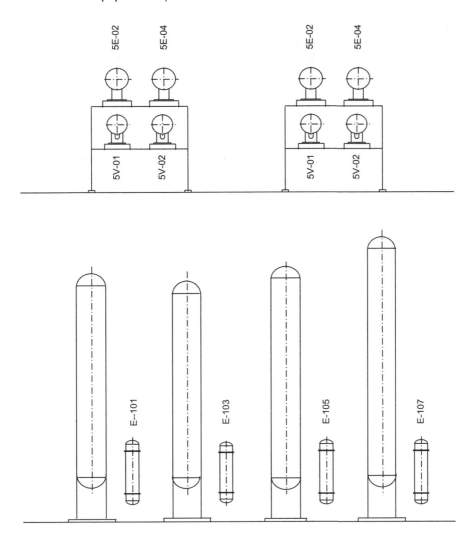

FIGURE 3.7(b)   Equipment layout – elevation.

Step 3: Locate column C-103 next to column C-102. Locate the reboiler E-105 adjacent to the column. Locate the column bottoms pump P-103 A/B close to the column, at grade level, below the pipe rack. Locate the overhead condenser E-106 above the reflux drum V-103, keeping in mind the NPSH requirements of the reflux pumps P-106 A/B. Locate the pumps P-106 A/B close to the reflux drum V-103, at grade level, below the pipe rack.

Step 4: Locate column C-104 next to column C-103. Locate the reboiler E-107 adjacent to the column. Locate the column bottoms pump P-107 A/B close to the column, below the pipe rack. Locate the overhead condenser E-108 above the reflux drum V-104, keeping in mind the NPSH requirements of the reflux pumps P-108 A/B. Locate the pumps P-108 A/B close to the reflux drum V-104, at grade level, below the pipe rack.

REFERENCES

1. Oil India Safety Directorate, "Layouts for Oil and Gas Installations", *OISD Standard*, 118, 2004.
2. www.pipingengineer.org
3. Jalnapurkar, K. M. and Amale, P. D., "Plant Layout – Doing it Economically", *Chemical Engineering World*, September 2001.
4. Bausbacher, E. and Hunt, R., *Process Plant Layout and Piping Design, Prentice Hall*, Englewood Cliffs, New Jersey, 1993.
5. Weaver, R., *Piping Process Design, Vol. 1*, Gulf Publishing Company, Houston, Texas, 1985.

4 Piping and Mechanical Considerations

4.1 INTRODUCTION

It is sometimes understood that the role of a chemical engineer centers around process simulation, heat and material balance, process flow diagrams, distillation columns, heat exchangers, pumps, vessels, etc. However, once the engineer begins to work with a project team on a chemical process plant, he realizes that a minimum knowledge of certain mechanical aspects is required, without which he would not be able to fully contribute in a live project. These mechanical aspects include knowledge of piping material specification, pipe classes, the concept of pound-rating, and some knowledge of piping and pipe routing. This chapter focuses on these issues.

4.2 PIPING MATERIAL SPECIFICATION

A piping material specification is a document prepared by the piping engineer and contains the definition of a pipe and all related components that are to be used under a specific pressure and temperature condition – including sometimes the service they are in. A typical definition consists of the pipe material, type, pound rating, and dimensional data.

Of all the piping components, flanges constitute the weakest point and these determine the pipe class. The rest of the piping components are given as thicknesses in the form of pipe schedules. The most common method of specifying a pipe class is that given by ASME 16.5 in the form of pound ratings, viz., 150#, 300#, 600#, 900#, 1500#, etc.

Flanges can withstand different pressures at different temperatures. As the temperature increases, the pressure which a flange can withstand decreases. For example, for carbon steel material, a 150# rating flange can withstand approximately 19.6 kg/cm^{2}g at ambient conditions, 13.8 kg/cm^{2}g at 200°C, 10.2 kg/cm^{2}g at 300°C, and 6.5 kg/cm^{2}g at 400°C. ASME 16.5 provides pressure-temperature ratings for various materials. Tables 4.1 and 4.2 illustrate such tables for carbon steel and stainless steel pipes, respectively.

While the piping material specification is prepared by the piping engineer, the inputs for preparing the same are provided by the process engineer. For every service in a process plant, viz., steam, water, fuel gas, hydrocarbon feed, cooling water, etc., the process engineer is required to provide the following information to the piping engineer:

- Design pressure
- Design temperature

DOI: 10.1201/9780429284656-4

TABLE 4.1
Pressure-Temperature Ratings for Carbon Steel
(Group 1.1 Materials)

Nominal Designation	Forgings	Castings	Plates
C-Si	A105	A216 Gr. WCB	A515 Gr. 70
C-Mn-Si	A350 Gr. LF2	.....	A516 Gr. 70
	A350 Gr. LF6 Cl.		
C-Mn-Si-V	1	.....	A537 Cl. 1
3½ Ni	A350 Gr. LF3	.....	.....

Working Pressures by Classes, Bar

Temp, °C	Class				
	150	300	600	900	1500
−29 to 38	19.6	51.1	102.1	153.2	255.3
50	19.2	50.1	100.2	150.4	250.6
100	17.7	46.6	93.2	139.8	233.0
150	15.8	45.1	90.2	135.2	225.4
200	13.8	43.8	87.6	131.4	219.0
250	12.1	41.9	83.9	125.8	209.7
300	10.2	39.8	79.6	119.5	199.1
325	9.3	38.7	77.4	116.1	193.6
350	8.4	37.6	75.1	112.7	187.8
375	7.4	36.4	72.7	109.1	181.8
400	6.5	34.7	69.4	104.2	173.6
425	5.5	28.8	57.5	86.3	143.8
450	4.6	23.0	46.0	69.0	115.0
475	3.7	17.4	34.9	52.3	87.2
500	2.8	11.8	23.5	35.3	58.8

Note: For further details, refer to ASME B16.5-2009.
Source: Reprinted from ASME B16.5-2009, by permission of the American Society of
Mechanical Engineers. All rights reserved.

- Base material
- Maximum line size
- Corrosion allowance

Table 4.3 illustrates typical information provided by the process engineer.

From the above information, the piping engineer calculates the pipe thicknesses for various pipe diameters and for various pressure ranges. Based on the calculations, standard pipe thicknesses are selected for each pipe diameter (refer to Table 4.4 for standard pipe thicknesses). The calculations are reproduced in a tabular form. Sometimes, the piping engineer produces such tables for a wider pressure and temperature range so that the same could be used for a broader range of applications. Based on the services listed by the process engineer in the form of Table 4.3, the piping engineer generates the pipe classes allotted to each of the services. Table 4.5 illustrates this information in the form of a table typically called, "Pipe Class Index." Table 4.6 provides further details of these applicable pipe classes.

It can be seen from Table 4.6 that each piping class is identified by two alphabetical characters and a two-digit figure, e.g., A01P, B09P, etc.

TABLE 4.2
Pressure-Temperature Ratings for Stainless Steel (Group 2.2 Materials)

Nominal Designation	Forgings	Castings	Plates
16Cr-12Ni-2Mo	A182 Gr. F316 A182 Gr. F316H	A351 Gr. CF3M A351 Gr. CF8M	A240 Gr. 316 A240 Gr. 316H
18Cr-13Ni-3Mo	A182 Gr. F317	. . . . .	A240 Gr. 317
19Cr-10Ni-3Mo	. . . . .	A351 Gr. CG8M	

Working Pressures by Classes, Bar

Temp, °C	Class				
	150	300	600	900	1500
−29 to 38	19.0	49.6	99.3	148.9	248.2
50	18.4	48.1	96.2	144.3	240.6
100	16.2	42.2	84.4	126.6	211.0
150	14.8	38.5	77.0	115.5	192.5
200	13.7	35.7	71.3	107.0	178.3
250	12.1	33.4	66.8	100.1	166.9
300	10.2	31.6	63.2	94.9	158.1
325	9.3	30.9	61.8	92.7	154.4
350	8.4	30.3	60.7	91.0	151.6
375	7.4	29.9	59.8	89.6	149.4
400	6.5	29.4	58.9	88.3	147.2
425	5.5	29.1	58.3	87.4	145.7
450	4.6	28.8	57.7	86.5	144.2
475	3.7	28.7	57.3	86.0	143.4
500	2.8	28.2	56.5	84.7	140.9

Note: For further details, refer to ASME B16.5-2009.

Source: Reprinted from ASME B16.5-2009, by permission of the American Society of Mechanical Engineers. All rights reserved.

The first alphabetical character indicates pressure rating of flange, i.e., ANSI classes as follows:

Character A   ANSI rating 150#
Character B   ANSI rating 300#
Character C   ANSI rating 600#

The two-digit figure indicates the process fluids being handled. For example:

01   Cooling Water, Nitrogen, Service Water, Waste Water
02   Low Pressure Steam, Low Pressure Condensate

The second alphabetical character indicates different material groups. For example:

P.   Carbon Steel
Q.   Killed Carbon Steel
R.   Carbon Steel (Galvanized)
S.   Stainless Steel

TABLE 4.3
Process Inputs for Generating Piping Classes

Sl. No.	Service	Maximum Line Size, inch	Design Temperature, kg/cm²g	Design Pressure, kg/cm²g	Base Material
1	Hydrocarbon Feed	10	180	10	carbon steel
2	Solvent	8	180	12	carbon steel
3	Hydrocarbon Product	8	180	10	carbon steel
4	Hydrocarbon Vapors	12	180	10	carbon steel
5	Atmospheric Vent	10	100	3.5	carbon steel
6	Dry Flare	16	(−) 45–200	3.5	killed carbon steel
7	Fuel Gas	8	80	10	carbon steel
8	Lubricated Oil	4	80	10	carbon steel
9	Seal Oil	4	80	10	carbon steel
10	Contaminated Oil	4	80	10	carbon steel
11	Cooling Water	18	80	8	carbon steel
12	Service Water	4	80	8	carbon steel
13	Waste Water	8	70	6	carbon steel
14	Nitrogen	6	70	8	carbon steel
15	Drinking Water	3	70	6	carbon steel (galvanized)
16	Instrument Air	4	70	8	carbon steel (galvanized)
17	Demineralized Water	4	70	8	SS 304
18	Low Pressure Steam	10	185	7	carbon steel
19	Low Pressure Condensate	8	185	7	carbon steel
20	Medium Pressure Steam	8	220	15	carbon steel
21	Medium Pressure Condensate	6	220	15	carbon steel
22	High Pressure Steam	8	255	30	carbon steel
23	High Pressure Condensate	6	255	30	carbon steel

For each set of temperature and pressure, keeping in mind the corrosion allowance, the thickness of the pipe is calculated. Once the theoretical thickness is calculated, the nearest standard thickness is considered.

Consider the pipe class represented by A01P in Table 4.6. It can be seen that at 38°C, the pipe can take a pressure of 20 kg/cm²g. At 350°C, it can withstand a

TABLE 4.4
Standard Pipe Thicknesses

Pipe Size, inch	Nominal OD, inch	Schedule 10		Schedule 40		Schedule 80		Schedule 160	
		Pipe ID inch	Wall Thk, inch	Pipe ID inch	Wall Thk, inch	Pipe ID inch	Wall Thk, inch	Pipe ID, inch	Wall Thk, inch
1/2	0.840	0.674	0.083	0.622	0.109	0.546	0.147	0.464	0.188
3/4	1.050	0.884	0.109	0.824	0.113	0.742	0.154	0.612	0.219
1	1.315	1.097	0.109	1.049	0.133	0.957	0.179	0.815	0.250
1 1/2	1.900	1.682	0.109	1.610	0.145	1.500	0.200	1.338	0.281
2	2.375	2.157	0.109	2.067	0.154	1.939	0.218	1.687	0.344
3	3.500	3.260	0.120	3.068	0.216	2.900	0.300	2.624	0.438
4	4.500	4.260	0.120	4.026	0.237	3.826	0.337	3.438	0.531
6	6.625	6.357	0.134	6.065	0.280	5.761	0.432	5.187	0.719
8	8.625	8.329	0.148	7.981	0.322	7.625	0.500	6.813	0.906
10	10.750	10.420	0.148	10.020	0.365	9.562	0.594	8.500	1.125
12	12.750	12.390	0.148	11.938	0.406	11.376	0.687	10.126	1.312
14	14.000	13.500	0.148	13.124	0.438	12.500	0.750	11.188	1.406
16	16.000	15.500	0.148	15.000	0.500	14.314	0.843	12.812	1.594
18	18.000	17.500	0.148	16.876	0.562	16.126	0.937	14.438	1.781
20	20.000	19.500	0.148	18.812	0.594	17.938	1.031	16.062	1.969
24	24.000	23.500	0.148	22.624	0.688	21.564	1.218	19.312	2.344

Note: ID = Inner diameter.

pressure of 8.5 kg/cm²g. The pipe thicknesses for this range of temperature and pressure can be seen from Table 4.6. Pipe classes in this range of pressure fall under the 150# rating category.

Now consider the pipe class represented by B09P in Table 4.6. It can be seen that at 38°C, the pipe can take a pressure of 49.7 kg/cm²g, and at 350°C, it can withstand a pressure of 37.6 kg/cm²g. In other words, this pipe class can withstand higher pressures. It can also be seen from Table 4.6 that the corresponding pipe thicknesses are higher, particularly for pipe sizes of 200 mm and above (at lower sizes, the wall thicknesses are high enough to accommodate high pressures). Pipe classes in this range of pressure fall under the 300# rating category.

Similarly, a look at pipe class C10P reveals that beyond a pipe size of 150 mm, the wall thicknesses are even higher than that of B09P. Pipe classes in this range of pressure fall under the 600# rating category.

4.3   INSULATION OF PIPING SYSTEMS

There are different types of insulation in piping systems. Pipes are insulated either for conservation of heat, minimizing the ingress of heat into a cold fluid or for personnel protection.

Insulation for Heat Conversation: Whenever pipes carry hot fluids, viz., steam or other process fluids which are meant to remain at high temperatures, the pipes are

TABLE 4.5

Pipe Class Index Generated by Piping Engineer

Pipe Class	Service	Temperature Min-Max, °C	Pound Rating	Base Material	Corrosion Allowance, mm
A01P	Cooling Water Nitrogen Service Water Waste Water	0–120	150#	carbon steel	1.5
A02P	Low Pressure Steam Low Pressure Condensate	0–260	150#	carbon steel	1.5
A03P	Fuel Gas Contaminated Oil Lubricated Oil Seal Oil	0–350	150#	carbon steel	1.5
A04P	Hydrocarbon Feed Solvent Hydrocarbon Product Hydrocarbon Vapors	0–350	150#	carbon steel	1.5
A05P	Hydrocarbon Feed Solvent Hydrocarbon Product Hydrocarbon Vapors	0–350	150#	carbon steel (sour service)	1.5
A06P	Atmospheric Vent	0–420	150#	carbon steel	1.5
A07Q	Dry Flare	(–) 45–200	150#	killed carbon steel	1.5
A08R	Drinking Water Instrument Air	0–75	150#	carbon steel (galvanized)	1.5
A09S	Demineralized Water	0–75	150#	SS 304	0.0
B10P	Medium Pressure Steam Medium Pressure Condensate	0–350	300#	carbon steel	1.5
C11P	High Pressure Steam High Pressure Condensate	0–420	600#	carbon steel	1.5

insulated to minimize loss of heat to the surroundings. Such type of insulation is called "insulation for heat conservation." While mathematical equations are beyond the scope of this book, Table 4.7 illustrates typical insulation thicknesses calculated for various pipe diameters at different temperature ranges for an assumed ambient temperature. Readers may note that the numbers are typical. The thicknesses may be different in countries having very low ambient temperatures. The numbers also depend on the material of the insulation. Exact thicknesses should be calculated for project specific cases.

Insulation for Personnel Protection: Pipes carrying hot fluids above 60°C and located within a height of 2.1 m from grade level or from platforms should be insulated to

TABLE 4.6
Pipe Classes

A01P (150# Rating)

Cooling Water, Nitrogen, Service Water, Waste Water
Basic Material: Carbon Steel, Corrosion Allowance: 1.5 mm

Pressure, kg/cm²g	20.0	19.6	18.0	16.1	14.2	12.3	10.4	8.5
Temperature, °C	38	50	100	150	200	250	300	350

DN, mm	15	20	25	40	50	80	100	150
Outside Dia., mm	21.3	26.7	33.4	48.3	60.3	88.9	114.3	168.3
Wall Thickness, mm	3.73	3.91	4.55	5.08	3.91	5.49	6.02	7.11
Sch. Number	80	80	80	80	40	40	40	40

DN, mm	200	250	300	350	400	450	500	600
Outside Dia., mm	219.1	273.0	323.9	355.6	406.4	457.0	508.0	610.0
Wall Thickness, mm	6.35	6.35	6.35	7.92	7.92	7.92	9.53	9.53
Sch. Number	20	20	20	20	20	20	20	20

B10P (300# Rating)

MP Steam, MP Condensate
Basic Material: Carbon Steel, Corrosion Allowance: 1.5 mm

Pressure, kg/cm²g	49.7	49.7	47.2	46.0	44.7	42.6	39.6	37.6
Temperature, °C	38	50	100	150	200	250	300	350

DN, mm	15	20	25	40	50	80
Outside Dia., mm	21.3	26.7	33.4	48.3	60.3	88.9
Wall Thickness, mm	3.73	3.91	4.55	5.08	5.54	5.49
Sch. Number	80	80	80	80	80	40

DN, mm	100	150	200	250	300	350
Outside Dia., mm	114.3	168.3	219.1	273.0	323.9	355.6
Wall Thickness, mm	6.02	7.11	8.18	9.27	10.31	11.12
Sch. Number	40	40	40	40	40	40

C11P (600# Rating)

HP Steam, HP Condensate
Basic Material: Carbon Steel, Corrosion Allowance: 1.5 mm

Pressure, kg/cm²g	60.4	60.4	60.4	60.4	60.4	60.4	60.4	59.9	52.0	45.6	0.033
Temperature, °C	38	50	100	150	200	250	300	350	400	420	420
DN, mm	15	20	25	40	50	80					
Outside Dia., mm	21.3	26.7	33.4	48.3	60.3	88.9					

(Continued)

TABLE 4.6 (Continued)

C10P (600# Rating)

Wall Thickness, mm	3.73	3.91	4.55	5.08	3.91	5.49
Sch. Number	80	80	80	80	40	40
DN, mm	100	150	200	250	300	350
Outside Dia., mm	114.3	168.3	219.1	273.0	323.9	355.6
Wall Thickness, mm	6.02	7.11	10.31	12.70	14.27	15.09
Sch. Number	40	40	60	60	60	60

TABLE 4.7
Insulation Thickness for Heat Conservation

DN, mm	Temperature, °C							
	50–125	126–150	151–200	201–250	251–300	301–350	351–400	401–450
20	30	40	55	70	80	80	90	100
25	30	40	55	70	90	90	100	120
40	30	50	60	80	100	120	130	140
50	30	50	60	80	100	120	130	140
80	50	50	70	90	90	110	130	150
100	50	70	100	110	120	130	140	150
150	70	70	100	110	130	140	160	180
200	70	90	100	120	130	140	170	180
250	70	90	110	120	140	160	180	200
300	70	100	110	120	140	160	180	200
350	70	100	120	120	150	160	180	200
400	70	110	120	130	150	180	190	210
500–700	90	110	140	150	160	180	200	210
>700	90	150	170	180	200	200	210	220

protect personnel against burns. Pipes located within a height of 2.1 m and up to 0.6 m outside the platforms should also be insulated. The significance of 2.1 m and 0.6 m is that personnel are liable to come in physical contact with the piping or equipment if they are located within such distances. Table 4.8 illustrates typical insulation thickness for personnel protection. Similarly, Table 4.9 illustrates typical insulation thickness for a cold medium.

TABLE 4.8
Insulation Thickness for Personnel Protection

DN, mm	Temperature, °C		
	60–250	251–450°C	451–550
20	25	40	60
25	25	40	60
40	25	50	70
50	30	50	70
80	30	70	90
100	30	70	90
150	40	100	120
200	40	100	120
250	40	100	140
300	50	120	140
350	50	120	160
400	60	140	160
500–700	60	140	180

TABLE 4.9
Insulation Thickness for a Cold Medium

DN, mm	Temperature, °C			
	+20 to 0	−1 to −30	−31 to −60	−61 to −90
15	30	80	100	140
20	30	80	120	140
25	40	90	120	160
40	40	90	130	160
50	50	100	140	180
80	50	100	150	180
100	60	120	160	200
150	60	120	180	220
200	70	130	200	240
250	70	140	210	260
300	80	150	220	270

4.4 PIPING CONSIDERATIONS FOR EQUIPMENT AND ACCESSORIES

4.4.1 PIPE RACK

The pipe rack acts as the main artery of a process plant and supports process and utility piping as well as electrical and instrument cables. The width of the pipe rack can be adequately estimated on the basis of line sizing of process and utility lines, including insulation requirements provided by the process engineer. It should also include cable tray requirements provided by the instrumentation and electrical engineers. Figure 4.1 shows a typical pipe rack arrangement [1].

While the pipe rack as such does not fall under the purview of a process engineer, it is important to have a bigger picture of a process plant as already explained in the preface of this book. The following are some general guidelines for pipe rack arrangements [2]:

- Inside the process plant, pipe racks are inevitable. However, interconnecting lines between the pipe racks of two units can run at grade level supported on sleepers, unless roads and walkways come on the way.

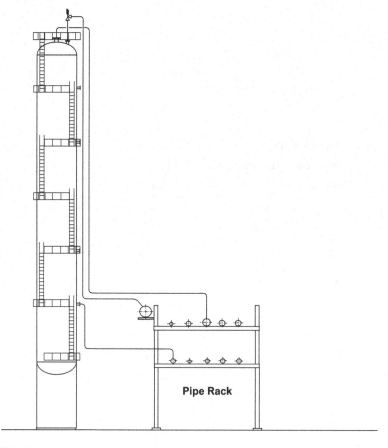

FIGURE 4.1 Typical pipe rack arrangement.

- Large process units require mostly a two-level pipe rack with width in the range of 6 to 12 m. If the total requirement exceeds 24 m, then an extra level is added.
- When using a double-deck pipe rack, it is usual to place the utility and service piping on the upper tier.
- Provide distribution space for future pipes.
- Group hot lines requiring expansion loops at one side for ease of support.

4.4.2 Centrifugal Pumps

Chapter 5 discusses pumps in detail; however, in this section, we will discuss piping aspects. The majority of the pumps in the chemical process industry are of the centrifugal type [2]. Centrifugal pumps have one important requirement: they need to have a flooded suction. To cater to this requirement, piping should be such that bubbles are avoided from entering the pump casing. Suction piping from an overhead suction source should run continuously down and should not at any point rise above the pump suction nozzle. Entry of vapor into the pump casing on account of a poorly routed suction piping may cause cavitation.

Generally, centrifugal pumps are provided with suction and discharge nozzles of diameter large enough to handle the full rated flow [2]. However, based on the densities and viscosities of the fluids handled, and keeping the NPSH in mind, pump suction pipes have invariably been found to be one or two sizes bigger than the suction nozzle. In Chapter 5, line sizing criteria are provided for various applications. It has been seen that once these line sizing criteria are followed, the above guideline on suction line sizes invariably fits in. The guideline also serves as a check for cases where the estimated suction line size becomes too large compared to the size of the suction nozzle [3].

In Chapter 2, a P&ID was presented for a distillation column system (Figure 2.9) where two centrifugal pump assemblies were illustrated, P-101 A/B and P-102 A/B. It can be seen that any centrifugal pump has a suction isolation valve followed by a suction strainer. In addition, a discharge isolation valve is also provided. The discharge piping has a pressure gauge followed by the isolation valve and finally a check valve to prevent possible reverse flow. Figure 4.2 illustrates typical isometrics for a centrifugal pump suction and discharge assembly.

4.4.3 Vessels and Columns

Vertical vessels, including reactors and particularly tray columns, are amongst the most complicated in terms of piping design. In Chapter 9, tray columns are discussed in detail; however, certain important features from the piping point of view are covered below.

The job of a designer here is orienting the column, which includes orientation of the tray downcomers, process nozzles, instrument connections, manholes, ladders and platforms [4].

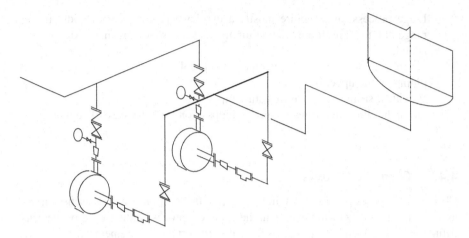

FIGURE 4.2   Suction and discharge line isometrics for a centrifugal pump.

Platforms and Ladders: For horizontal vessels, platforms and ladders are provided if the vessel centerline is located 4.5 m above grade. For columns, platforms are provided at each manhole. Further, additional platforms may need to be provided for access to critical instruments or for operating certain valves. Ladders should be caged if they are more than 6 m above grade [4]. Ladder lengths are restricted to 9–12 m. If platforms are further apart compared to the maximum permissible height of ladder, a small intermediate platform needs to be provided [2]. The reader may refer to the following video link for more on ascending a distillation column, platform-by-platform using the ladder provided.

YouTube link: SDENG - Oil Refinery Tower Ascending - YouTube

Manhole: It is common practice to provide manholes after every 10–12 trays in a tray column. The purpose of manholes is to gain access into the column, first for installation of the internals and thereafter for inspection of internals. Manholes are also typically provided at all feed trays. Thus, a compromise in terms of raising or lowering of manholes is often done to provide the most economical ladder and platform arrangement. Manholes should be oriented such that access is above the column tray and not at the downcomer. Figures 4.3(a) and 4.3(b) show possible locations of manholes for 1-pass and 2-pass trays, respectively.

Process Nozzles: Figure 4.4(a) shows the orientation of the reflux nozzle as well as the details of feeding the reflux for the case of reflux nozzle at an orientation of 90°. At an orientation of 270°, the feeding arrangements are different, i.e., Figure 4.4(b). Similarly, for feed to the column, the nozzle could be located based on piping convenience. However, a suitable feed pipe should direct the feed to the appropriate location on the tray.

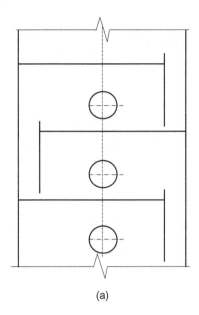 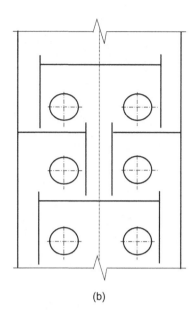

(a)   (b)

FIGURE 4.3 (a) Location of manholes for 1-pass trays. (b) Location of manholes for 2-pass trays.

Reboiler Return Nozzles: Reboiler return nozzles are very critical in distillation columns. The reboiler return nozzle feeds a mixture of liquid and vapor (only vapor in case of kettle type reboilers) back to the column. The liquid drops to the column sump and the vapor travels up through the column.

In a once-through thermosiphon reboiler, the liquid from the last tray flows directly to the reboiler. The two-phase fluid from the reboiler enters the column. The liquid portion of this two-phase fluid drops to the column sump. This liquid portion does not enter the reboiler again and is transferred as column bottoms product. That is why it is called a "once-through reboiler," i.e., there is no intermixing of the fluid streams.

In contrast, in a circulating reboiler, some of the liquid from the reboiler outlet will always recirculate into the reboiler feed [5].

In Chapter 15, Section 15.3, there is an interesting illustration of the importance of reboiler return nozzles. Figure 4.5 shows typical nozzle orientations for reboiler return nozzles.

4.4.4 HEAT EXCHANGERS

Piping for shell-and-tube heat exchangers should be routed to meet both economy as well as access to operation and maintenance. Piping should be such that it allows adequate space for the removal of channel heads and shell covers. Piping connected to channel head nozzles should be furnished with break flanges (Figure 4.6).

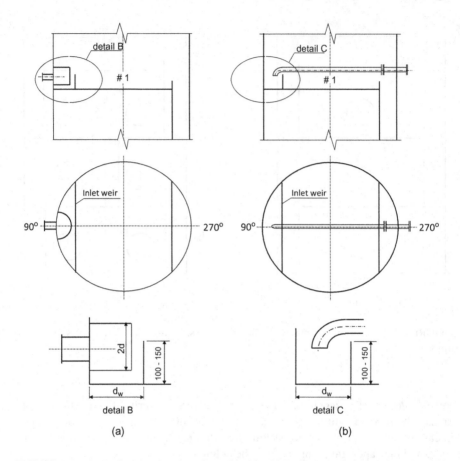

FIGURE 4.4   (a) Reflux nozzle – for orientation 90°. (b) Reflux nozzle – for orientation 270°

Figure 4.7 shows a typical piping arrangement for cooling water connection from an underground system [1].

4.4.5 AIR COOLERS

Increased scarcity of process water today has led to the increase in the use of air coolers in place of shell-and-tube heat exchangers, especially in process plants in cases where process integration is no longer possible. The heat removed by a cross-current steam of air is rejected to the atmosphere. Air coolers are normally supported on the pipe rack and can overhang the pipe rack equally on either side. For example, an air cooler with a bundle length of 12 m can be adequately supported on a pipe rack of 10.5 m width [1].

Piping considerations: Similar to the case with heat exchangers, the lines being cooled in air coolers should flow down. Accordingly, the piping to the air cooler bundles should enter from the nozzle on top of the bundle and exit from the nozzle at the bottom.

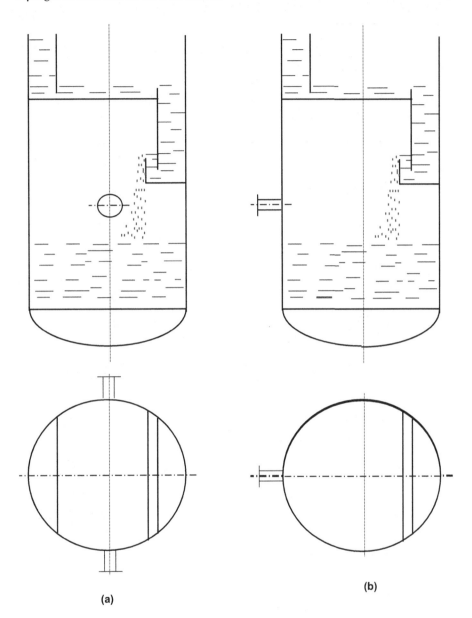

FIGURE 4.5 Nozzle orientations for reboiler return.

Air coolers are also used to condense column overhead vapors. This leads to the existence of two-phase flow, and piping becomes critical. Lines must be routed to ensure that stretch between the column top and the air cooler inlet must be devoid of any pockets. In case of large air coolers with several tube bundles, care must be taken to ensure that piping does not contribute to unequal pressure drop through the bundles [3]. In such cases, the symmetrical piping or the tree type piping becomes

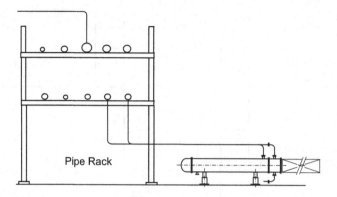

FIGURE 4.6 Piping connections to channel head nozzles.

Source: Adapted from *Process Plant Layout and Piping Design*, Prentice Hall

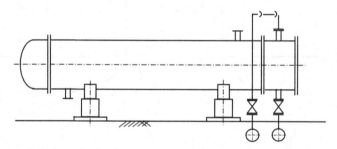

FIGURE 4.7 Cooling water piping.

Source: Adapted from *Process Plant Layout and Piping Design*, Prentice Hall

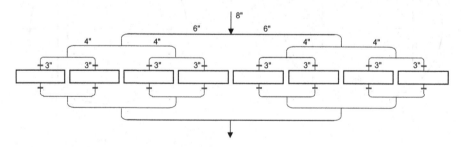

FIGURE 4.8 Tree type piping configuration for air coolers.

necessary (Figure 4.8). However, in cases where the fluid is a liquid, the comb type piping is sufficient (Figure 4.9). In this latter configuration, the branching of pipes is such that equal quantities of fluid flow are ensured through each of the branch pipes.

The nozzle locations of air coolers can have a significant effect on the piping configurations. A single-pass arrangement can make return piping of an overhead condenser cumbersome. If, however, the single-pass arrangement becomes mandatory, the option would be to go for an alternate location of the reflux drum. However, reorienting the air cooler or changing to a two-pass arrangement could greatly improve the piping configuration. Figure 4.10 illustrates a comparison between the

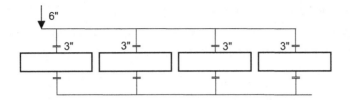

FIGURE 4.9 Comb type piping configuration for air coolers.

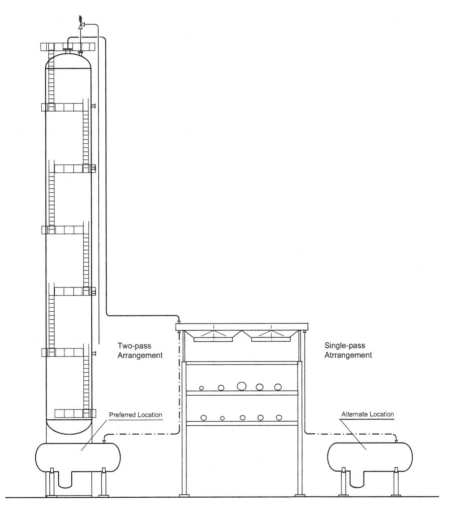

FIGURE 4.10 Air cooler piping.

location of the reflux drum for a single-pass arrangement and a two-pass arrangement. Although the piping engineer does not have the freedom to independently relocate the nozzles, suggestions could be provided to the process engineer in the interest of improving piping arrangements [1].

4.4.6 STEAM TRACING OF PIPING

Steam tracing is one of the commonly used ways of keeping lines warm. Steam tracing is done for the following important reasons:

- Prevent the fluid inside from solidifying due to low ambient temperatures
- Maintain the fluidity of viscous fluids
- Prevent freezing of water-containing process fluids
- Prevent corrosive compounds from forming in case of condensation
- Keep inlet of pressure relief valves free from any solidified material

A steam tracing system typically consists of tracer lines that are fed from a steam header. Each tracer runs through a certain pipe length and finally terminates with a separate steam trap. It is common practice to trace horizontal pipes along the bottom by a single tracer. Multiple-traced pipes with more than two tracers are not normally used. A steam tracing system could be a close system in which the condensate is collected and re-used, or an open system where the condensate is discharged into the drain [2, 6].

Steam Pressures: Steam used for tracing could be of low pressure (LP), medium pressure (MP) or high pressure (HP) with condensing pressures of 150–180°C, 200–270°C and 350–400°C, respectively. If available steam pressure is too high, it is reduced by means of control valve [2, 6].

Steam Tracer Headers: The easiest way to calculate a steam tracer sub-header is to estimate the total cross-sectional area of all the individual tracers and then calculate the header size providing the same cross-sectional area.

Tracer Lengths: The length of the tracer in contact with the pipe is determined by the rate at which condensate forms and fills the line. There are too many variables which determine maximum tracer lengths. Most companies have their own guidelines based on experience [2].

Pipes, Tubes and Fittings for Tracing: Carbon steel pipe, copper or stainless steel tubing are normally used for tracers. The selection is based on the pressure of steam and the required size of tracer. In practice 1/2" or 3/8" size tracers are used. Smaller sizes lead to too much pressure drop, while larger materials do not bend well enough for customary field installations. It has been found that 1/2" copper tubes are the most economical material for tracing straight pipes and can be used for pressures up to 10 kg/cm^{2}g [2].

Figure 4.11 illustrates a typical steam side configuration. Similarly, Figure 4.12 illustrates a typical condensate side configuration.

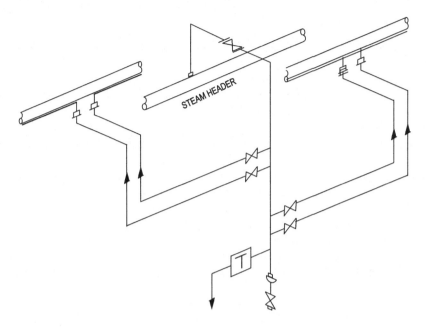

FIGURE 4.11 Steam tracing – stream side configuration.

Source: Adapted from *The Piping Guide – for the Design and Drafting of Industrial Piping Systems*, Syentek Books Company Inc.

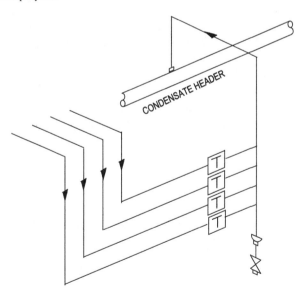

FIGURE 4.12 Steam tracing – condensate side configuration.

Source: Adapted from *The Piping Guide – for the Design and Drafting of Industrial Piping Systems*, Syentek Books Company Inc.

4.4.7 RELIEF SYSTEM PIPING

Pressure relief valves are located to ensure that the inlet piping lengths are short and are self-draining. This, however, does not apply to cases where the inlet pipe comes from a reciprocating compressor in which case the flow is likely to be pulsating and the valve may need to be located in a comparatively stable pressure zone. Pressure relief valves are located at an elevation higher than the respective flare header.

Piping from the outlet of the relief valves should drain towards the unit flare header, so that any liquid condensed on the way is separated in the unit flare knock-out drum. Therefore, the unit flare header is one of the highest located lines in the pipe rack of the respective unit to keep any condensed liquid flowing towards the unit flare knock-out drum. Similarly, the main flare header should drain towards the main flare knock-out drum. A typical outlet piping system for the main relief header is illustrated in Figure 4.13. If for some reason a particular pressure relief valve needs to be lower than the flare header, the piping arrangement should be equipped with a manual drain valve to prevent build-up of liquid downstream of the relief valve. Figure 4.14 illustrates such a configuration.

Further details on relief system piping are discussed in Chapter 12.

4.4.8 UTILITY STATIONS

Utility stations are generally provided around the plant at various locations to supply utility services like plant air, plant water, steam and nitrogen. The number of utility stations shall be such that all equipment are approachable from at least one utility station. Such stations are distributed in a way that any equipment can be supplied with these utilities using a 15 m hose pipe. Such stations may also be provided at elevations and operating platforms to cater to equipment located on structures [7].

Utility stations are normally located adjacent to pipe rack columns for support. Figure 4.15 illustrates a typical utility station provided close to the column of a two-tier pipe rack.

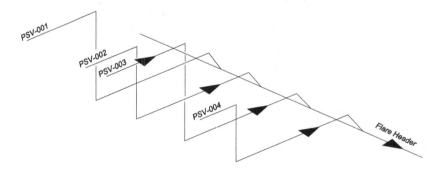

FIGURE 4.13   Main relief header.

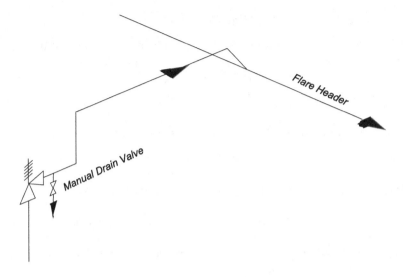

FIGURE 4.14   Relief valve located below the flare header.

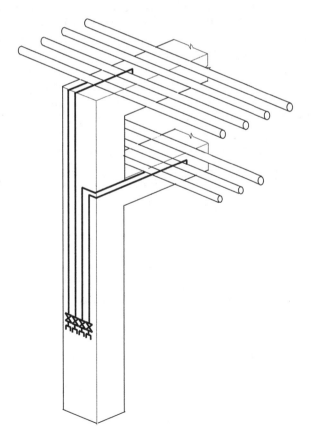

FIGURE 4.15   Utility station.
Source: Adapted from *Process Plant Layout and Piping Design*, Prentice Hall.

4.5   MECHANICAL CONSIDERATIONS

In the course of daily engineering activities, there are two terms that the process engineer comes across quite regularly, i.e., the design pressure and the design temperature. Vessels are mechanically designed for pressures and temperatures higher than those at which they are meant to operate.

4.5.1   DESIGN PRESSURE, DESIGN TEMPERATURE AND MAWP

The process simulation a process engineer carries out specifies pressures and temperatures in various sections of the plant, viz., top of the column, outlet of the heat exchanger, inside the vessel, etc. It should be noted that these are operating conditions. Operating conditions are of interest to the process engineer. However, when the process engineer generates a process data sheet of an equipment and issues it to the mechanical engineer for further engineering, the latter has no specific interest in the operating conditions. The mechanical engineer needs to know the design pressure and the temperature of the equipment.

Design Pressure: Incidentally, the above design conditions are also to be furnished by the process engineer since he is the best person to judge the same. The process engineer adds a suitable margin above the most severe pressure expected in the equipment during normal operation. This pressure is called the "design pressure." This concept is also applicable to piping. Equipment downstream of a centrifugal pump are specified for a design pressure equal to that at the shut-off condition of the pump. This shut-off condition arises if there is a chance that a control valve downstream of the equipment could completely close. Every company has guidelines for such margins.

Table 4.10 illustrates recommended design pressures for various operating pressures. By using pressure relief devices, it is ensured that operating pressures during normal plant operation will not exceed these recommended design pressures. The set pressures of these pressure relief valves should be selected such that they do not exceed the design pressure values.

TABLE 4.10

Recommendations for Design Pressures

Maximum Operating Pressure	Recommended Design Pressure
kg/cm²g	kg/cm²g
<atm	full vacuum
atm	3.5
atm –6	OP x 1.4
6–24	OP + 2.4
24–80	OP x 1.1
> 80	OP + 8

TABLE 4.11
Recommendations for Design Temperatures

Maximum Operating Temperature	Recommended Design Temperature
°C	°C
0–35	65
35–150	OT + 30
150–420	OT + 20
>420	OT + 10

Design Temperature: Similar to design pressure, design temperature is arrived at by adding a suitable margin to the highest operating temperature possible in an equipment. Table 4.11 illustrates the recommended design temperatures for various operating temperatures.

Maximum Allowable Working Pressure: Let us now come to maximum allowable working pressure (MAWP). Suppose for a vessel of a given diameter and for a design pressure of 6 kg/cm^{2}g the mechanical engineer calculates a shell thickness of 9.2 mm. The standard commercially available being 10 mm, the same thickness is selected. It is now apparent that although the vessel is designed for a pressure of 6.0 kg/cm^{2}g it can withstand a *higher pressure* since there is a margin of 0.8 mm on the thickness. This higher pressure is known as the MAWP. Theoretically speaking, pressure relief valves could be set at the MAWP. However, in reality this is rarely the case and the design pressure is considered for all practical purposes. The design pressure is sometimes called the *soft limit* and the MAWP, the *hard limit*. The extra thickness sometimes is useful during revamp scenarios, when revised process simulations may call for slight increase in operating (and hence design pressures) in some cases.

REFERENCES

1. Bausbacher, E. and Hunt, R., *Process Plant Layout and Piping Design, Prentice Hall*, Englewood Cliffs, New Jersey, 1993.
2. Sherwood, D. R. and Wheistance, D. J., *The Piping Guide – for the Design and Drafting of Industrial Piping Systems*, Syentek Books Company Inc., 1991.
3. Weaver, R. *Process Piping Design, Vol. 2*, Gulf Publishing Company, Houston, Texas, 1995.
4. Weaver, R. *Process Piping Design, Vol. 1*, Gulf Publishing Company, Houston, Texas, 1995.
5. Liebermann, N. and, Liebermann, E. T., *A Working Guide to Process Equipment*, McGraw-Hill Inc., 1997.
6. www.enggcyclopedia.com/2012/05/steam-tracing-pipeline/
7. www.pipingengineer.org/utility-stations/

5   Flow of Fluids

5.1   INTRODUCTION

Equipment in a chemical process plant can be broadly classified as:

Performing equipment: include distillation columns, reactors, heat exchangers, air coolers, fired heaters, etc. Each of these equipment performs a duty whether it is to facilitate mass transfer from a gas to liquid and (or) exchange heat between two steams at different temperatures, or facilitate a chemical reaction.

Storage equipment: include tanks and vessels.

Transporting equipment: include pumps, compressors ejectors, etc. The importance of transporting equipment cannot be underestimated because without these the plant would not operate. Whether it is transporting a feed stream to the feed surge drum, routing a column top product to its destination, or conveying column bottoms product to the next column, the importance of pumps needs no further elaboration. Similarly, while conveying gases, e.g., make-up gas to a hydroprocessing unit, or a recycle gas to a reactor, a feed air stream to an air separation unit, compressors do enjoy a dominant position.

While the transporting equipment have their own importance in a process plant, the associated piping, valves and fittings also play a role in completing the loop.

This chapter deals, in detail, first with the flow of fluids in general. It then discusses the various tools used for calculations and guidelines used to facilitate the design. Finally, it discusses the various transport equipment along with solved examples.

5.2   FLOW OF FLUIDS

5.2.1   General Equations

Flow in pipes is always accompanied by friction of fluid particles rubbing against each other, resulting in loss of energy available for work. In other words, there occurs a pressure drop in the direction of flow.

The general equation for pressure drop in cylindrical pipes, running full is given by the Darcy-Weisbach equation written as

$$h_f = f \frac{L}{D} \frac{V^2}{2g} \qquad (5.1)$$

DOI: 10.1201/9780429284656-5

The above equation may be written to express pressure drop in kg/m² as follows:

$$\Delta P = f\rho \frac{L}{D} \frac{V^2}{2g} \tag{5.2}$$

The Darcy-Weisbach equation is valid for laminar or turbulent flow of any liquid in a pipe. In laminar flow, the friction factor f is given by:

$$f = \frac{64}{N_{Re}} \tag{5.3}$$

In turbulent flow, $N_{Re} > 3000$, Colebrook [1] has developed the following correlation for friction factor f:

$$\frac{1}{\sqrt{f}} = -2log\left(\frac{\varepsilon}{3.7D} + \frac{2.51}{N_{Re}\sqrt{f}}\right) \tag{5.4}$$

Colebrook's equation is the basis of Moody's diagram [2], which is the most widely used and accepted data for friction factor for use in the Darcy-Weisbach equation. Moody's diagram (Figure 5.1) presents friction factors as a function of the relative roughness and Reynolds number. Values of relative roughness are also obtained from Figure 5.1.

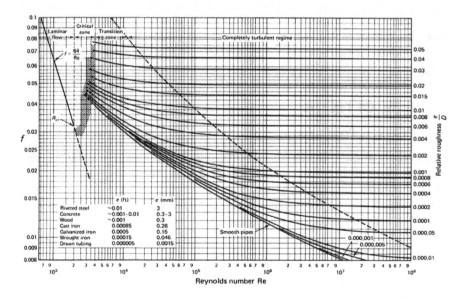

FIGURE 5.1   Moody's diagram.

The Darcy-Weisbach equation is valid for laminar or turbulent flow of any liquid in a pipe. However, when extreme velocities occurring in a pipe cause the downstream pressure to fall to the vapor pressure of the liquid, cavitation occurs and the calculated flowrates will be inaccurate.

When dealing with compressible fluids such as air, steam, etc., certain restrictions should be observed while applying the Darcy-Weisbach formula. If the calculated pressure drop is less than about 10% of the inlet pressure, it could be considered as reasonably accurate if the density used in the formula was based upon either the upstream or downstream conditions. If the calculated pressure drop is in the range of 10–40% of the inlet pressure, the results may be considered reasonably accurate if the density was based on the average of upstream and downstream conditions [3]. For higher pressure drops, other methods are applicable, which are beyond the scope of this book.

Head Losses through Fittings and Valves: In addition to the head loss for flow in a pipe expressed by Equation 5.1, there are additional losses in the pipe due to valves, bends, elbows, etc. There are a number of methods to estimate the additional losses due to fittings. However, the two-K method [4] predicts these losses fairly accurately over a wide range of pipe diameters and Reynolds Numbers. According to this method, the additional losses due to pipe fittings are expressed as:

$$h_f = K \frac{V^2}{2g} \tag{5.5}$$

K is expressed as:

$$K = \frac{K_1}{N_{Re}} + K_\infty \left(1 + \frac{1}{39.37D} \right) \tag{5.6}$$

where
$K_1 = K$ for the fitting at $N_{Re} = 1$
$K_\infty = K$ for a large fitting at $N_{Re} = \infty$

Table 5.1 provides a list of K_1 and K_∞ for various types of fittings.
Equation (5.5) may be written to express pressure drop in kg/m^2 as follows:

$$\Delta P = K\rho \frac{V^2}{2g} \tag{5.7}$$

5.2.2 LIQUID AND GAS PIPE SIZING CRITERIA

In a chemical plant, the capital investment in process piping is in the range of 25–40% of the total plant investment, and the power consumption for pumping, which depends on the pipe size, is a substantial fraction of the total cost of utilities. Accordingly, economic optimization of pipe size is a necessary aspect of plant design. As the diameter of a pipe increases, its cost goes up, but it is accompanied by decreased

TABLE 5.1
Constants for Two-K Method

Fitting Type			K1	K∞
Elbows	90°	Standard (R/D = 1), screwed	800	0.40
		Standard (R/D = 1), flanged/welded	800	0.25
		Long-radius (R/D = 1.5), all types	800	0.20
		Mitered Elbows (R/D=1.5) — 1 Weld (90° angle)	1000	1.15
		2 Weld (45° angle)	800	0.35
		3 Weld (30° angle)	800	0.30
		4 Weld (22½° angle)	800	0.27
		5 Weld (18° angle)	800	0.25
	45°	Standard (R/D = 1), all types	500	0.20
		Long-radius (R/D = 1.5), all types	500	0.15
		Mitered, 1 weld, 45° angle	500	0.25
		Mitered, 2 weld, 22½° angle	500	0.15
	180°	Standard (R/D = 1), screwed	1000	0.60
		Standard (R/D = 1), flanged/welded	1000	0.35
		Long-radius (R/D = 1.5), all types	1000	0.30
Tees	Used as elbow	Standard, screwed	500	0.70
		Long-radius, screwed	800	0.40
		Standard, flanged or welded	800	0.80
		Stub-in-type branch	1000	1.00
	Run through tee	Screwed	200	0.10
		Flanged or welded	150	0.50
		Stub-in-type branch	100	0.00
Valves	Gate Ball Plug	Full line size, ß = 1.0	300	0.10
		Reduced trim, ß = 0.9	500	0.15
		Reduced trim, ß = 0.8	1000	0.25
	Globe, standard		1500	4.00
	Globe, angle or Y-type		1000	2.00
	Diaphragm, dam type		1000	2.00
	Butterfly		800	0.25
	Check	Lift	2000	10.00
		Swing	1500	1.50
		Tilting-disk	1000	0.50

Note: Use R/D = 1.5 values for R/D = 5 pipe bends, 45 to 180°. Use appropriate tee values for flow through crosses.

Source: Reprinted by special permission from *CHEMICAL ENGINEERING*, Copyright© August 1981, by Access Intelligence, Rockville, MD 20850.

electrical power consumption because of a decrease in the frictional losses. Somewhere, there is an optimum balance between operating cost and capital cost.

For small capacities and short lines, near optimum pipe sizes may be obtained on the basis of typical velocities or pressure drops detailed in Tables 5.2 through 5.3. Where large capacities are involved, pipes are long, or expensive materials are needed, the selection of pipe sizes may need elaborate economic analysis. In addition, when handling highly viscous materials, it may be worth considering heating the material to reduce the viscosity since this may lead to a reduced power requirement for pumping.

TABLE 5.2
Pipe Sizing Criteria for Liquids

Service	Pipe Size, mm	Velocity, m/s	Pressure Drop, kg/cm²/km	
			Minimum	Maximum
Pump Suction, at bubble point	≤50	0.3–0.6		
	80–150	0.4–0.8		
	200–250	0.7–1.0	0.1	0.6
	300–450	0.9–1.2		
	≥500	1.4–2.5		
Pump Suction, sub-cooled	≤50	0.4–0.8		
	80–150	0.6–1.1		
	200–250	1.0–1.5	0.5	2
	300–450	1.2–2.0		
	≥500	1.6–3.0		
Pump Discharge	≤50	0.8–1.5		
	80–150	1.0–2.5		
	200–250	1.5–3.0	2	5.0
	300–450	2.0–4.0		
	≥500	3.0–4.0		
Cooling Water		2.0–4.0	2.3	3.5
Gravity Flow		0.3–0.6	0.3	0.5
Slurry Flow	≥50	1.0–2.7		

TABLE 5.3
Pipe Sizing Criteria for Gases and Steam

Service	Pipe Size, mm	Velocity, m/s	Pressure Drop, kg/cm²km Recommended
Gases, General			
vacuum		40–60	0.1–0.6
atm - 7 kg/cm²g		20–50	0.6–2.0
7–70 kg/cm²g		15–40	2.0–5.0
>70 kg/cm²g		10–35	7.0% of operating pressure
Steam (saturated)		10–60	1.2–3.8
Steam (superheated)		15–70	2.0–4.8
Flare Header		0.5 Mach	

5.3   PUMPS

The two main categories of pumps as defined by their basic principle of operation are dynamic pumps and positive displacement pumps. While, in dynamic pumps, energy is converted by hydrodynamic means, positive displacements pumps generate

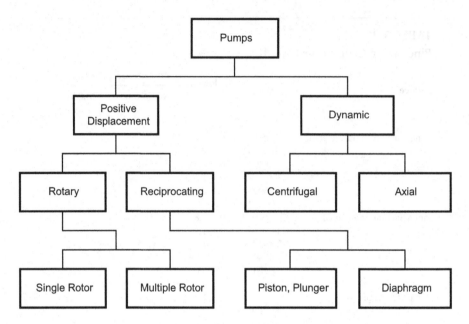

FIGURE 5.2 Classification of pumps.

pressure hydrostatically. Pumps are further classified under various sub-categories. Figure 5.2 provides detailed classifications of these two categories of pumps.

5.3.1 SELECTION OF PUMP TYPE

Except for special purpose pumps for specific services, most of the process pumps are made in standard sizes, and hence the main problem is to select the size and type that most nearly fits the service in question. Though the final selection of the pump will be made in close cooperation between the vendor and the design engineer, a preliminary selection of the type of pump required is made by the process engineer. Selection of the pump type can be made on the basis of head-capacity requirements or fluid properties, e.g., viscosity, solid content and corrosive or erosive nature. Figure 5.3 can be used for section type based on head-capacity requirements.

Centrifugal pumps are the most widely used pumps in the process industry and are suitable for conveying various types of liquids. They are available in capacities ranging from 0.5 to 20,000 m^3/hr, and for discharge heads of a few meters to 480 kg/cm². While the head developed by centrifugal pumps is determined by the speed at which it rotates, in the case of positive-displacement pumps, the head will be impressed upon them by restrictions in the downstream piping. The maximum head attainable by a positive-displacement pump will therefore depend upon the power of the driver provided the pump parts have the strength to bear that head. Positive-displacement pumps can be either rotary or reciprocating [5]. Further details of pumps are provided in the subsequent sections.

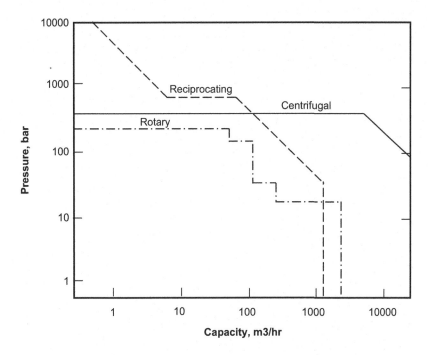

FIGURE 5.3   Limits of pressure and capacity for various types of pumps.

Source: Reproduced from *Pump Handbook,* 3rd Edition, with permission from Mc-Graw Hill Inc.

5.3.2   DEFINITION OF TERMS

Pump Suction Pressure: This is the absolute pressure at the liquid surface h_s (also known as "source pressure"), plus the static head at the suction flange of the pump h_{st}, minus the friction losses on the suction side h_{sf} (Figure 5.4):

$$H_s = h_s + h_{st} - h_{sf} \tag{5.8}$$

Pump Discharge Pressure: This is the absolute pressure at the liquid surface h_d (also known as "destination pressure"), plus the static head at the discharge flange of the pump h_{dt}, plus the friction losses on the discharge side h_{df}:

$$H_d = h_d + h_{dt} + h_{df} \tag{5.9}$$

Pump Differential Pressure: The differential head of a pump H is the total discharge head H_d minus the total suction head H_s:

$$H = H_d - H_s \tag{5.10}$$

5.3.3   SUCTION CONDITIONS AND NPSH

Improper suction conditions are the largest source of pump troubles. Careful attention should be given to net positive suction head (NPSH). NPSH is the net

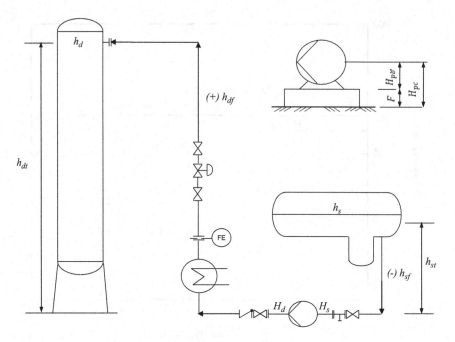

FIGURE 5.4   Pump head calculation sketch.

remaining pressure at the suction flange of the pump after all negative forces that restrict liquid from getting into the pump are subtracted from all the positives forces that assist liquid in getting into the pump. Two terms of NPSH are referred to as follows:

NPSH$_a$: Net positive suction head available in the system expressed as meters of liquid.
NPSH$_r$: Net positive suction head required by the pump expressed as meters of liquid.

NPSH$_a$ is calculated as:

  NPSH$_a$ = Terminal pressure in the vessel (in gauge)

(+) Static head of fluid above pump centerline (see note below).
(+) Atmospheric pressure
(-) Vapor pressure of liquid at pumping temperature
(-) Friction loss in suction piping up to pump centerline consisting of the following:

  1. Entrance and exit losses
  2. Loss in suction strainer
  3. Loss in control valves, instruments, exchangers, etc., if any
  4. Line losses

Note:

1. Height of liquid in the vessel should be taken to be at the vessel bottom tangent line.
2. Height of the pump centerline is calculated as follows:

$$H_{pc} = H_{pb} + F$$

where

H_{pc} = Height of pump centerline above finished floor level
H_{pb} = Height of pump centerline above pump baseplate
F = Height of pump foundation above finished floor level

Normally, as per standard practice, F is 300 mm.

H_{pb} is estimated from vendor information. In the absence of vendor data, the following can be assumed as a first guess:

For pumps having rated flow of <200 m³/hr, H_{pb} = 400 mm
For pumps having rated flow of 200–300 m³/hr, H_{pb} = 600 mm
For pumps having rated flow of 300–400 m³/hr, H_{pb} = 700 mm

Calculate $NPSH_a$ carefully considering all conditions, i.e., start-up, original fill of lines, winter operation, all control valve pressure drops, summer operation, piping conditions, exit and entrance losses. Provide $NPSH_a$ at 1 meter over worse conceived $NPSH_r$ curve by the manufacturer. In case of submersible pumps, the available submergence should also be considered.

NPSH is a function of flow; therefore, it should always be determined at rated capacity regardless of the total head required. Quite often the $NPSH_r$ curve turns upwards at low flows. Figure 5.7 illustrates a typical case. Low flow effects can often be amplified by selecting too large a pump which forces recirculation within the pump.

When liquids at their bubble points are pumped from closed vessels, $NPSH_a$ is only the static liquid head above the pump center minus the friction losses in the suction piping. This is due to the fact that the vapor pressure equals the vessel pressure. In such cases, it is usual to elevate the vessels suitably (or sometimes cool the liquid before it enters the pump) to get the margin of $NPSH_a$ over $NPSH_r$. Sometimes, booster pumps which have low head and high capacity pumps and also require a low NPSH could be used upstream of the main pumps to improve the $NPSH_a$ for the main pump which has high capacity and high head requirements. However, such cases are rarely seen in practice.

5.4   CENTRIFUGAL PUMPS

Centrifugal pumps are the most widely used pumps in the industry for transferring all types of liquids, viz., inorganic and organic liquids, volatile fluids, slurries, viscous liquids, water supply, boiler feed water circulation, etc. These pumps are available for a wide range of sizes and capacities.

The primary advantages of a centrifugal pump are simplicity, low first cost, non-pulsating flow, small floor space, low maintenance cost, quiet operation and adaptability for use with a motor or a turbine drive [5].

5.4.1 CENTRIFUGAL PUMP CHARACTERISTICS

Figures 5.5, 5.6 and 5.7 illustrate typical characteristic curves of centrifugal pumps for different impeller diameters. The nature of head-capacity curve, efficiency and NPSH curves can be different depending upon the type of pump. At a given speed, the pump will operate along this curve and at no other point. Figure 5.8 shows a typical characteristic curve of a centrifugal pump at different speeds.

It is important to note that the head (in terms of meters of liquid) will be the same for any clean liquid of the same viscosity. The pressure (in terms of kg/cm^2) developed will however vary in proportion to the specific gravity of the liquid.

The pump model should be so chosen that the rated point falls close to the best efficiency point. This will be demonstrated subsequently in an example later in the chapter.

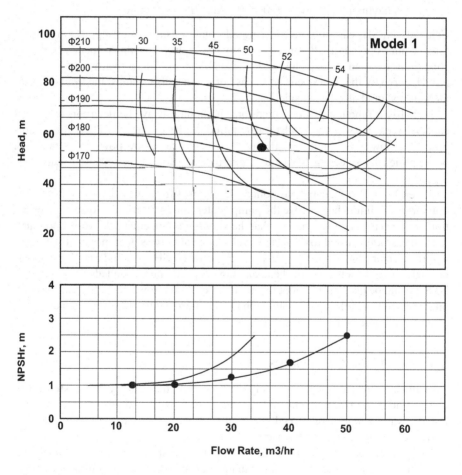

FIGURE 5.5 Pump characteristic curve (Model 1).

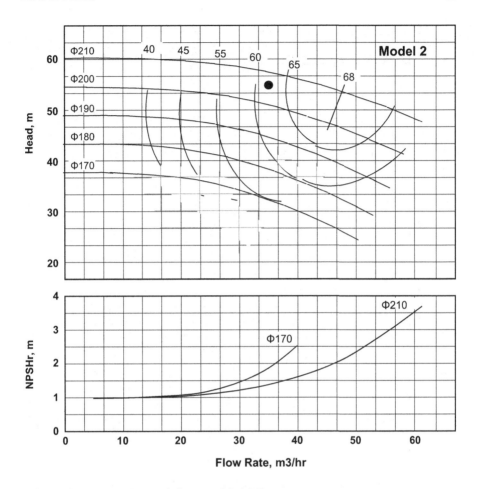

FIGURE 5.6 Pump characteristic curve (Model 2).

Manufacturers supply characteristic curves of pumps based on water as test fluid used in the testing shop. While handling viscous fluids in actual practice, certain corrections need to be made for viscosity to get the actual head, capacity and efficiency. Charts are available for making these corrections. A detailed discussion on these charts is beyond the scope of this book.

5.4.2 CENTRIFUGAL PUMP HEAD CALCULATIONS

Pump head calculations are carried out in the following steps:

1. The system sketch is drawn in the pump calculation sheet
2. Calculations are to be carried out for all the destinations to which the pump delivers
3. Fouled condition of piping/equipment is considered
4. Computations are carried out for the maximum flow Q_{max} for each case

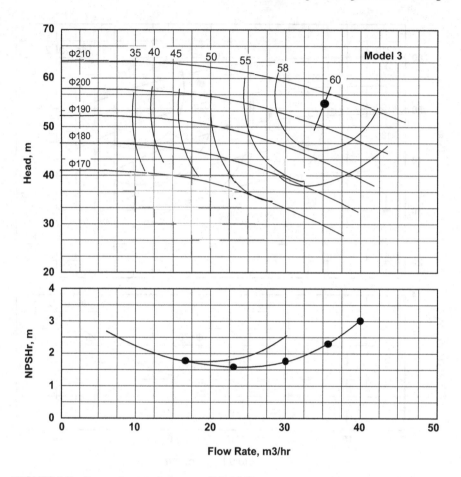

FIGURE 5.7 Pump characteristic curve (Model 3).

5. Computations are carried out by the Pump Calculation Program (nowadays manual calculations are rarely carried out) which reproduces data in the form of a pump calculation sheet (refer to Table 5.4)
6. Process specifications are reproduced in the form of a pump process data sheet illustrated in Table 5.5
7. The differential head so calculated at the maximum flow is specified in pump process data sheet as differential head at Q_{max}

A centrifugal pump will produce high pressures on the discharge side under the following conditions:

1. Downstream blockage which leads to the pump shut-off condition
2. Upstream upset when the suction vessel may see the design pressure.

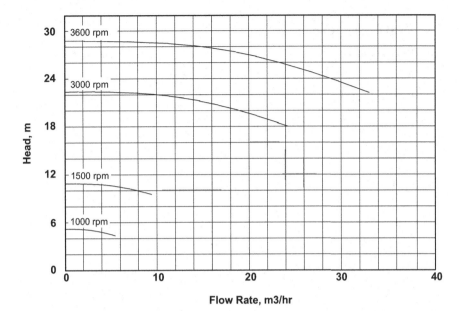

FIGURE 5.8   Centrifugal pump characteristics at different speeds.

TABLE 5.4
Pump Calculation Sheet

Suction Side

Parameter	Unit	Min	Nor	Max
Flow	m³/hr	15	30	36
Density	kg/m³	969	969	969
Viscosity	m-Pas	0.334	0.334	0.334
Source Pressure	kg/cm²a	1.033	1.033	1.033
Vapor Pressure	kg/cm²	0.57	0.57	0.57
Static Height	m	2.0	2.0	2.0
Frictional Losses	kg/cm²	0.01	0.06	0.08
Static Pressure		1.22	1.17	1.15
NPSH$_a$	m			5.95

Discharge Side		Destination 1			Destination 2		
Parameter	**Unit**	**Min**	**Nor**	**Max**	**Min**	**Nor**	**Max**
Flow	m³/hr	10	20	24	5	10	12
Static Height	m³/hr	30	30	30	15	15	15
Static Pressure	kg/cm²	2.78	2.78	2.78	1.35	1.35	1.35
Destination Pressure	kg/cm²	4.0	4.0	4.0	3.0	3.0	3.0
Line Losses	kg/cm²	0.05	0.19	0.28	0.10	0.38	0.55
Flowmeter Losses	kg/cm²	0.03	0.14	0.20	0.03	0.14	0.20
Control Loss	kg/cm²	2.59	1.23	0.70	4.97	3.47	2.86
Discharge Pressure	kg/cm²a	8.48	8.29	7.96	9.45	8.34	7.96
Differential Pressure	kg/cm²	7.27	7.12	6.81	8.24	7.17	6.81

(Continued)

TABLE 5.4   (Continued)

Discharge Side		Destination 1			Destination 2		
Parameter	Unit	Min	Nor	Max	Min	Nor	Max
Differential Head	m	75	73.5	70.3	75	73.5	70.3

Power Calculations

Parameter	Unit	Max
Flow	m³/hr	36
Pump Efficiency	%	51
Shaft Power	kW	13.1
Motor Rating	kW	16.4
Standard Motor Rating		18.5
Motor Efficiency	%	90
Power Consumption	kW	14.6

Note: Min, Nor, Max stand for Minimum, Normal and Maximum flow respectively.

TABLE 5.5
Centrifugal Pump – Process Data Sheet

Tag No.		P-101
Quantity		2 (1 Opn + 1 Std.by)
Service		Water
P&ID No.		40054-03-006
Line No.		4"-P-A01P-010
Pumping temperature	°C	75
Density of liquid	kg/m³	992
Dynamic viscosity	cP	0.8
Vapor pressure	kg/cm²a	0.37
Corrosive components		None
Rated condition definition		Normal flow
Maximum flow	m³/hr	43
Normal flow	m³/hr	36
Minimum flow	m³/hr	18
Differential pressure at Q_{rated}	kg/cm²	5.52
Differential head at Q_{rated}	m	55.6
NPSH available	m	4.3
Material casing		CI
Material impeller		CS

Hence, all piping, valves, exchangers and vessels downstream of the pump should be designed for the greater of the following pressures:

1. Suction vessel operating pressure + normal liquid static head + 1.25 times the pump differential pressure (pump shut-off condition)
2. Suction vessel design pressure + maximum liquid static head + pump differential pressure (upstream upset condition).

***

EXAMPLE 5.1   HEAD CALCULATIONS AT NORMAL FLOW

A centrifugal pump delivers water from a feed surge drum to the top of a water wash column via a heat exchanger. Refer to Figure 5.9 for the vessel elevations. Consider pressure drops of 1.0 and 0.5 kg/cm^2 for the control valve and the heat exchanger, respectively. The process data are as follows:

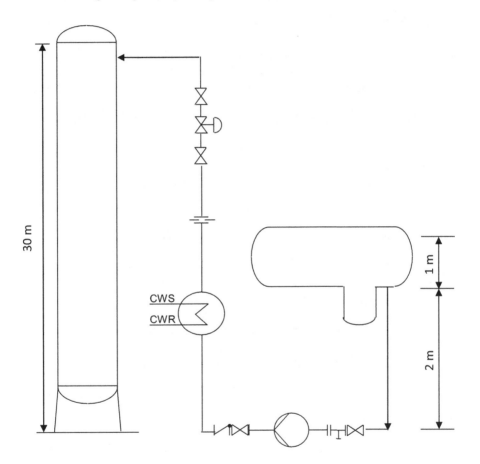

FIGURE 5.9   Pump head calculation sketch (Examples 5.1 and 5.2).

Fluid Properties

Fluid		Water
Temperature	°C	30
Flowrate	m³/hr	36.0
Density	kg/m³	992.0
Viscosity	cP	0.7
Vapor Pressure at 30°C	kg/cm²a	0.042

Suction Conditions

Static Height	m	2.00
Liquid Level in Reflux Drum	m	1.00
Pressure on Suction Drum	kg/cm²g	0.00
Suction Pipe Size	m	0.15
Straight Length of Suction Piping	m	3.00
Bends		2
Gate Valves		1
Strainers		1

Discharge Conditions

Static Height	m	30.00
Pressure on Column Top	kg/cm²g	1.00
Discharge Pipe Size	m	0.10
Straight length of Discharge Piping	m	40.00
Bends		3
Gate Valves		2
Control Valves		1
Check Valves (swing type)		1

Suction Side Calculations

Suction vessel pressure, h_s	0.0 kg/cm²g + 1.033 = 1.033 kg/cm²a
Suction static head, h_{st}	((2.0 + 1.0) × 992) kg/m
	= 2976 kg/m²a
	= 0.298 kg/cm²a
Velocity	(36 × 4) / (3600 × 3.142 × 0.15 × 0.15)
	= 0.565 m/s
Reynolds number	0.150 × 0.565 × 992 / 0.0007
	= 120103
Roughness	0.0457 mm
Friction factor (Equation (5.4))	0.0189
For bends	$K_1 = 800$ (refer to Table 5.1)
	$K_\infty = 0.25$ (refer to Table 5.1)

From Equation (5.6),

$$K = \frac{800}{120103} + 0.25 \left(1 + \frac{1}{39.37 \times 0.15}\right)$$

$$= 0.2990$$

For gate valves $\quad K = 300$ (refer to Table 5.1)
$K_\infty = 0.10$ (refer to Table 5.1)

From Equation (5.6)

$$K = \frac{300}{120103} + 0.10 \left(1 + \frac{1}{39.37 \times 0.15}\right)$$

$$= 0.1194$$

Thus, overall K for fittings $\quad (2 \times 0.2990) + (1 \times 0.1194)$
$= 0.7174$

Pressure drop through fittings
(Equation (5.7))

$$= K\rho \frac{V^2}{2g} = 0.7174 \times 992 \frac{0.565^2}{2 \times 9.81}$$

$$= 11.58 \text{ kg/m}^2$$
$$= 0.0012 \text{ kg/cm}^2$$

Pressure drop through straight pipe (Equation 5.2)	$= f\rho \dfrac{L}{D} \dfrac{V^2}{2g} = 0.0189 \times 992 \times \dfrac{3.0}{0.15} \dfrac{0.565^2}{2 \times 9.81}$
	$= 6.10 \text{ kg/m}^2$ $= 0.0006 \text{ kg/cm}^2$
Frictional line losses	$= 0.0012 + 0.0006$ $= 0.0018 \text{ kg/cm}^2$
Pressure drop through suction strainer	0.05 kg/cm^2
Total friction losses, h_{sf}	$(0.0018 + 0.05) \text{ kg/cm}^2$ $= 0.0518 \text{ kg/cm}^2$
Pump suction pressure, H_s	$(1.033 + 0.298 - 0.0518) \text{ kg/cm}^2\text{a}$ $= 1.279 \text{ kg/cm}^2\text{a}$

Discharge Side Calculations

Destination vessel pressure, h_d	$1.0 \text{ kg/cm}^2\text{g} + 1.033 = 2.033 \text{ kg/cm}^2\text{a}$
Discharge static head, h_{dt}	$(30 \times 992) \text{ kg/m}^2$ $= 29760 \text{ kg/m}^2$ $= 2.98 \text{ kg/cm}^2$
Velocity	$(36 \times 4) / (3600 \times 3.142 \times 0.10 \times 0.10)$ $= 1.273 \text{ m/s}$
Reynold's Number	$0.100 \times 1.273 \times 992 / 0.0007$ $= 180402$
Roughness	0.0457 mm
Friction Factor (Equation (5.4))	0.0188
For bends	$K_1 = 800$ (refer to Table 5.1) $K_\infty = 0.25$ (refer to Table 5.1)
From Equation (5.6),	$K = \dfrac{800}{180402} + 0.25 \left(1 + \dfrac{1}{39.37 \times 0.10}\right)$
	$= 0.3179$
For gate valves,	$K_1 = 300$ (refer to Table 5.1) $K_\infty = 0.1$ (refer to Table 5.1)
From Equation (5.6),	$K = \dfrac{300}{180402} + 0.10 \left(1 + \dfrac{1}{39.37 \times 0.10}\right)$
	$= 0.1271$
For check valve,	$K_1 = 1500$ $K_\infty = 1.5$
From Equation (5.6)	$K = \dfrac{1500}{180402} + 1.5 \left(1 + \dfrac{1}{39.37 \times 0.10}\right)$
	$= 1.8893$
Thus, overall K for all fittings	$((3 \times 0.3179) + (2 \times 0.1271) + (1 \times 1.8893))$ $= 3.0972$

Pressure drop through fittings (Equation (5.7))	$= K\rho \dfrac{V^2}{2g} = 3.0972 \times 992 \dfrac{1.273^2}{2 \times 9.81}$
	$= 253.77 \text{ kg/m}^2$
	$= 0.0254 \text{ kg/cm}^2$
Pressure drop through straight pipe (Equation 5.2)	$= f\rho \dfrac{L}{D} \dfrac{V^2}{2g} = 0.0188 \times 992 \times \dfrac{40.0}{0.10} \dfrac{1.273^2}{2 \times 9.81}$
	$= 616.15 \text{ kg/m}^2$
	$= 0.0616 \text{ kg/cm}^2$
Frictional line losses	$0.0254 + 0.0616$
	$= 0.0870 \text{ kg/cm}^2$
Pressure drop through exchanger	0.5 kg/cm^2
Pressure drop through orifice	0.2 kg/cm^2
Total friction losses, h_{df}	$(0.0870 + 0.5 + 0.2) \text{ kg/cm}^2$
	$= 0.787 \text{ kg/cm}^2$
Pressure drop through control valve	1.0 kg/cm^2
Pump discharge pressure, H_d	$(2.033 + 2.98 + 1.0 + 0.787) \text{ kg/cm}^2 \text{a}$
	$= 6.80 \text{ kg/cm}^2 \text{a}$
Pump differential pressure, H	$(6.80 - 1.279) \text{ kg/cm}^2$
	$= 5.52 \text{ kg/cm}^2$
	$= 5.52 \times 10000 / 992$
	$= 55.6 \text{ m}$

***

5.4.3 CHOOSING THE PUMP MODEL

The normal flow of the pump in this example is a flow of 36 m³/hr and a differential head of 55.6 m of liquid. This is the flow at which the pump is expected to operate for most of the time. Now, refer to the three pump models, i.e., Figures 5.5 (Model 1), 5.6 (Model 2) and 5.7 (Model 3). For Model 1, the desired operating point is achieved with an impeller diameter of 183 mm at a pump efficiency of 49%. For Model 2, the condition is achieved with an impeller diameter of 206 mm at a pump efficiency of 62%. For Model 3, the requisites are 207 mm impeller diameter at 60% pump efficiency. We therefore choose Model 2 since this gives the highest efficiency at normal flow, i.e., the flow in which the pump is expected to operate most of the time. Figure 5.10 shows the selected pump model (Model 2) and the pump curve at an impeller diameter of 206 mm. For the selected pump model, the differential pressures are summarized as follows:

59.5 m liquid (5.90 kg/cm²) at a flowrate of 18 m³/hr
55.6 m liquid (5.52 kg/cm²) at a flowrate of 36 m³/hr
53.0 m liquid (5.26 kg/cm²) at a flowrate of 43 m³/hr

Now, let us work out the calculations for maximum flow and minimum flow.

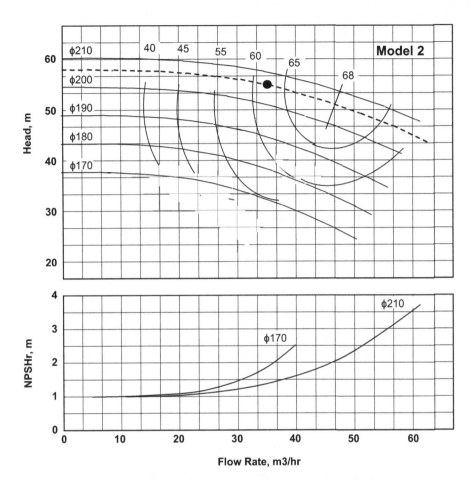

FIGURE 5.10   Pump characteristic curve – the selected pump model.

***

EXAMPLE 5.2   HEAD CALCULATIONS AT MAXIMUM FLOW

The maximum process flow for the pump illustrated in Example 5.1 is 43 m³/hr. The over- design of 20% takes care of an inherent margin in the production capacity of the plant. The pressure drop through the heat exchanger at maximum flow is 0.72 kg/cm² and that through the strainer is 0.072 kg/cm². Calculate the head loss through the control valve.

Calculations for maximum flow are summarized below and are similar to those of normal flow illustrated in Example 5.1 above. Since the procedure is repetitive, only the results are reproduced.

Suction Side Calculations

Suction vessel pressure, h_s 0.0 kg/cm²g + 1.033 = 1.033 kg/cm²a
Suction static head, h_{st} 0.298 kg/cm²

Friction line losses	0.0026 kg/cm^2
Pressure drop through strainer	0.07 kg/cm^2
Total friction losses, h_{sf}	0.0726 kg/cm^2
Pump suction pressure, H_s	1.258 kg/cm^{2}a

Discharge Side Calculations

Destination vessel pressure, h_d	1.0 kg/cm^{2}g + 1.033 = 2.033 kg/cm^{2}a
Discharge static head, h_{dt}	2.98 kg/cm^2
Friction line losses,	0.124 kg/cm^2
Pressure drop through exchanger	0.72 kg/cm^2
Pressure drop through orifice	0.29 kg/cm^2
Total friction losses, h_{df}	1.134 kg/cm^2

Pump differential pressure, H

= 5.26 kg/cm^2
= 53 m liquid
(from pump Model 2 at 43.2 m^3/hr and impeller diameter of 206 mm)

Pump discharge pressure, H_d

= 5.26 kg/cm^2 + 1.258 kg/cm^{2}a
= 6.52 kg/cm^{2}a

Pressure drop through control valve

6.52-2.033-2.98-0.124-0.72-0.29

= 0.373 kg/cm^2

***

EXAMPLE 5.3   HEAD CALCULATIONS AT MINIMUM FLOW

Calculations for minimum flow are summarized below and are similar to those of normal and maximum flows illustrated in Examples 5.1 and 5.2, respectively, above. Since the procedure is repetitive, only the results are reproduced.

Suction Side Calculations

Suction vessel pressure, h_s	0.0 kg/cm^{2}g + 1.033 = 1.033 kg/cm^{2}a
Suction static head, h_{st}	0.298 kg/cm^2
Friction line losses	0.0005 kg/cm^2
Pressure drop through strainer	0.0125 kg/cm^2
Total friction losses, h_{sf}	0.0130 kg/cm^2
Pump suction pressure, H_s	1.318 kg/cm^{2}a

Discharge Side Calculations

| Destination vessel pressure, h_d | 1.0 kg/cm^{2}g + 1.033 = 2.033 kg/cm^{2}a |
| Discharge static head, h_{dt} | 2.98 kg/cm^2 |

Friction line losses	0.021 kg/cm²
Pressure drop through exchanger	0.125 kg/cm²
Pressure drop through orifice	0.05 kg/cm²
Total friction losses, h_{df}	0.1960 kg/cm²
Pump differential pressure	5.90 kg/cm² = 59.5 m liquid (from pump Model 2 at 18 m³/hr and impeller diameter of 206 mm)
Pump discharge pressure, H_d	5.90+1.318 kg/cm² = 7.22 kg/cm²a
Pressure drop through control valve	7.22-2.033-2.98-0.021-0.125-0.05 = 2.011 kg/cm²

***

5.4.4 ANALYSIS OF THE DATA

Table 5.6 summarizes the above three cases. The first three columns provide data in kg/cm². The last three columns provide the same in meters of liquid column.

Table 5.7 represents a summary of the calculations for the differential pressure of the pump which has three components. The analysis is as follows:

Pump suction pressure =	Suction vessel pressure + Suction static head – Friction losses (line + strainer)
Pump discharge pressure =	Destination vessel pressure + Discharge static head + Friction losses (line + HE + FE) + Control valve pressure drop
Pump differential pressure =	Destination vessel pressure – Suction vessel pressure + Discharge static head – Suction static head + Variable friction losses (discharge line losses + suction line losses + pressure drop across heat exchanger + pressure drop across orifice + pressure drop across strainer) + control valve losses

or

Pump differential pressure =	Fixed portion losses + Variable portion losses + Control valve losses

Figure 5.11 illustrates the results. There are three operating points at flowrates of 18, 36 and 43 m³/hr. It may be noted that irrespective of the flowrate the fixed portion is a straight line, since the parameters of suction and discharge vessel pressures as

TABLE 5.6
Analysis of Calculations in Examples 5.1, 5.2 and 5.3

	Min	Nor	Max	Min	Nor	Max
Liquid flowrate, m³/hr	18	36	43	18	36	43
Pressure	kg/cm²a	kg/cm²a	kg/cm²a	m, liquid	m, liquid	m, liquid
Source pressure	1.033	1.033	1.033	10.41	10.41	10.41
Suction static head	0.298	0.298	0.298	3.00	3.00	3.00
Frictional line loss (suction)	0.0005	0.0018	0.0026	0.005	0.018	0.026
Pressure drop across strainer	0.0125	0.050	0.070	0.126	0.504	0.706
Total Suction Pressure	**1.318**	**1.279**	**1.258**	**13.3**	**12.9**	**12.7**
Destination pressure	2.033	2.033	2.033	20.5	20.5	20.5
Discharge head	2.98	2.98	2.98	30.0	30.0	30.0
Discharge FE loss	0.050	0.20	0.290	0.50	2.02	2.92
Discharge CV loss	2.01	1.00	0.37	20.3	10.1	3.8
Discharge HE loss	0.125	0.500	0.720	1.260	5.040	7.258
Discharge line loss	0.021	0.087	0.124	0.21	0.88	1.25
Total Discharge Pressure	**7.22**	**6.80**	**6.52**	**72.8**	**68.5**	**65.7**
Differential Pressure	**5.90**	**5.52**	**5.26**	**59.5**	**55.6**	**53.0**

TABLE 5.7
Summary of Calculations

Flowrate	m3/hr	18	36	43
Fixed Portion Losses	m	37.09	37.09	37.09
Variable Portion Losses	m	2.11	8.46	12.16
Line losses	m	0.341	1.402	1.982
Pressure drop across heat exchanger	m	1.26	5.04	7.258
Pressure drop across flow meter	m	0.50	2.02	2.92
Control Valve Losses	m	20.3	10.1	3.8
Pump Differential Head	m	59.5	55.6	53.0

well as suction and discharge heads are fixed values irrespective of the flowrates. Further, this portion is calculated based on the difference between the discharge vis-à-vis suction conditions, since the suction conditions of source pressure and static head contribute to reducing the total head.

The friction losses however are additive, since irrespective of elements being located either on the suction or on discharge lines, the pump has to overcome these losses. The frictional losses follow a curve that increases with increasing flowrate. This in line with the theory that the frictional losses vary as the square of the flowrate.

What is interesting is that the control valve losses follow a reverse trend, i.e., the losses decrease with increasing flowrate. This phenomenon is explained as follows:

At the normal flowrate of 36 m³/hr, at a pump differential head of 55.6 m of liquid, we have selected a control valve loss of 1 kg/cm² (i.e., 10.1 m of liquid). We had

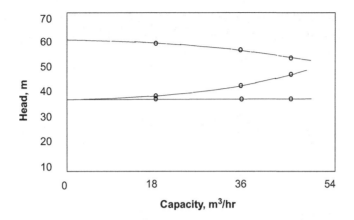

FIGURE 5.11   Fitting data into the pump curve.

selected Model 2. Now, at the maximum flowrate of 43 m³/hr, the differential head is
53.0 m liquid. Out of this, the fixed portion is 37.09 m. The variable losses constitute
12.16 m. This means that the remaining portion of 3.8 m of liquid has to be absorbed
by the control valve. The control valve will thus open to such an extent that the losses
across the valve would constitute 3.8 m. By a similar analogy, a loss of 20.3 m would
occur across the control valve at the minimum flowrate of 18 m³/hr.

This leads to the so-called "slogan" for control valves, i.e., high flow --> low
pressure drop, low flow --> high pressure drop.

5.4.5   ANALYSIS OF NPSH

Liquid temperature well below its boiling point: In the example illustrated above, let
us analyze the NPSH requirements. Let us consider the maximum flow case which
gives the lowest $NPSH_a$ value because of higher suction side frictional losses. The
liquid being water, and the temperature 30°C, is therefore well below its boiling
point.

We have the following data:

Pump suction pressure   = 1.258 kg/cm²a
                          = 1.258 × 10,000 / 992 = 12.68 m of liquid

Vapor pressure (at 30°C) = 0.042 kg/cm²a
                          = 0.042 × 10,000 / 992 = 0.42 m of liquid

Available $NPSH_a$ is   12.68 m – 0.42 m = 12.26 m

The selected pump model is Model 2 (Figure 5.10).

The $NPSH_r$ for the maximum flow of 43 m³/hr max flow case and at an impeller
diameter of 206 mm is 2 m (refer to Figure 5.10).

That is, $NPSH_r$ is   2 m

It is a standard engineering practice to keep the $NPSH_a$ 1.0 m higher than the
$NPSH_r$. This criterion is therefore satisfied here.

Liquid temperature at its boiling point: Assume a case where the liquid is at its boiling temperature at the vessel operating pressure (atmospheric in this case).

The vapor pressure of the liquid would thus be 1.033 kg/cm^{2}a

$$= 1.033 \times 10{,}000/992 = 10.41 \text{ m}$$

In this case,

Available NPSH$_a$ is   12.68 m – 10.41m = 2.27 m
NPSH$_r$ is   2 m

The difference between NPSH$_a$ and NPSH$_r$ is only 0.27 m and thus falls short of the industry practice of 1.0 m.

In order to fulfil this criterion, the elevation of the suction vessel would need to be increased by at least 0.73 m.

In most cases, liquids at their boiling points create such challenges for NPSH$_a$. In such cases, the suction vessels need to be suitably placed to meet the NPSH$_a$ requirements.

5.4.6   MINIMUM FLOW REQUIREMENTS OF CENTRIFUGAL PUMPS

Centrifugal pumps have a minimum flow requirement for stable operation. In case the minimum required flow specified in the pump process data sheet is below the required pump minimum flow for stable operation, the difference should be recirculated through a restriction orifice back to the suction vessel (refer to Figure 5.12(a) for a typical configuration). Let us take Examples 5.1 through 5.3 illustrated above. The process minimum required flow specified is 18 m^3/hr. Suppose the pump minimum flow for stable operation calls for 20 m^3/hr. The configuration should be such that 2 m^3/hr flows under recirculation through a restriction orifice. In this way, even if the process draws the minimum flow of 18 m^3/hr, the pump would operate at 20 m^3/hr, since the shortfall of 2 m^3/hr would be drawn through the bypass line. Thus, pump stable operation is ensured. In this configuration, a flow of 2 m^3/hr continuously flows through the bypass line irrespective of the demands of the process, Thus, while selecting the pump model, a flow of 2 m^3/hr is added to the process maximum flow. It is important to note that the recirculation line is needed only if the minimum required flow specified in the pump process data sheet is below the required pump minimum flow for stable operation. The process data sheet of the pump would thus be specified with flowrates as follows:

Minimum flow: 20 m^3/hr (instead of 18 m^3/hr)
Normal flow: 38 m^3/hr (instead of 36 m^3/hr)
Maximum flow: 45 m^3/hr (instead of 43 m^3/hr)

The drawback of this configuration (i.e., recirculation through restriction orifice) is that pumps get oversized by the addition of recirculation. Therefore, such configurations are normally deployed for pumps of smaller capacities whereby a small increase in the flow does not substantially add to the operating cost. Also, the pump model may still remain unchanged in most cases.

For pumps of larger sizes, such configurations may result in pumps being significantly oversized. Apart from this, there is an additional penalty of higher operating cost in view of a significant flow being routed through the bypass line. Imagine a pump with a capacity of 150 m³/hr having a pump minimum flow for stable operation specified at 65 m³/hr. In such cases, the usual practice is to add a control valve on the recirculation line which opens once the flow through the pump goes below the pump minimum flow (refer to Figure 5.12(b)) for a typical configuration).

The control valve option, although avoids a comparatively high recurring cost of recirculation, is also cost intensive in the sense that it accounts for the cost of a flow element, flow transmitter and control valve, although incurred once. This brings the third option and possibly the most attractive one. It consists of the addition of an automatic recirculation (ARC) valve through which recirculation takes place. Such valves are located in the place of the main check valve. Normally, there is no flow through this valve; however, once the process flow reaches close to the pump minimum flow for stable operation, the pump head (and thereby the discharge pressure) also moves up the curve. At the set pressure, the ARC valve opens and recirculation takes place through this valve back to the suction vessel. Similarly, as the flow in the main line increases beyond the pump minimum flow, the pump head (and hence the discharge pressure) also decreases. At the set pressure, the spring-loaded disc returns to its seat closing the bypass line. In this way, no additional flow is added to the process flow requirements while selecting the pump model. Such valves combine the contributions of the check valve, flow element, flow transmitter and control valve, all in one piece of equipment (refer to Figure 5.12(c)) for a typical configuration).

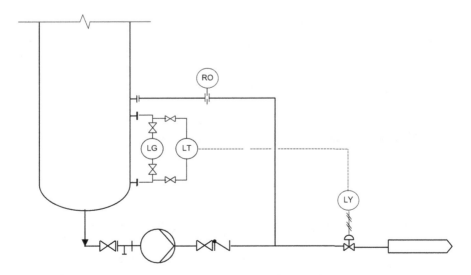

FIGURE 5.12(a) Minimum flow recirculation in centrifugal pumps – through restriction orifice.

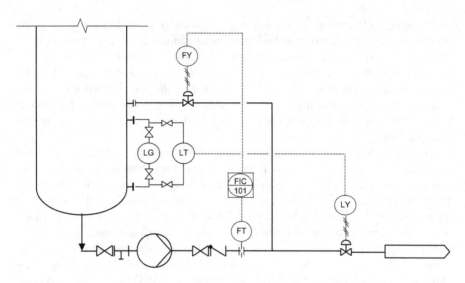

FIGURE 5.12(b)   Minimum flow recirculation in centrifugal pumps – through control valve.

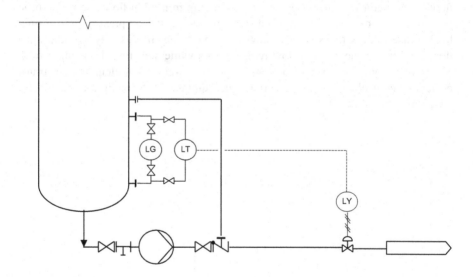

FIGURE 5.12(c)   Minimum flow recirculation in centrifugal pumps – through ARC.

5.5   RECIPROCATING PUMPS

Positive displacement pumps could be reciprocating or rotary. In all such pumps, cavities are alternately filled and emptied of the pumped liquid by the action of the pump.

A reciprocating pump is a positive displacement machine consisting of one or more cylinders, each containing a piston or plunger driven by slider-crank mechanisms and a crankshaft from an external source. The pump capacity is determined by the rotational speed of the crankshaft. In the third type of reciprocating pump, the driving member is a flexible diaphragm fabricated of metal, rubber or plastic [5].

Unlike a centrifugal pump, a positive displacement pump does not develop pressure; it only produces a flow of fluid which is independent of the pressure in the discharge piping system. The downstream equipment and piping system induce a resistance to this flow, thereby developing a pressure in the discharge section of the pump. The flow generated varies in proportion to the pump speed and number of cylinders.

All positive displacement pumps are capable of operating over a wide range of speeds. Therefore, they are capable of operating at various capacities when coupled to a variable speed drive. Such pumps are capable of generating very high pressures, provided the pump driver has enough power to generate that and the pump body is designed to withstand it [6]. Therefore, the discharge systems of positive displacement pumps are equipped with pressure relief devices to limit the pressure in the discharge system to its design pressure. These devices must be located between the discharge connection on the pump and the first isolation valve in the piping system.

5.6   POWER CONSUMPTION FOR PUMPS

The power consumption for the pump at shaft can be theoretically computed from the following equation:

$$kW_{shaft} = \frac{QH}{36.7\eta} \qquad (5.11)$$

where,

kW_{shaft} = Shaft Power (kW)
Q = Pump Flowrate (m³/hr)
H = Pump Differential Head (kg/cm²)
η = Pump Efficiency

Having arrived at the shaft power, the electrical motor power is estimated as follows. Refer to API 610 for details [7]:

- For shaft power less than 22 kW, multiply the shaft power by 1.25
- For shaft power between 22 and 55 kW, multiply the shaft power by 1.15
- For shaft power more than 55 kW, multiply the shaft power by 1.10

Having arrived at the electrical motor power, the next higher motor rating that is commercially available is specified. Table 5.8 lists the standard motors ratings (in kW) available commercially.

TABLE 5.8

Standard Motor Ratings, kW

0.18	0.25	0.37	0.55
0.75	1.1	1.5	2.2
3.7	5.5	7.5	9.3
11.0	15.0	18.5	22.0
30.0	37.0	45.0	55.0
75.0	90.0		

***

EXAMPLE 5.4   ESTIMATE THE SHAFT POWER AND THE MOTOR RATING OF THE PUMP ILLUSTRATED IN EXAMPLE 5.2

From Example 5.2, we have:

Q = 43 m³/hr
H = 5.26 kg/cm² (53 m of liquid)

From Figure 5.10, the pump efficiency is 67%

$$\eta = 0.67$$

Hence, kW_{shaft} works out to 9.2 kW

As explained above, since the shaft power is below 22 kW, the electrical motor power is calculated as:

Electrical motor power

$$= 1.25 \times 9.2$$

$$= 11.5 \text{ kW}$$

From Table 5.8, we select the next standard motor rating of 15 kW.

Inexperienced engineers sometimes get confused between motor rating and electric power consumption. Electric power consumption is arrived at by dividing the shaft power by the motor efficiency. Table 5.9 provides typical motor efficiencies. Thus, the electric power consumption is calculated as:

Shaft power	= 9.2 kW
From Table 5.9, motor efficiency	= 89% (assuming that the motor would fall in the 11–22 kW range)
	(i.e., η_{motor} = 0.89)
Electric power consumption	= 9.2 / 0.89 = 10.3 kW

Keep in mind that electric power consumption (in this case, 10.3 kW) will always be less than the motor rating (in this case, 15 kW).

***

TABLE 5.9
Typical Motor Efficiencies

Motor Rating, kW	Motor Efficiency, %
150–400	95
110–150	94
55–90	93
30–45	91
11–22	89
3.7–7.5	87
1.1–2.2	81
<0.75	72

Source: Adapted from National Electrical Manufacturer's Association, NEMA MG 1-2014.

5.7   CHOICE OF DRIVER FOR PUMPS

Electric motor is by far the most common drive for pumps in the process industry. Occasionally, special considerations such as reliability of power, safety consider-ations and criticality of service require turbine drives to be used. Sometimes, the plant utility balance makes turbine drives necessary on large pumps.

Steam turbines are effective for standby for a few larger, vital pumps such as charge pumps, cooling water pumps, unit pump-out pumps or flushing-out pumps. These considerations have to be firmed up during the design basis stage.

Steam turbines can be condensing or non-condensing. The choice of non-con-densing or condensing operation is affected by the demand for the exhaust steam. Generally, pump drives are not made of the condensing type without a rigorous review, as the increased complexity of condensing is rarely worth the small savings achieved in the utilities.

In cases like offshore platforms, oil and gas processing terminals, gas turbines are used as drivers for large pumps.

5.8   PIPING AND INSTRUMENTATION DIAGRAMS FOR PUMPS

Figure 5.13 illustrates a typical P&ID for a centrifugal pump. Isolation valves are provided at inlet and outlet lines. A check valve is provided downstream to prevent reverse flow when the pump is not in operation. The inlet line is equipped with a strainer for protecting the pump against solid matter which might inadvertently be stuck in the suction loop. A pressure gauge is provided at the discharge. As explained in Section 5.4.6, the minimum flow recirculation line is provided. Since centrifugal pumps are constant pressure machines in the sense that the pressure developed is fairly constant in the range between minimum and maximum flows; hence, pressure relief valves are normally not installed on the discharge lines of such pumps as long as the discharge line is selected for the proper pound rating.

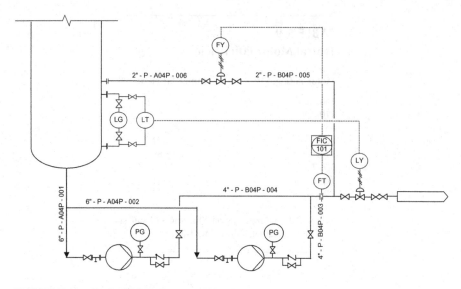

FIGURE 5.13 Typical P&ID for a centrifugal pump.

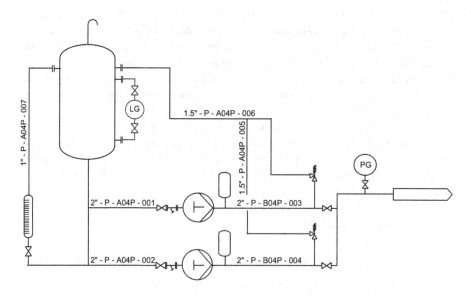

FIGURE 5.14 Typical P&ID for a reciprocating pump.

Figure 5.14 illustrates a typical P&ID for a reciprocating pump. A reciprocating pump is a constant flow machine, in the sense that flow is constant while the discharge pressure can vary depending upon the resistance in the downstream loop. In case there is a valve closure downstream, the discharge pressure can shoot up. The application of reciprocating pumps is thus limited to low flow applications, viz., dosing of chemicals, etc. For bulk flow of process fluids, reciprocating pumps have limited application and for cases which require high pressure, one has to resort to multi-stage centrifugal pumps. The P&ID is also configured in accordance with this requirement. Apart from the isolation valves and suction strainer, there is a pressure relief valve on the downstream line. Further, flow in a reciprocating pump is pulsating as a result of the oscillating motion of the plungers. Pulsation dampeners are thus installed at the discharge of such pumps. The suction line is equipped with a metering device which can be used for calibration of the pump by isolating it from the suction vessel.

5.9 COMPRESSORS

Compressors find important applications in chemical process plants. They could be deployed to recycle gas in a closed circuit in hydroprocessing units in refineries, for supplying feed gas to reactors, for transporting associated gases from oil fields, as main air compressors in cryogenic air separation units, as booster compressors or as main air blowers to supply air to fluid catalytic cracking units.

The machines could be driven either by steam or by electric power.

5.9.1 TYPES OF COMPRESSORS

Compressors broadly fall under two categories: positive displacement and dynamic. Positive displacement compressors are further classified as reciprocating and rotary compressors. Dynamic compressors are further classified as centrifugal and axial (refer to Figure 5.15).

Reciprocating Compressors: Reciprocating compressors are widely used in the chemical process industry. The flowrates for such compressors fall on the lower end of the range, typically 1000–5000 m³/hr. Discharge pressures are typically 140–350 kg/cm²g. Low reliability and high maintenance cost contribute to the drawbacks of these types of compressors [8].

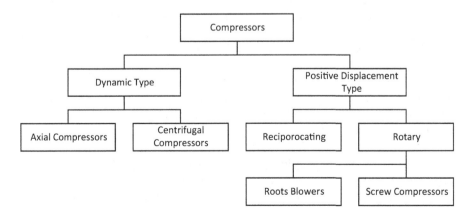

FIGURE 5.15 Classification of compressors.

Rotary Compressors: These compressors use two intermeshing helical and spiral lobes to compress the gas between the lobes and the rotor chamber of the casing. Such compressors are attractive in the range of 5100–10200 m³/hr and for medium discharge pressures, typically 20–28 kg/cm²g. In terms of application range, these compressors bridge the gap between the reciprocating compressor and the centrifugal compressor [8, 9].

Compressors with helical lobes are also called "screw compressors" in the process industry. Straight lobe compressors are similar to the helical lobes machines but are less sophisticated. They have two straight lobe rotors that intermesh as they rotate. Straight lobe compressors have a higher flowrate range but are limited in discharge pressures [9].

Screw compressors come in two versions: dry and oil-flooded. In the dry type, no oil is injected into the screws. However, in the oil-flooded type, lubrication oil is injected into the screw mesh. This acts both for sealing the screw and for removal of heat of compression through an external oil cooler. On account of heat removal, the oil-flooded screw compressors can carry out a given compression duty in a fewer number of stages compared to the dry running type. Conversely, high compression ratios (15–18) can be achieved even for low molecular weight gases in a single stage [8].

Dynamic Compressors: In such compressors, the energy is transferred in the form of velocity from a set of moving blades. The velocity is then converted to pressure in the outlet gas. Based on the direction of flow through the machine, dynamic compressors are sub-divided primarily into two categories: radial and axial flow.

Radial or centrifugal compressors use an impellor consisting of radial or backward leaning blades. As the impeller rotates, gas is moved between the rotating blades and radially outward towards the discharge. Centrifugal compressors can handle flows as high as 255,000 m³/hr. A common low-pressure compressor can deliver discharge pressures of 0.7–0.8 kg/cm²g. In higher models, pressure ratios of 3 are possible [9].

5.9.2 Capacity Control of Compressors

Reciprocating Compressors: In many applications, the availability of compressed gas is intermittent. To handle such cases, there are ways of controlling the output flow from the compressor. The type of mechanism chosen has a bearing on the power consumption. Some of the control mechanisms are covered below [10].

Automatic On-Off Control: The on-off control mechanism starts or stops the compressor through a pressure activated switch. In this method, the motor idle-running losses are eliminated since the motor gets switched off once the set pressure is reached. This is thus an efficient system but is suitable only for small compressors.

Load and Unload: In this two-step control, the compressor is loaded when there is a demand and unloaded when there is no demand. During unloading, the compressor can consume as low as 30% of the rated power, depending upon the configuration.

Multi-step Control: For large capacity reciprocating compressors, a multi-step control is the common configuration. In this scheme, the unloading is achieved in a series of steps such as 0, 25, 50, 75 and 100%. In such a scheme, at full load, the power consumption would be 100%, at 50% load, the power may be 52–55%, and at no load the power consumption could be as low as 10–14%.

Rotary Compressors: Capacity control in rotary screw compressors is achieved in three ways by suction throttling or modulation, variable displacement and speed control [9, 11].

Suction Throttling: Before the slide valve came into picture, suction throttling was the only method that was deployed in flooded screw compressors. The method could be used in dry-type compressors as well if provision is made for lubrication of the slide. However, this system is not applicable for cases of 70% of the load.

Variable Speed Control: In this method, the speed of the motor is varied, thus controlling the output. It provides the best performance among the three types of capacity control. With a variable speed driver, energy efficient speed control is achieved over a wide range of speeds.

Slide Valve: For small compressors, the slide valve offers an economic alternative to speed control. In this system, the slide valve moves parallel to the rotor axis and changes the area of the opening in the bottom of the rotor casing. In this way, the region of the compression of the rotor lengthens or shortens.

Dynamic Compressors: For centrifugal compressors, throttling range is limited on account of the damaging effect called "surge." In such compressors, therefore, capacity control is achieved either by using adjustable inlet guide vanes or by varying the speed. At low flowrates, the capacity is normally controlled by using inlet guide vanes. Such a scheme is recommended for volumetric flowrates below 40%. The other way to meet the load requirement is by speed control. This scheme is deployed at flowrates above 40% [10].

5.9.3   ANTISURGE CONTROL AND CHOKING IN CENTRIFUGAL COMPRESSORS

The performance curves of centrifugal compressors are similar to those of centrifugal pumps. At a particular point, the head has a maximum value and the flow a minimum. Surge is the lowest flow at which the compressor still achieves stable operation. Below this point, the operating becomes unstable and the instability is accompanied by a characteristic noise called "surge."

   Figure 5.16 illustrates a typical curve for centrifugal compressors. The inlet flowrate is on the X-axis and the discharge pressure is on the Y-axis. With an increasing flowrate, the discharge pressure developed by the compressor decreases. The flowrate versus the discharge pressure curves are shown at different speeds of the compressor. The surge points at different speeds are shown as a surge line. The surge control line is also shown. The antisurge flow is considered at 10% higher than the surge flow. The choking or the stonewall condition occurs when the compressor operates at very high flowrate and flow through the compressor reached close to the sonic

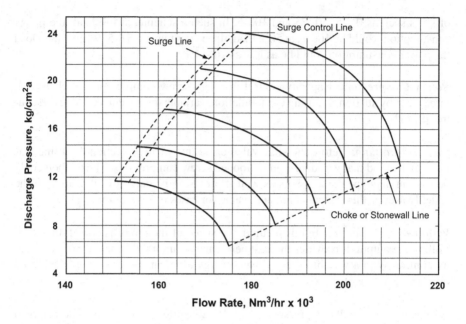

FIGURE 5.16 Characteristic curves of centrifugal compressors.

velocity and therefore cannot be increased further. The choking or the stonewall line is also shown.

Compressor surge causes the machine to vibrate and may cause damage to the compressor parts. Dynamic compressors should therefore be equipped with an anti-surge control system, wherein a minimum flow through the machine is maintained such that the surge condition is never reached. This system consists of a line from the discharge of the aftercooler to the suction knock-out drum or to the atmosphere, equipped with an antisurge valve. When the compressor approaches surge condition (low flow, high differential head), the antisurge valve opens up and bleeds part of the flow either to the atmosphere or recirculates it back to the suction [12].

Technically speaking, surge control can be achieved in different ways. In the conventional approach, only the differential pressure and the flow through the compressor are measured. In another approach, the differential pressure is neglected, while the flow and the rotational speed are considered. In more sophisticated applications, in addition to flow and differential pressure, inlet temperature and pressure are measured to accurately calculate the operating point [12, 13]. Figure 5.17 illustrates a typical antisurge control configuration in a centrifugal compressor.

5.9.4 POWER CONSUMPTION IN COMPRESSORS

In the process of compression, the absolute pressure p and the volume V are related by the formula:

$$pV^n = C = constant \qquad (5.12)$$

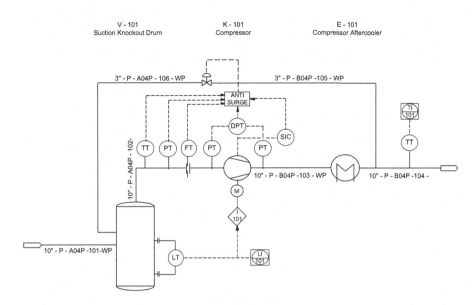

FIGURE 5.17 Typical P&ID for a centrifugal compressor.

The plot of pressure versus volume for each value of exponent n is known as the "polytropic curve." For $n = 1$, the compression is isothermal. For adiabatic compression (i.e., no heat is added or taken away), $n = k$ (i.e., the ratio of specific heat at constant pressure to that at constant volume). Since most compressors operate along a polytropic path approaching the adiabatic, compressor calculations are generally based on the adiabatic curve. The adiabatic power is expressed as [5]:

$$kW_{ad} = 0.0273 \frac{k}{k-1} Q p_1 \left[\left(\frac{p_2}{p_1} \right)^{(k-1)/k} - 1 \right]$$
(5.13)

The adiabatic temperature rise is given by:

$$T_2 = T_1 \left(p_2 / p_1 \right)^{(k-1)/k}$$
(5.14)

Table 5.10 provides a quick estimation of the k values knowing the gas molecular weight [14].

The temperature of the gas after compression needs to be maintained within a limit. This ensures adequate life of packing and avoids possible degradation of lube oil [15]. At temperatures beyond 149°C, oxygen, if present, may cause ignition of the lube oil. In some cases, this limit is followed to protect against polymerization as in olefin or butadiene plants [16]. At high temperatures, problems of sealing and casing growth start to occur. High temperature calls for more expensive machines. This explains why in cases of high compression ratios, the compression is split into stages. It is ensured that the exit temperature of each stage does not exceed a limit, typically 120–140°C. Interstage coolers are installed between stages.

TABLE 5.10

Approximate Heat Capacity Ratios of Hydrocarbon Gases

Temperature, °C	Mol. Wt, kg/kgmol	k	Temperature, °C	Mol. Wt, kg/kgmol	k
40	80	1.070	95	80	1.055
	70	1.073		70	1.061
	60	1.090		60	1.074
	50	1.108		50	1.090
	40	1.135		40	1.114
	30	1.178		30	1.158
	20	1.248		20	1.220
65	80	1.059	120	80	1.052
	70	1.070		70	1.059
	60	1.082		60	1.060
	50	1.100		50	1.088
	40	1.128		40	1.108
	30	1.169		30	1.146
	20	1.235		20	1.218

Source: Adapted from *GPSA Engineering Data Book.*

For reciprocating compressors, typical compression ratios range from 1.2 to 4.0. However, for centrifugal compressors, flowrates are much higher. In such cases, power consumption becomes important, and lower compression ratios save power. Assume a gas to be compressed from 5 to 20 kg/cm²a. There could be two options, i.e., compression from 5 kg/cm²a directly to 20 kg/cm²a, or compression first from 5 kg/cm²a directly to 10 kg/cm²a, and then from 10 kg/cm²a directly to 20 kg/cm²a with an interstage cooler in between. In the latter case, installing a second stage, one can cool the gas discharged from the first stage. Thus, the gas entering the second stage is at a lower temperature and has a lower volume than what it would have had it continued to be compressed in a single stage. The net result is savings in power consumption. Centrifugal compressors normally work with compression ratios of 2–3.

***

EXAMPLE 5.5   COMPRESSOR POWER CALCULATIONS

Natural gas, at 150,000 m³/hr, is to be compressed from 1.033 to 30 kg/cm²a. Going by the capacity, a centrifugal machine is necessary. Two options are worked out, i.e., a four-stage machine with a compression ratio of 2.3215, and a three-stage machine with a compression ratio of 3.074. Let us first consider a three-stage machine.

Calculations for a three-stage machine:
First stage

Volumetric flowrate Q, m³/hr	150,000 m³/hr
Ratio of specific heats, k (Table 5.10)	1.30

Pressure at inlet p_1, kg/cm^{2}a	1.033
Compression ratio, p_1/p_2	3.074
$(k-1)/k$	0.2308
$k/(k-1)$	4.333

The adiabatic power works out to $0.0273 \times 4.333 \times 150,000 \times 1.033 \times$
$(3.074^{0.2308} - 1) = 5423$ kW

Assuming an efficiency of 80%, the power at shaft works out to 6779 kW.

Second stage

Volumetric flowrate Q, m^3/hr	48,796 m^3/hr
Ratio of specific heats, k (Table 5.10)	1.30
Pressure at inlet p_1, kg/cm^{2}a	3.175

The adiabatic power works $0.0273 \times 4.333 \times 48,796 \times 3.175 \times (3.074^{0.2308} - 1)$
out to $\quad\quad\quad\quad = 5423$ kW

Assuming an efficiency of 80%, the power at shaft works out to 6779 kW.

Third stage

Volumetric flowrate Q, m^3/hr	15,874 m^3/hr
Ratio of specific heats, k (Table 5.10)	1.30
Pressure at inlet p_1, kg/cm^{2}a	9.760

The adiabatic power works $\quad 0.0273 \times 4.333 \times 15,874 \times 9.760 \times (3.074^{0.2308} - 1)$
out to $\quad\quad\quad\quad = 5423$ kW

Assuming an efficiency of 80%, the power at shaft works out to 6779 kW.
It can be seen that all three stages have identical power consumption.
The total power at shaft for all three stages sums up to 20337 kW (or 20.34 mW).
Assuming a motor efficiency of 96%, the power consumption works out to 21.18 mW.

Calculations for a four-stage machine:
First stage

Volumetric flowrate Q, m^3/hr	150,000 m^3/hr
Ratio of specific heats, k (Table 5.10)	1.30
Pressure at inlet p_1, kg/cm^{2}a	1.033
Compression ratio, p_1/p_2	2.3215
$(k-1)/k$	0.2308
$k/(k-1)$	4.333

The adiabatic power works out to $0.0273 \times 4.333 \times 150{,}000 \times 1.033 \times$
$(2.3215^{0.2308} - 1)$
$= 3933 \text{ kW}$

Assuming an efficiency of 80%, the power at shaft works out to 4916 kW.

Second stage

Volumetric flowrate Q, m³/hr	64,613 m³/hr
Ratio of specific heats, k (Table 5.10)	1.30
Pressure at inlet p_1, kg/cm²a	2.398

The adiabatic power works out to $0.0273 \times 4.333 \times 64{,}613 \times 2.398 \times$
$(2.3215^{0.2308} - 1)$
$= 3933 \text{ kW}$

Assuming an efficiency of 80%, the power at shaft works out to 4916 kW.

Third stage

Volumetric flowrate Q, m³/hr	27,832 m³/hr
Ratio of specific heats, k (Table 5.10)	1.30
Pressure at inlet p_1, kg/cm²a	5.567

The adiabatic power works out to $0.0273 \times 4.333 \times 27{,}832 \times 5.567 \times$
$(2.3215^{0.2308} - 1)$
$= 3933 \text{ kW}$

Assuming an efficiency of 80%, the power at shaft works out to 4916 kW.

Fourth stage

Volumetric flowrate Q, m³/hr	11,989 m³/hr
Ratio of specific heats, k (Table 5.10)	1.30
Pressure at inlet p_1, kg/cm²a	12.924

The adiabatic power works out to $0.0273 \times 4.333 \times 11{,}989 \times 12.924 \times$
$(2.3215^{0.2308} - 1)$
$= 3933 \text{ kW}$

Assuming an efficiency of 80%, the power at shaft works out to 4916 kW.
Total power at shaft sums up to 19664 kW (or 19.7 mW).
Assuming a motor efficiency of 96%, the power consumption works out to 20.48 mW.
The difference in electrical power between the two cases is significant. The machine with four stages demands a lower shaft power. It could be argued that a

four-stage machine would incur a higher capital expenditure in terms of the cost towards the extra stage including the intercooler, and additional piping and instrumentation. However, in a period of 15–20 years, the savings in power for a four-stage machine could be significant. Let us consider the following calculation:

A rough estimate of the price of such a compressor would be in the range of $12,000,000.

Using a very conservative estimate, let us take the cost difference between a three-stage machine and a four-stage as 15% of the cost of the machine, i.e., $1,800,000.

The difference in power consumption between the four-stage and the three-stage machines is 0.7 mW.

Let us take the cost of electric power at $0.04 per kWh.

For a 15-year period, the savings in cost towards power consumption work out to:

$$\$0.04 \times 0.7 \times 1000 \times 8000 \times 15 = \$3,360,000.$$

***

The above calculations give a clear feel that for such large machines, in the long term, the savings in electrical power consumption using a four-stage machine are likely to outweigh the additional expenditure towards the extra compression stage.

5.9.5 CHOICE OF DRIVES FOR COMPRESSORS

Motor Drive: In the same line as that for pumps, motor is the most commonly used driving mechanism. Motors call for less maintenance requirements than their gas-fired counterparts, in addition to their production of less noise and vibration. Therefore, for small to medium power requirements, if electricity is readily available, motor drives are the usual choice. However, in remote locations, viz., onshore/offshore operations, electrical power may not be readily available or reliable. In such cases, fuel fired engines may need to be used.

The motor kilowatt requirement to drive a compressor is determined by the operating conditions. The maximum operating condition, along with a proper judgement on the margin to be provided, gives the motor rating. As a guide, the American Petroleum Institute (API) recommends a 10% margin on the highest power requirement of the compressor to arrive at the motor rating.

Turbine Drive: For large drives, typically of the order of 5 mW or above, if there is a reliable source of steam at relatively cheap rates, it may be worthwhile to go for a steam turbine drive. In principle, there are broadly two types of steam turbines, viz., back pressure and condensing [17].

In a back pressure turbine, high pressure steam, typically 80 kg/cm^{2}g enters the turbine. The steam expands and a part of its thermal energy is converted to mechanical energy which drives the compressor or the pump. The outlet steam leaves the turbine as a low pressure steam. The back pressure type turbines are normally preferred, particularly in cases where the low pressure steam could be used in the process plant suitably. They have a lower capital cost, are more reliable and are simple in construction.

In the condensing type turbine, high pressure steam enters the turbine, expands to below atmospheric pressure and condenses using cooling water. At the turbine outlet, the pressure of the steam is so low that it has no practical value. Condensing type steam turbines require less steam because the enthalpy drop for such machines is higher. They have a higher capital cost and incur additional cost for condenser, ejector and other auxiliary equipment. In spite of these drawbacks, such turbines are still selected for very large compressor drives.

***

<h3>EXAMPLE 5.6</h3>

Take Example 5.5 illustrated above. We have seen that for a four-stage machine we calculated a shaft power of 19.66 mW. We compare two cases: motor as the driver and steam turbine as the driver.

First let us consider a motor drive. With a motor efficiency of 96% efficiency, we arrive at an electrical power consumption of 20.48 mW. For a cost of power at $0.10 per kWh, we arrive at the expenditure towards electric power as:

$$\text{Electric power}(\text{for motor drive}) = \$0.10 \times 22.48 \times 1000 = \$2248 \text{ per hour.}$$

Now we consider turbine as the driver. We consider a turbine efficiency of 80. We consider steam at the following conditions:

$$\text{Inlet}(\text{high pressure steam}) : 80\,\text{kg/cm}^2\text{g at } 500°\text{C}$$

$$\text{Outlet}(\text{low pressure steam}) : 5\,\text{kg}/\text{cm}^2\text{g}$$

The quantity of steam works out to 134,100 kg/hr (134.1 tonnes/hr)

Considering the cost of steam at $20 per tonne, we arrive at the expenditure towards steam as,

$$\text{Steam}(\text{for turbine drive}) = \$20 \times 134.1 = \$2682 \text{ per hour.}$$

***

Thus, with the above costs of steam and electric power, it is advisable to go ahead with a motor drive.

However, if the steam price goes down below $17 per tonne, the expenditure towards steam equals that for electric power. Thus, for steam at a cost lower than this, it could be advisable to go for a turbine drive.

5.9.6 Piping and Instrumentation Diagrams for Compressors

Figure 5.17 illustrates a typical P&ID for a centrifugal compressor. A knock-out drum is located upstream of the compressor to remove possible entrained liquids and liquid droplets. Sometimes, such knock-out drums are further equipped with demister pads. Downstream of the compressor, an aftercooler is installed to cool down the gas. Pressure transmitters are provided at the suction and discharge lines. The suction line is also equipped with a flow transmitter. Temperature transmitters are also provided, if required. Signals from these transmitters are sent to the anti-surge control system. Depending upon the flow and differential pressure, the anti-surge valve operates to prevent any possible surge condition (refer to Section 5.9.3). Level instruments are provided on the knock-out drum to control the liquid level. A compressor shutdown is activated in case of a high liquid level in the knock-out drum.

Figures 5.18 and 5.19 illustrate typical P&IDs for reciprocating and screw compressors, respectively. These are positive displacement machines, and therefore spill-back lines are provided for pressure control. In addition, pressure relief valves are also provided. Since these are positive displacement machines, a possibility exists for a pressure build-up on the discharge side and subsequently a temperature rise. To take care of this, compressor trips are initiated for cases of high suction pressure, high suction temperature, and high discharge temperature. Compressor trips are also initiated for high liquid levels in the knock-out drum.

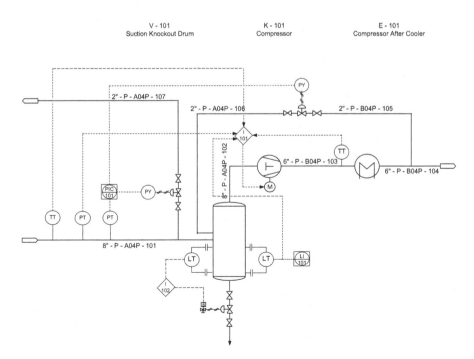

FIGURE 5.18 Typical P&ID for a reciprocating compressor.

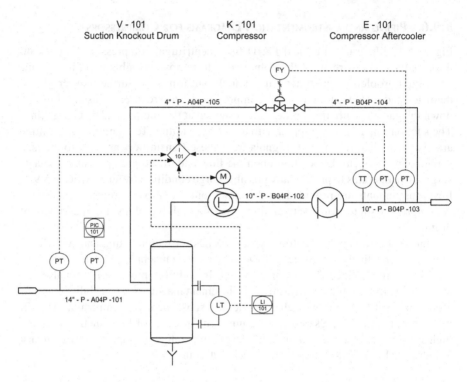

FIGURE 5.19 Typical P&ID for a screw compressor.

Symbols

Q Pump flowrate, m^3/hr

H_s Pump suction pressure, kg/cm^2a

H_d Pump discharge pressure, kg/cm^2a

H Pump differential head, kg/cm^2

h_f Pressure drop, m of liquid

h_s Source pressure, kg/cm^2a

h_{st} Static head at suction flange, kg/cm^2a

h_{sf} Friction losses on suction side, kg/cm^2a

h_d Destination pressure, kg/cm^2a

h_{dt} Static head at discharge flange, kg/cm^2a

h_{df} Friction losses on discharge side, kg/cm^2a

H_{pc} Height of pump centerline above finished floor level

H_{pb} Height of pump centerline above finished pump baseplate

f Friction factor

L Length of pipe, m

D Diameter of pipe, m

g Acceleration due to gravity, m/s^2

V Velocity though pipe, m/s

ρ Liquid density, kg/m^3

p_1 Absolute suction pressure, kg/cm^2a

p_2	Absolute discharge pressure, kg/cm^{2}a
ΔP	Pressure drop, kg/m^2
K	Dimensionless factor (for fittings)
K_1	K for the fitting at $N_{Re} = 1$
K_∞	K for a large fitting at $N_{Re} = \infty$
N_{Re}	Reynolds number
ε	Roughness height, m
η	Pump efficiency
η_{motor}	Motor efficiency
kW_{shaft}	Shaft power

REFERENCES

1. Colebrook, C. F., *J. Inst. Civ. Eng. London,* Vol. 11, pp. 133–156, 1938–1939.
2. Moody, L. F., "Friction Factors for Pipe Flow", *Trans. A.S.M.E.*, November 1944.
3. Crane Co., "Flow of Fluids through Valves, Fittings and Pipe", 1982.
4. Hooper, W. B., *Chemical Engineering*, August 24, 1981.
5. Perry, R. H. and Green, D., *Perry's Chemical Engineers' Handbook*, McGraw Hill International Editions, 1984.
6. Karrasic, I. J., Joseph, P. M., Cooper, P. and Heald, C. C., *Pump Handbook*, McGraw Hill, 2001.
7. American Petroleum Institute, "Centrifugal Pumps for Petroleum, Petrochemical and Natural Gas Industries", Eleventh Edition, July 2011.
8. Barnett, J. M. and Schranke, T. M., "Cost-effective Compressor Selection and Specification", *Chemical Engineering*, September 2000.
9. Brown, R. N., *Compressors – Sizing and Selection*, Gulf Professional Publishing, 2nd edition, Houston, Texas, 1997.
10. Bureau of Energy Efficiency: https://beeindia.gov.in/sites/default/files/3Ch3.pdf
11. https://www.plantengineering.com/articles/controls-increase-compressor-efficiency/
12. Hanlon, P. C., *Compressor Handbook*, McGraw Hill, 2001.
13. Kvangardsnes, T., *Anti-surge Control – Control Theoretic Analysis of Existing Anti-surge Control Strategies"*, Master's Thesis – Norwegian University of Science and Technology, 2009.
14. *GPSA Engineering Data Book*, Gas Processors Suppliers Association, Vol. 1, 10th Ed, 1987.
15. Montemayor, A., *Reciprocating Compressors Compression Ratios*, www.cheresources.com, September 12, 2013.
16. Branan, C., *Rules of Thumb for Chemical Engineers*, Gulf Professional Publishing, 2005.
17. Liebermann, N.P., Liebermann, E.T., *A Working Guide to Process Equipment*, McGraw Hill Inc., 1997.

6 Vessels and Tanks

6.1 INTRODUCTION

Vessels are of two types: pressure vessels and atmospheric vessels. Pressure vessels operate at pressures above atmospheric and have dished heads. Examples of pressure vessels include feed surge drums which are located upstream of a process unit. The pressure vessel maintains a relatively constant flow into the unit even if the feed from the upstream unit varies a bit. Other such examples are the reflux drums which serve as a reservoir of the condensed reflux from a column overhead. Such vessels have residence times of 10–15 minutes. Other examples of pressure vessels include steam condensate pots, closed blowdown drums, instrument air receivers, vapor liquid separators, etc. which are part of a process plant. In addition, there are large pressure vessels which store liquid hydrocarbons with storage times of several days. Such vessels are typically the spheres and bullets and form part of offsite facilities.

Most atmospheric vessels fall under the category of tanks. Tanks operate at atmospheric pressures and come with different designs as described later in the chapter. There are small tanks called "day tanks" which hold a day's inventory of a product and are located within the process unit. However, large tanks are located outside a process unit and form a part of the offsite facilities. Such tanks could hold a week's or more supply of a feed or product. Typical diameters of such tanks would be 30–60 m with heights in the range of 10–20 m.

In a typical refinery, petrochemical or chemical plant, process vessels constitute about 30–40% of the total equipment, and their contribution to the total unit investment varies between 10 and 30%.

6.2 ATMOSPHERIC STORAGE TANKS

Atmospheric storage tanks are large vessels with hold-ups of up to several hours and are used to store feed/products to and from process units. In cases, where the feed/products are in liquid state at atmospheric pressures, the tanks store such materials under atmospheric or low pressures. Such tanks normally consist of a cylindrical shell with a fixed roof that can be flat, conical or dome-shaped, or a floating roof (discussed below). In cases where the feed/products are in vapor form at atmospheric pressures, viz., propane, butane, etc., such vessels store the materials at high pressures such that they remain in liquid state. In such cases, the vessels are normally designed as spheres or bullets.

6.2.1 CLASSIFICATION OF STORED LIQUIDS

While tanks can be used to store any kind of liquid, it is the hydrocarbon liquids which deserve the utmost care because of their flammable or combustible properties. According to the National Fire Protection Association (NFPA), a flammable liquid is

TABLE 6.1
Classification of Flammable and Combustible Liquids

Classification of Liquid	Type of Liquid	Characteristics
Class IA	Flammable	Liquids that have a flash point below 22.8°C, and boiling point below 37.8°C
Class IB	Flammable	Liquids that have a flash point below 22.8°C, and boiling point at or above 37.8°C
Class IC	Flammable	Liquids that have a flash point at or above 22.8°C but below 37.8°C
Class II	Combustible	Liquids that have a flash point at or above 37.8°C and below 60°C
Class IIIA	Combustible	Liquids that have a flash point at or above 60°C but below 93°C
Class IIIB	Combustible	Liquids that have a flash point at or above 93°C

Source: Reproduced from Flammable and Combustible Liquid Code, NFPA 30, National Fire Protection Association, 2003.

that which has a a closed-cup flash point below 37.8°C. Similarly, a combustible liquid is that which has a closed-cup flash point at or above 37.8°C. The NFPA has published several volumes on recommended practices on fire protection. Guidelines pertaining to storage of hydrocarbon liquids are also covered in these volumes. For detailed classification of flammable and combustible liquids, refer to Table 6.1 [1].

6.2.2 DEFINITIONS

According to NFPA [1], atmospheric storage tanks are permitted to be operated at pressures between atmospheric and 0.07 kg/cm^{2}g (6.9 kPag) measured at the top of the tank. The design pressures for such tanks shall not exceed 0.176 kg/cm^{2}g (17.2 kPag). However, it is mandatory to design and construct such tanks in accordance with Appendix F of API Standard 650. All other tanks shall be limited in operation between atmospheric pressure and 0.036 kg/cm^{2}g (3.5 kPag), unless engineering analysis is performed to determine that the tank can withstand the elevated pressure. Such tanks are built in two basic designs: fixed roof design where the roof remains fixed and external floating roof design where the roof floats on top of the liquid, rising and falling with the liquid level.

6.2.2.1 Fixed Roof Design

Fixed roof storage tanks are those that consist of a cylindrical shell with a permanently welded roof that could be flat, conical or dome-shaped. Such tanks are used to store materials with a true vapor pressure of less than 0.105 kg/cm^{2}a (10.3 kPa(a)) [2]. From a construction point of view, fixed roof tanks are the least expensive and are in general considered as a minimum acceptable equipment for liquid storage. Such tanks are equipped with a breather valve which allows operation at slight internal pressure or vacuum [3].

6.2.2.2 External Floating Roof Design

External floating roof storage tanks consist of an open-topped cylindrical shell equipped with a roof that rests on the stored liquid and is free to move as the level goes up and down in the tank. Such tanks are preferred for storage of petroleum products which have a true vapor pressure range from 0.105 kg/cm²a (10.3 kPa (a)) to 0.77 kg/cm²a (75.5 kPa (a)) [2]. External floating roof tanks are equipped with a rim seal, which is attached to the perimeter of the roof and contacts the tank wall. The rim seal system slides against the tank wall as the roof moves up and down. In such tanks, the evaporative losses from the stored liquid are limited to losses from the rim seal system and deck fittings [3].

There are, in principle, three different types of floating roofs:

1. The pan type roof is a single-deck roof having full contact with the surface of the liquid. The roof has no buoyancy other than provided by the deck; hence, any leak through the deck will cause it to sink. Further, rain or snow could cause it to deform. It is the cheapest among the three types of floating roofs. Pan roofs are therefore generally not used since they are unsafe. The trend is more towards use of pontoon type or double-deck type floating roofs.
2. In the pontoon type, the roof has an annular pontoon around the edge and a deck of single thickness. The annular pontoon is divided into several chambers by radial bulkheads. The ratio of pontoon area to the total area varies according to the size of the tank, storing product and accessories and is typically about 20–40% of the roof area. The upper surface of the pontoon is slanted down towards the center for drainage purposes [4]. This type of roof has a better buoyancy and stability.
3. The double-deck roof comprises upper and lower decks separated by bulkheads and trusses. The space between the decks is divided into liquid tight compartments. They have a superior loading capacity and are recommended for tank diameters below 12 m and above 60 m.

Figures 6.1(a–c) illustrate typical floating roofs under each category.

6.2.2.3 Internal Floating Roof Design

Internal floating storage tanks are basically cone roof tanks with an inside floating deck that travels up and down along with the liquid level. Such tanks are preferred in areas of heavy snowfall where accumulation of snow or water on the floating roof may affect buoyancy. In such tanks, the vapor space is normally blanketed with an inert gas. Figure 6.1(d) illustrates a typical internal floating roof.

6.2.3 Design and Engineering Aspects

Tank Capacity: There are three different definitions of capacity of a storage tank: nominal capacity, gross capacity and net capacity.

In case of fixed roof tanks, the nominal capacity is the geometric volume of the tank from the bottom of the tank up to the curb angle. For floating roof tanks, the nominal

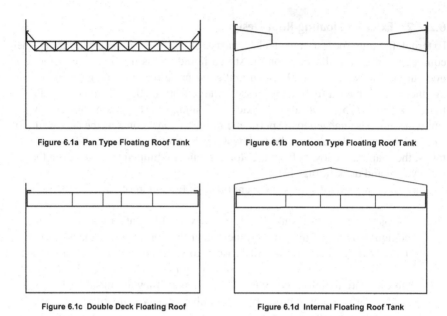

Figure 6.1a Pan Type Floating Roof Tank   Figure 6.1b Pontoon Type Floating Roof Tank

Figure 6.1c Double Deck Floating Roof   Figure 6.1d Internal Floating Roof Tank

FIGURE 6.1(a–d)   Floating roof tanks.

Source: Reprinted by special permission from *CHEMICAL ENGINEERING*, Copyright © April 2006, by Access Intelligence, Rockville, MD 20850.

capacity is the volume calculated from the bottom of the tank to the underside of the roof deck up to the maximum floating position of the roof in case of floating rook tanks.

The gross capacity is the volume of the tank up to the maximum safe filling height of the tank. This is also sometimes referred to as the "total capacity," or the "actual capacity."

The net capacity is the volume of the tank contents between two datum points, namely the low liquid level (LLL) and the high liquid level (HLL).

Tank Dimensions: The height of a tank should not normally exceed one and a half times its diameter. However, for large diameter tanks, the criteria for setting the tank dimensions change. Soil conditions, seismic activity and the cost of land govern the selection of tank dimensions. In windy or seismically active areas, reduced tank heights and larger diameters are preferred [2]. As the tank height increases, the wall thickness also increases. In addition, higher tank heights would result in a greater load on the soil. If the load on the soil exceeds the soil's allowable bearing pressure, pile-supported foundations are necessary and these become expensive. This is particularly of concern for soils poor in load-bearing capacity. Therefore, in cases where availability of land is not a constraint, it is preferable to go for tanks with larger diameters and smaller heights. In general, tanks of height beyond 15–20 m are not commonly used in the chemical process industry [5].

Roof Ladders, Staircases, Handrails and Walkways: While these items may not be of direct concern to a process engineer and are important only from a mechanical point of view, nevertheless, it is important for the engineer to have an overall understanding of this subject. Individual tanks should be provided access to the roof. Generally, tanks 6 m or less in height must have a ladder without a cage. Tanks taller than 6 m require a spiral stairway [6]. Staircases should have landing platforms at every 5 m height.

Floating roof tanks are also normally provided with a ladder that automatically adjusts to any position, so that access to the roof is always provided. The ladder is designed for full roof travel. These are illustrated in Figures 6.2 and 6.3 for typical external and internal floating tanks and are self-explanatory.

Roof Drains: As per API 650 [7], primary roof drains shall be of the hose, jointed or siphon type. In each case, a check valve shall be provided near the roof end of the hose to prevent stored product from flowing back in case of a leak. The maximum hourly rainfall rate should be considered for designing the number of sizes of drains for floating roof tanks. The primary roof drain system shall be the closed type using

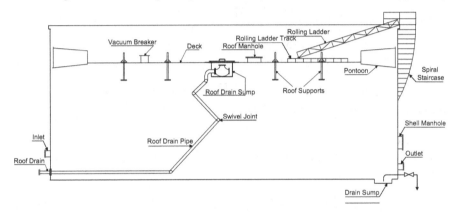

FIGURE 6.2   External floating roof tank showing various appurtenances.

Source: Reprinted by special permission from *CHEMICAL ENGINEERING*, Copyright © April 2006, by Access Intelligence, Rockville, MD 20850.

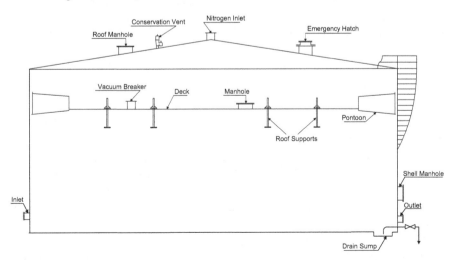

FIGURE 6.3   Internal floating roof tank showing various appurtenances.

Source: Reprinted by special permission from *CHEMICAL ENGINEERING*, Copyright © April 2006, by Access Intelligence, Rockville, MD 20850.

pipe and swing joints. The inlet to these drains should have a swing check valve [7]. Refer to Figure 6.2 for details.

Bottom Drains: Bottom drains, sometimes also referred to as "water draw-off sumps," should be provided in all tanks for draining water and also for storage tanks handling petroleum products to enable emptying the tank out for cleaning. API 650 [6] provides dimensional details of such drains.

Refer to Figures 6.2 and 6.3 also for tank appurtenances.

Manholes: Shell Manholes are normally 600 mm in diameter. The minimum number of manholes depends on the diameter of the tank. Typical guidelines are:

- One manhole for tanks less than 12 m in diameter
- Two manholes (180° apart) for tanks of diameter 12–44 m
- Three manholes (120° apart) for tanks of diameter 45–59 m
- Four manholes (90° apart) for tanks of diameter greater than 60 m

For tanks less than 12 m in diameter, one roof manhole is enough. For tanks larger than 12 m, two roof manholes are recommended.

Tank Instrumentation: Instrumentation in tanks is an important factor, especially for large tanks subjected to frequent filling and emptying. The following guidelines could be useful while selecting tank instrumentation:

- *Level:* It is recommended to have at least two level instruments in a tank, one having local indication and the other having indication in the control room. In several cases, two level instruments both having local and remote indications are also used. The level instruments should have high/low level alarms.
- *Temperature:* As a minimum, there should be a local indication for temperature. In case of high storage temperatures, it is recommended to have remote indication with alarm.
- *Pressure and Flow*: For tanks provided with inert gas blanketing, a flow and pressure indicator with alarms is recommended so that an alarm is triggered in case of problems in the blanketing gas line.

6.2.4 TANK BLANKETING

Blanketing of Tanks: If tanks contain liquid vapors which are harmful or hazardous to health, they require blanketing with an inert gas. This inert gas blanket cover exists on the vapor space over the liquid. The blanketing also helps in avoiding contact of tank vapors with atmospheric air, which may result in possible formation of explosive compounds or product degradation. Blanketing pressures are typically in the range of 100–150 mm water column.

Blanketing of tanks is maintained by means of a blanketing valve that senses the pressure in the vapor space of the tank and controls the flow of inert gas (normally nitrogen) into the vapor space so that the tank pressure can be maintained within the range. In the event the tank sees vacuum due to tank inbreathing (discussed later), the

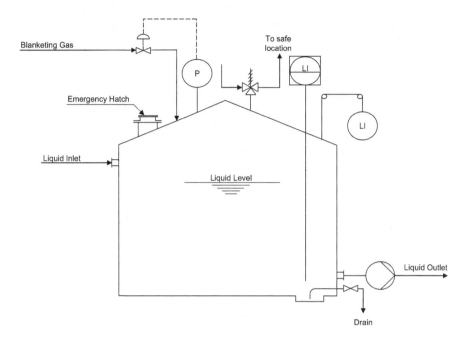

FIGURE 6.4 P&ID for a blanketed tank.

Source: Adapted from *CHEMICAL ENGINEERING*, Copyright © April 2006, by Access Intelligence, Rockville, MD 20850.

blanketing valve opens and supplies inert gas to the vapor space. When the vapor space pressure increases (outbreathing), the valve reseals. In other words, the blanketing valve provides primary vacuum relief to the tank. The secondary relief is provided by pressure/vacuum vents described later in this chapter. Figure 6.4 illustrates a typical P&ID of a blanketed tank containing a hydrocarbon mixture.

Tank Venting Calculations: A tank breathes in or out due to the following conditions:

- Inbreathing due to emptying of liquid from the tank
- Inbreathing due to contraction or condensation of vapors as a result of a drop in the atmospheric temperature (this is also known as "thermal inbreathing")
- Outbreathing due to filling of liquid into the tank
- Outbreathing due to expansion or vaporization of vapors as a result of a rise in the atmospheric temperature (this is also called "thermal outbreathing")
- Outbreathing due to external fire leading to vaporization

The following section deals with tank inbreathing and outbreathing rates at various conditions [5]. Such requirements are applicable to storage thanks that meet the following criteria:

- Tank volumes are less than 30,000 m³
- Maximum operating temperatures do not exceed 48.9°C
- Tanks are uninsulated

- Temperatures of the tank contents are less than the boiling temperature at the maximum operating pressure

For tanks that do not satisfy the above conditions, the reader may refer to Section 3.3.2 of API 2000 [8] for venting requirements.

Inbreathing: According to API 2000 [8], the inbreathing rate for maximum liquid draw-off from a tank should be equivalent to 0.94 Nm³/hr of air per m³/hr emptying rate of liquids of any flash point. In other words, for an emptying rate of Q_e m³/hr, the venting capacity shall be $0.94Q_e$ Nm³/hr.

In addition, there are thermal inbreathing guidelines for a given tank capacity for liquids of any flash point. API 2000 [8] furnishes these as a function of tank capacity in the form of tables. The values in such tables can also be expressed in the form of equations, wherein the thermal venting is expressed as a function of tank capacity.

The total venting capacity expressed as a sum of inbreathing due to liquid movement and inbreathing due to thermal effects can be reproduced as:

For tanks up to a capacity of 3500 m³:

$$V_{ibr} = 0.94Q_e + 0.178C \tag{6.1}$$

For tanks larger than a capacity of 3500 m³:

$$V_{ibr} = 0.94Q_e + 3.20C^{0.651} \tag{6.2}$$

The first component on the right-hand side represents venting capacity due to liquid movement out of the tank. The second component represents that due to thermal inbreathing.

The reason for adding the venting requirements on account of both liquid movement and thermal venting is to make sure that the venting arrangement (discussed later) is designed for the most severe case.

Outbreathing (Case-1): This case applies to liquids with a flash point above 37.8°C or a normal boiling point above 148.9°C. For this case, the outbreathing rate for maximum liquid movement into a tank should be equivalent to 1.01 Nm³/hr of air per m³/hr filling rate of liquids. In other words, for a filling rate of Q_f m³/hr, the venting capacity shall be $1.01Q_f$ Nm³/hr.

In addition, there are requirements for thermal outbreathing. The total venting capacity, expressed as a sum of liquid movement and thermal outbreathing can be expressed as:

For tanks up to a capacity of 3500 m³:

$$V_{obr} = 1.01Q_f + 0.107C \tag{6.3}$$

For tanks larger than a capacity of 3500 m³:

$$V_{obr} = 1.01Q_f + 1.92C^{0.651} \tag{6.4}$$

For such liquids where the flash points of the liquids are high, the thermal outbreathing is approximately 60% of the thermal inbreathing requirement. The reason is that the roof and shell temperatures cannot rise as rapidly as they can fall, for example during a sudden rain shower.

Outbreathing (Case-2): This case applies to liquids with a flash point below 37.8°C or a normal boiling point below 148.9°C. For this case, the outbreathing rate for maximum liquid movement into a tank should be equivalent to 2.02 Nm³/hr of air per m³/hr filling rate of liquids. In other words, for a filling rate of Q_f m³/hr, the venting capacity shall be $2.02Q_f$ Nm³/hr.

In addition, there are requirements for thermal outbreathing. The total venting capacity, expressed as a sum of liquid movement and thermal outbreathing can be expressed as:

For tanks up to a capacity of 3500 m³:

$$V_{obr} = 2.02Q_f + 0.178C \tag{6.5}$$

For tanks larger than a capacity of 3500 m³:

$$V_{obr} = 2.02Q_f + 3.20C^{0.651} \tag{6.6}$$

Means of Venting: Venting in storage tanks is of three types: open vents, pressure/vacuum vents and emergence vents.

Atmospheric Vents: In the case of open venting, tanks are vented to the atmosphere. Such type of venting is used when the stored products comprise harmless or non-toxic liquids such as firewater, service water, etc. Such type of tanks operate at atmospheric pressures and the vent pipe is usually provided either with some type of weather hood, or the pipe is provided with a goose neck that prevents rain or snow from entering the tank. When the tank is under filling, it breathes out through the vent. When the tank is under emptying, it breathes in through the vent.

An earlier version of API, i.e., API 2000 (1992), provides a clear definition of when to use open vents in tanks.

Open vents with a flame arrester may be used in tanks where:

- The stored product has a flash point below 37.8°C
- The temperature of the stored product may exceed its flash point.

Likewise, open vents without flame arresters may be used under the following circumstances:

- The stored product has a flash point of 37.8°C or above
- In heated tanks, where the temperature of the stored product is below its flash point
- Tanks having a capacity of less than 9.46 m³ used for storing any product

Needless to say, open vents without a flame arrester may also be used in tanks that do not contain a flammable vapor space.

Pressure/Vacuum Vents: Pressure/vacuum vents are usually used to protect blanketed tanks. Such tanks are used to store petroleum products with flash points below

37.8°C or petroleum products where the temperature could exceed their flash point. With such venting systems, flame arresters are not necessary. In situations where the blanketing valve fails and gets stuck in the open position, the tank gets pressurized by continuous inflow of blanketing gas. The pressure vent opens to protect the tank from rupture. Likewise, in situations where the tank is under emptying and the blanketing gas fails, the tank sees vacuum conditions. The vacuum valve opens, thus preventing a tank collapse.

Pressure/vacuum vents may be weight loaded or spring loaded. The set pressures depend upon the design pressure of the tanks. Typical settings on the pressure side are in the range of 1.5–2.5 kPag. Likewise, settings on the vacuum side are in the range of (–)0.25 to (–)0.5 kPag. The pressure setting of the vent is kept slightly above the tank blanketing pressure, but below the maximum pressure the tank can withstand. Similarly, the vacuum setting is kept higher than what the normal operating conditions can bring about, but below the maximum vacuum the tank can withstand.

Pressure/vacuum vents also conserve the product, in the sense that they minimize evaporation losses and fugitive emissions. This is achieved by preventing the release of vapors during minor variations in temperature, pressure or level, which would otherwise occur in case of open vents. The vents are designed to remain closed until they must open to protect the tanks. Thus, vapors are contained and not released into the atmosphere. In other words, the difference between the blanketing pressure and the set pressure acts as a cushion for pressure fluctuations arising out of liquid movement into or out of the tank or due to atmospheric temperature changes. Figure 6.5 illustrates a pressure-vacuum vent assembly which is self-explanatory.

Refer to the following video link for better understanding:

YouTube link:

PRESSURE VACUUM RELIEF VALVE (TANKS SAFETY EQUIPMENT) Finekay® - YouTube

Emergency Vents: Emergency vents are used when the storage tanks are exposed to fire. In such cases, the venting rate may exceed the rate resulting from a combination

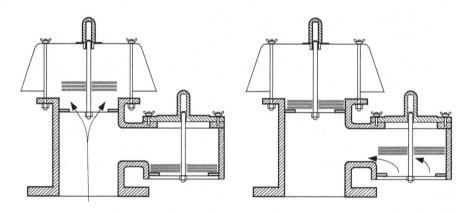

FIGURE 6.5   Pressure-vacuum venting device.

of thermal effects and liquid movement. Emergency vents can be in the form of a gauge hatch (Figure 6.4), that permits the cover to lift under emergency venting loads, or a manhole cover that lifts when exposed to emergency venting loads. For tanks subjected to fire exposure, the required venting capacities are provided in API 2000 [8].

6.2.5 TANK LAYOUT, REMOTE IMPOUNDING AND DIKING

In order to prevent any accidental discharge of Class I, II or IIIA liquids from endangering adjoining property, appropriate facilities should be provided in the tankage area. National Fire Codes [1] describe these requirements. There are two possible facilities that could be provided:

Remote Impounding: In remote impounding, the adjoining property is protected by drainage of the discharge to a remote impounding area. In such a case, the impounded liquid is not held against the tanks. The following guidelines, however, need to be followed for this option:

- A minimum slope of 1% needs to be provided for at least 15 m towards the impounding area.
- The impounding area should at least have a capacity equal to that of the largest tank that can drain into it.
- In case it is not practically possible to have a full capacity remote impounding, partial remote impounding may be permitted. The remainder of the impounding volume shall be accommodated by diking as discussed below.

Impounding by Diking: In cases where the adjoining property is protected by providing diked enclosures with roads all around the enclosure, the following guidelines should be followed:

- The height of the dike wall should be between 1.0–1.8 m from the internal grade. The minimum distance between the tank shell and the toe of the interior dike wall shall be 1.5 m.
- The diked enclosure should be able to contain the contents of the largest tank in the tank farm in case of any emergency. This is calculated as follows:
  1) Calculate the total volume of the diked area (say x).
  2) Add the volume of all the tanks below the height of the dike except the largest tank (say y).
  3) Volume of the largest tank is, say, z.

Now, $x - y$ should be at least equal to z. Accordingly adjust the dimensions of the diked area.

Some companies add a specified volume of fire water that is possible to accumulate in one hour to this diked enclosure. In any case, it would ultimately be the local regulations that govern the type of containment area.

For further details, refer to [1].

TABLE 6.2
Minimum Tank Spacing (shell-to-shell)

	Floating Roof Tanks	Fixed or Horizontal Tanks	
		Class I or II Liquids	Class IIIA Liquids
Tanks under 45 m in diameter			
	1/6 sum of adjacent tank diameters but not less than 0.9 m	1/6 sum of adjacent tank diameters but not less than 0.9 m	1/6 sum of adjacent tank diameters but not less than 0.9 m
Tanks larger than 45 m in diameter			
If remote impounding is provided (in accordance with NFPA 30, 2003)	1/6 sum of adjacent tank diameters	1/4 sum of adjacent tank diameters	1/6 sum of adjacent tank diameters
If diking is provided (in accordance with NFPA 30, 2003)	1/4 sum of adjacent tank diameters	1/3 sum of adjacent tank diameters	1/4 sum of adjacent tank diameters

Source: Reproduced from Flammable and Combustible Liquid Code, NFPA 30, National Fire Protection Association, 2003.

Layout Considerations: NFPA provides detailed guidelines for layout and installation of aboveground storage tanks. Tanks storing Class I, II or IIIA stable liquids shall be separated by distances specified in Table 6.2 [1]. However, in case of crude petroleum storage tanks having individual capacities not exceeding 476 m^3 and located at production facilities in isolated areas, the separation distance need not be more than 0.9 m.

6.2.6 TANK DATA SHEET

Having done the basic tank calculations, the process engineer now generates the tank process data sheet. A typical tank data sheet illustrates all process features of a tank including the type, dimensions, physical properties of the stored liquid, operating and design conditions, material of construction, corrosion allowances and insulation, to name a few. In addition, the data sheet consists of a nozzle table providing the details of all nozzles in the tank, their designations, sizes and pound ratings. As part of the data sheet, a process sketch is also provided. In addition, there are also certain other items which form part of the tank system, although not specifically a part of the tank specification. These particularly include the conservation vents and blanketing valves, in addition to instrumentation items like pressure, temperature and level instruments. These items are also specified in the form of instrument data sheets. This completes the tank system design.

Table 6.3 illustrates a typical tank process data sheet.

TABLE 6.3
Tank Data Sheet

Item Number				T - 301		
Number required				One		
Designation				Naphtha Tank		
Type				Floating Roof		
Design						
Type of roof				Single Deck Pontoon		
Capacity	Nominal / Net Working			1020	/	865
Inside diameter			mm	10000		
Cylindrical height			mm	15000		
Product stored				Naphtha		
Corrosive components				None		
Hazardous class						
Temperature	Operating / Design		°C	25	/	−5/55
Pressure	Operating / Design		mmWC	Atmospheric /	−25/FOW	
Flash point			°C	15		
Liquid density at operating temperature			kg/m³	755–770		
Viscosity at operating temperature			cP	0.49–0.51		
Inert gas blanketing required / Inert gas medium				No	/	none
Corrosion allowance	Roof / Shell		mm	1	/	2
	Bottoms / Internal		mm	2	/	-
Filling rate	nor/max		m³/hr	45/50		
Emptying rate	nor/max		m³/hr	90/100		
Construction						
Roof shape				refer type		
Bottom shape				flat		
Bottom sump				yes (as per API 620)		
Agitator				No		
Insulation	type (roof)	thickness	mm	none	nil	
	type (shell)	thickness	mm	none	nil	
Material (tank)				CS		
Material (internals)				CS		

6.3 PRESSURE STORAGE VESSELS

While atmospheric storage tanks are low-pressure vessels as explained above with a fixed roof or a floating roof, chemical process plants also have pressurized storage vessels which store material having high vapor pressures. Such vessels include spheres and bullets and are typically used to store C3, C4 and LPG which can be readily liquefied under pressure.

Spheres: Spheres come with their uniquely rounded construction in capacities in the range of 1600–12000 m³. One of the advantages of spheres is that to hold a given volume of liquid they require a relatively small plot area. The footprint required for a sphere is substantially lower than that required by a number of storage bullets to store a given volume of liquid.

However, the drawback of spheres is the time required for fabrication, which could be as high as 12–18 months in some cases. Further, due to logistic limitations, such spheres are fabricated in sections (petals and crown) in the workshop and then assembled at the site. The process in addition calls for proper coordination and monitoring at the site to ensure proper sequencing of assembly and welding of the sections [9].

Bullets: In contrast to spheres, bullets can be fabricated entirely in the workshop with fabrication time as low as 8–12 weeks. Multi-bullet installations can easily accommodate the needs of large-scale projects. Further, in case of a multi-bullet installation, if there is a problem or a need for repair or maintenance of one of the bullets, it can be isolated and the contents can be transferred to another bullet in the same facility. However, in the case of a sphere, if there is such a problem, the entire plant may come to a standstill.

Cost Considerations: Spheres have the smallest surface area for a given volume. Therefore, the overall cost of painting per unit volume is also lower for spherical vessels compared to spheres. Further, in view of the fact that the size of a sphere is large compared to that of a bullet, fewer spheres are required compared to bullets for a particular service. This gives rise to a low overall piping quantity for interconnecting of spheres, thus reducing the overall costs [9].

However, having said this, the cost of fabrication for spheres increases substantially since the sections (petals and crown) need to be fabricated in the workshop, shipped to the site, assembled and welded piece by piece in proper sequence, heat treated, and radiographed at the site. This leaves little or no advantage for spheres with regard to the weight saved as a result of the small surface area for a given volume and the reduced piping [10].

Cryogenic Tanks: There is a third category of pressure vessels, namely, the cryogenic vessels. Such vessels store liquefied gases, viz., oxygen, nitrogen natural gas, etc. Such tanks store liquids at cryogenic temperatures. The tanks typically consist of two vessels. The inner vessel is made of stainless-steel or other material having the required strength to withstand cryogenic temperatures. The outer vessel is made of carbon steel and is not exposed to cryogenic temperatures. The space between the inner and the outer jacket is under vacuum and is filled with an insulating material. The inner vessel is typically designed for a working pressure of 30 kg/cm²g. The tanks could be constructed as vertical or horizontal installations. Sizes range from a few cubic meters to more than 1000 cubic meters.

6.4   MISCELLANEOUS PROCESS VESSELS

There is a third category of vessels. These are part of the main process and are not meant for large-scale storage of process fluids. They are, therefore, located within the

TABLE 6.4
Normal Liquid Hold-up in Process Vessels

Service	Residence Time, min
Feed Surge Drum	15
Distillate Reflux Drum	5
Condensate Flash Drum	5
Compressor Interstage Knock-out Drum	5
Flare Knock-out Drum	20–30
Gas-Liquid Separator	5
Liquid Feed to a Furnace	10–20

process plant and not in the utilities and offsites area. Such vessels normally operate at pressures higher than atmospheric, are either horizontal or vertical and are also sometimes called "drums." Unlike tanks, such vessels have hold-up times of a few minutes. Table 6.4 summarizes the hold-up times in this category of vessels. Such vessels can be located depending upon their purpose. The following are some examples of such purposes:

- As feed surge drum in front of a unit.
- Between equipment as accumulator to take care of fluctuations in the process, viz., reflux drums.
- As knock-out drum upstream of compressor to prevent liquid free vapor going to the compressor.
- As a pulsation dampener downstream of a reciprocating compressor to smooth out pressure surges.

The length to diameter ratios of such drums are in the range of 2.5–5.0. At higher pressures, the vessel thickness increases, and it becomes more economical to increase the length rather than the diameter. At pressures up to 17 kg/cm²g, one can go with a length/diameter ratio of 3. Between 18 and 34 kg/cm²g, it is common to have the ratio at 4. Beyond a pressure of 34 kg/cm²g the length/diameter ratio is typically kept at 5 [11].

6.4.1 Reflux Drums

Reflux drums are process vessels used to store overhead products from distillation columns (Figure 6.6). They provide one or more of the following functions [11]:

- Surge volume to protect the column reflux pump.
- Smoothing out of fluctuations of distillate product going to downstream unit or to reflux.
- Separation of vapor from liquid.
- Settling time for separating the water from the hydrocarbon phase.

6.4.2 Vapor-Liquid Separators

Typically, separation of vapor and liquid in a process vessel is accomplished by density difference aided by gravitational force, by impingement separation, or both.

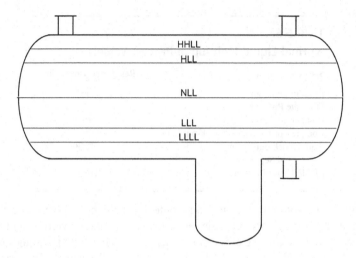

FIGURE 6.6 Reflux drum.

In gravitational separation, whenever a particle falls through a surrounding medium, the surrounding medium exerts a drag force upon the particle opposing its motion. The drag force is given as:

$$F_d = C_d A_p \frac{\rho_v U_s^2}{2} \qquad (6.7)$$

where A_p is the projected area of the particle on a plane normal to the flow and is given by $A_p = \pi D_p^2 / 4$

Rearranging, (Equation 6.7) can be written as,

$$F_d = C_d \frac{\pi \rho_v U_s^2 D_p^2}{8} \qquad (6.8)$$

There will also be a gravitational force on the particle given by:

$$F_g = \frac{(\rho_l - \rho_v) \pi D_p^3 g}{6} \qquad (6.9)$$

The liquid drops falling under the action of gravity will continue to accelerate until the drag force just balances the gravitational force, after which the fall will be at a constant velocity. Under these conditions, $F_d = F_g$. Rearranging for U_s, we have:

$$U_s = \left(\frac{4 D_p g (\rho_l - \rho_v)}{3 \rho_v C_d} \right)^{0.5}. \qquad (6.10)$$

Two-Phase Vertical Separator: Design of such vessels is based on the principle that the allowable vapor velocity should always be less than the settling velocity of the droplets. Consider a compressor interstage knock-out drum or a flare knock-out drum. While sizing such vessels, Figures 6.7 is commonly used to calculate the drag coefficients. Figure 6.8 provides guidelines to arrive at the dimensions.

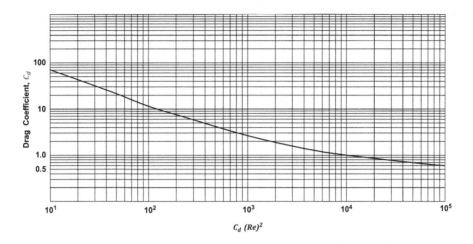

FIGURE 6.7   Determination of drag coefficients.

Source: Developed with permission using data from Table 5.22, Page 5.67, "Perry's Chemical Engineers' Handbook" 6th Edition.

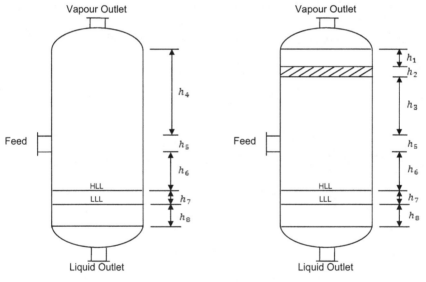

h_1	200 mm
h_2	100 - 150 mm
h_3	greater of 900 mm or 0.45D
h_4	greater of 900 mm or 0.7D
h_5	Feed Nozzle diameter
h_6	500 mm
h_7	As per hold-up time
h_8	200 mm

FIGURE 6.8   Vessel sizing – vertical separators.

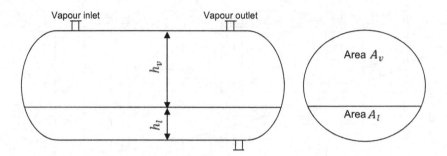

FIGURE 6.9   Vessel Sizing – Horizontal Separators

The following is the step-wise procedure for sizing of the vessel:

1. The drag coefficient C_d is expressed as

$$C_d (Re)^2 = 0.13 \times 10^8 \rho_v D^3 (\rho_l - \rho_v) / \mu_v^2 \tag{6.11}$$

Figure 6.7 provides values of C_d as a function of $C_d (Re)^2$.
2. Calculate C_d from Figure 6.7.
3. Calculate settling velocity U_s from (Equation 6.10).
4. Calculate maximum allowable vapor velocity U_v as follows:

$U_v = 0.8U_s$ in case there is no mist eliminator
$U_v = 1.7U_s$ in case there is a mist eliminator. If specific information is available
    from the mist eliminator vendor, the same should be used.

5. Calculate normal liquid hold-up volume V_l using guidelines in Table 6.4.
6. Calculate drum diameter as follows:

$$D = \left(\frac{4Q}{\pi U_v} \right)^{0.5} \tag{6.12}$$

7. Estimate height of drum section h_7 between HLL and LLL as follows:

$$h_7 = \frac{4}{\pi} \frac{V_l}{D^2} \tag{6.13}$$

8. Establish vessel length L using Figure 6.8 which serves as a guideline.
9. Check that L/D is between 2.5 and 3.0.

***

EXAMPLE 6.1

Vapors are discharged from the first stage of a reciprocating compressor to a
compressor interstage knock-out drum. Consider a liquid droplet size of 75
microns. Size the knock-out drum as per the following data:

Fluid Properties

Temperature	65°C
Vapor flowrate	10,000 kg/hr
Amount of liquid	110 kg/hr
Density of vapor, ρ_v	7.0 kg/m³
Density of liquid, ρ_l	800.0 kg/m³
Viscosity of vapor, μ_v	0.022 cP

Calculations

D_p

= 75 microns
= 0.000075 m

μ_v

= 0.022 cP

$C_d (Re)^2$

$= 0.13 \times 10^8 \rho_v D^3 (\rho_l - \rho_v) / \mu_v^2$

$= 0.13 \times 10^8 \times 7.0 \times (0.000075)^3$
$(800-7.0)/(0.022)^2$

= 62.9

C_d (from Figure 6.7)

= 14

Settling velocity, U_s (from
(Equation 6.10)

$$= \left(\frac{4 \times 0.000075 \times 9.81(800 - 7.0)}{3 \times 7 \times 14} \right)^{0.5}$$

= 0.089 m/sec

Maximum allowable vapor velocity,
U_v

= $1.7 U_s$ (assuming a demister)
= 0.15 m/sec

Normal liquid hold-up volume, V_l
(Table 6.4)

= 110 × 5/(800 × 60)
= 0.011 m³ (based on 5-minute
hold-up)

Vapor flowrate, Q

= 10000/(7.0 × 3600)
= 0.397 m³/sec

Vessel diameter, D (Equation 6.12)

= [(4 × 0.397)/(π × 0.15)]$^{0.5}$
= 1.84 m

Height of vessel section, h_7
(Equation 6.13)

= 4 × 0.011/π × 1.84^2

= 0.004 m (i.e., 4 mm)

From the mass flowrate and vapor density, we get the volumetric flowrate of
vapor as 1429 m³/hr. Assuming a velocity of 20 m/s, we get an inlet nozzle
diameter of 159 mm. We take a 6" pipe. Therefore, h_5 = 150 mm.

From Figure 6.8,

h_1

= 200 mm

h_2

= 150 mm

h_3	= 900 mm (which is greater than 0.45D)
h_5	= 150 mm
h_6	= 500 mm
h_7 (calculated above)	= 4 mm
h_8	= 200 mm
Therefore, height of vessel	= 2104 ≃ 2100 mm

***

Two-Phase Horizontal Separator:

Design of such vessels is based on the principle that the time in which the droplets get carried horizontally from one end to the other, they must fall onto the liquid surface.

1. Assume a vessel diameter D and a length L, keeping L/D between 2.5 and 3.0.
2. Calculate total vessel cross-sectional area as

$$A_t = \frac{\pi}{4}D^2 \qquad (6.14)$$

3. Estimate the liquid hold-up volume V_l corresponding to HLL.
4. The cross-sectional area corresponding to the liquid hold-up is given as

$$A_l = \frac{V_l}{L} \qquad (6.15)$$

The corresponding liquid height is given by h_l (Figure 6.9).
5. The cross-sectional area remaining for the vapor flow is given as

$$A_v = A_t - A_l \qquad (6.16)$$

The corresponding vapor height is given by h_v (Figure 6.9).
6. The residence time of vapor in the vessel is given by

$$\theta = \frac{A_v L \times 3600}{Q} \qquad (6.17)$$

7. To avoid carryover of droplets, during this period θ, the droplets should fall through the vapor space into the liquid. That is, they should traverse a vertical distance of h_v given by (Figure 6.9).

$$h_v = U_s \theta \qquad (6.18)$$

8. During this period, the droplet also gets carried by a horizontal distance L. L is calculated by Equation 6.17.
9. The calculated value of L should be less than the assumed value. If the calculated value is greater, then a new vessel diameter D is assumed and the calculations repeated.

***

EXAMPLE 6.2   SIZE A HORIZONTAL FLARE KNOCK-OUT
DRUM WITH THE FOLLOWING PROCESS DATA

Fluid Properties

Temperature	70°C
Flare load	135,000 kg/hr
Liquid load	5,000 kg/hr
Density of vapor, ρ_v	3.9 kg/m³
Density of liquid, ρ_l	750.0 kg/m³
Viscosity of vapor, μ_v	0.015 cP
Droplet diameter	400 microns

Calculations

Volumetric flowrate of vapor is given by

Q	= (135000-5000)/3.9
	= 33333.3 m³/hr
	= 9.26 m³/s
D_p	= 400 microns
	= 0.0004 m
μ_v	= 0.015 cP

From Equation 6.11,

$$C_d (Re)^2 = 0.13 \times 10^8 \rho_v D^3 (\rho_l - \rho_v) / \mu_v^2$$

$$= 0.13 \times 10^8 \times 3.9 \times (0.0004)^3 \times (750\text{-}3.9)/(0.015)^2$$

$C_d (Re)^2$	= 10760
C_d (from Figure 6.7)	= 0.9

$$\text{Settling velocity } U_s \text{ (from Equation 6.10)} = \left(\frac{4 \times 0.0004 \times 9.81(750 - 3.9)}{3 \times 3.9 \times 0.9} \right)^{0.5}$$

$$= 1.05 \text{ m/sec}$$

Trial 1

Assume D = 2.5 m and L = 5 m for the separator.

Assume a liquid hold-up of 30 minutes in the separator.

Liquid collected in the separator	= (5000 × 0.5)750
	= 3.3 m³.

From Figure 6.9,	$A_l L$ = 3.3 m³.
Since L is assumed as 5.0 m,	A_l = 0.66 m².
With D = 2.5 m and A_l = 0.66 m²,	A_v = 4.25 m².

From geometry, $h_l = 0.48$ m and $h_v = 2.02$ m.

The 400-micron droplet falls a
distance of 2.02 m.

From Equation 6.18, $h_v = U_s\theta$

Or $\theta = h_v/U_s$

$$= 2.02/1.05 = 1.9 \text{ s}$$

During the time θ, the droplet covers a vertical distance of h_v and also gets
carried by a distance of L.

From Equation 6.17, L is given by

$$L = \frac{Q\theta}{A_v \times 3600}$$
$$= (33333.3 \times 1.9)/(4.25 \times 3600)$$
$$= 4.14 \text{ m}.$$

The calculated value of L is still well below 5.0 m so we have chance to
optimise on the diameter

Trial 2

Assume $D = 2.3$ m and $L = 5$ m.
Assume a liquid hold-up of 30 minutes in the separator.

Liquid collected in the separator $= (5000 \times 0.5)750$

$$= 3.3 \text{ m}^3.$$

From Figure 6.9, $A_lL = 3.3$ m^3

Since L is assumed as 5.0 m, $A_l = 0.66$ m^2

With $D = 2.3$ m and $A_l = 0.66$ m^2, $A_v = 3.50$ m^2

From geometry, $h_l = 0.50$ m and $h_v = 1.80$ m.

The 400-micron droplet falls a distance of 1.80 m.

From Equation 6.18, $h_v = U_s\theta$

Or $\theta = h_v/U_s$

$$= 1.80/1.05 = 1.71 \text{ s}$$

During the time θ, the droplet covers a vertical distance of h_v and also gets
carried by a distance of L.

From Equation 6.17, L is given by

$$L = \frac{Q\theta}{A_v \times 3600}$$
$$= (33333.3 \times 1.710)/(3.50 \times 3600)$$
$$= 4.52 \text{ m}$$

The calculated value of L looks reasonable while being slightly lower than 5.0 m.
Hence we converge at $D = 2.3$ m and $L = 5.0$ m

***

Symbols

A_p Projected area of the particle, m^2
A_t Vessel total cross-sectional area, m^2
A_l Vessel cross-sectional area of the liquid hold-up, m^2
A_v Vessel cross-sectional area of the vapor portion, m^2
C Tank capacity, m^3
C_d Drag coefficient, dimensionless
D Vessel diameter, m
D_p Particle/Drop/Bubble diameter, m
F_d Drag force, N
F_g Gravitational force, N
g Acceleration due to gravity, m/s^2
h_l Distance along the diameter occupied by liquid, m
h_v Distance along the diameter occupied by vapor, m
L Length of vessel, m
U_s Settling velocity, m/s
U_v Maximum allowable vapor velocity, m/s
V_l Liquid hold-up volume, m^3
Q Volumetric flowrate of vapor, m^3/hr (for horizontal vessels)
Q Volumetric flowrate of vapor, m^3/s (for vertical vessels)
Q_e Tank emptying rate, m^3/hr
Q_f Tank filling rate, m^3/hr
θ Residence time of vapor, s
ρ_l Density of liquid, kg/m^3
ρ_v Density of vapor, kg/m^3
V_{ibr} Inbreathing, Nm^3/hr
V_{obr} Outbreathing, Nm^3/hr
μ_v Viscosity of vapor, cP

REFERENCES

1. NFPA 30, Flammable and Combustible Liquid Code, National Fire Protection Association, 2003.
2. Amrouche, Y. et al., "General Rules for Aboveground Storage Tank Design and Operation", *Chemical Engineering Progress*, pp. 54–58 (December 2002).
3. http://www.wermac.org/equipment/storage_tanks_vessels_general.html
4. http://www.s-tank.com/eng/products/pro03.htm
5. Mukherjee, S., "Understanding Atmospheric Storage Tanks", *Chemical Engineering*, April 2006.
6. Welded Steel Tanks for Oil Storage, American Petroleum Institute, API Standard 650, 10th edition, November 1998.

7. Oil Industry Safety Directorate, Government of India, Recommended Practice on Oil Storage and Handling, OISD-Standard-108, 1997.

8. Venting Atmospheric and Low-Pressure Storage Tanks, API Standard 2000, *American Petroleum Institute*, March 2014.

9. Ezzell, G., "LPG Storage Bullet Tanks vs. LPF Storage spheres/Hortonspheres", *Transenergy*, November 14, 2016, https://www.transtechenergy.com/lpg-ngl-storage-news/lpg-storage-bullet-tanks-vs.-lpg-storage-spheres-/-hortonspheres

10. BNH Gas Tanks, http://www.lpgstoragetanks.com/

11. Wallas, S., *Chemical Process Equipment – Selection and Design*, Butterworth-Heinemann, 1990.

7 Heat Exchangers

7.1 INTRODUCTION

Heat exchangers constitute a major part of performing equipment in the chemical process industry, especially in oil refineries and petrochemical units. Whether it is a feed preheater, overhead condenser, trim cooler or column bottoms reboiler, heat exchangers play a major role in a chemical process plant.

This chapter deals with shell-and-tube exchangers, plate exchangers and double pipe exchangers. It is not the intention to cover theory or mathematical equations in this chapter. Rather, the focus is to provide an insight into the practical details of heat exchangers, construction features, choice of fluids, etc. The finer aspects of design such as the Tubular Exchanger Manufacturers Association (TEMA) type and their criteria for selection are also explained. The concept of leakage streams in shell-and-tube heat exchangers and how they affect performance is also discussed.

7.2 SHELL-AND-TUBE HEAT EXCHANGERS

Components of Shell-and-Tube Heat Exchangers: The main components of a typical shell-and-tube heat exchanger are covered by the following:

- Shells and Shell Covers
- Channels and Channel Covers
- Tubes and Tubesheets
- Baffles
- Nozzles

There are, however, several other components including impingement plates, pass partition plates, longitudinal baffles, tie rods, spacers, sealing strips, supports and foundation [1, 2].

A shell-and-tube heat exchanger is divided into three sections: front head, shell and rear head. Figure 7.1 illustrates the TEMA nomenclature for the various construction options. Shell-and-tube heat exchangers are described by a combination of three letters each of the three sections. For example, an AEL type exchanger has a removable channel and cover, a shell, and fixed tubesheet rear head.

7.2.1 CLASSIFICATION OF SHELL-AND-TUBE HEAT EXCHANGERS

Fixed Tubesheet: In a fixed tubesheet heat exchanger, there are straight tubes which are secured at both ends to the respective tubesheets. The construction may be removable channel covers (e.g., AEL), or bonnet-type channel covers (e.g., BEM) or others (Figure 7.2).

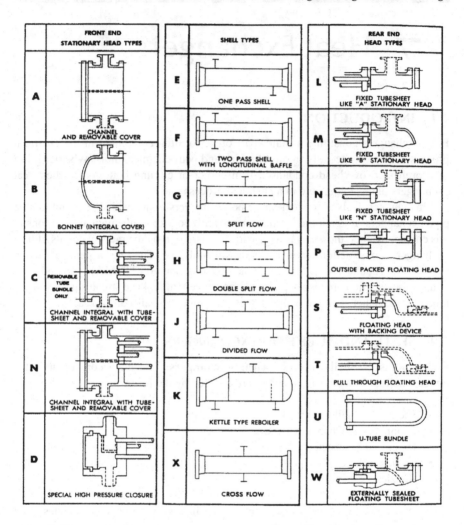

FIGURE 7.1 TEMA construction types.

Source: TEMA.

The fixed tubesheet exchanger has a simple construction; therefore, it is relatively cheap and this is its main advantage. In addition, in such exchangers, the tubes can be mechanically cleaned after removal of the channel cover or bonnet.

However, these exchangers suffer a drawback. Since the bundle is fixed to the shell and cannot be removed, it is not possible to clean the outside of the tubes by mechanical means. The application of such exchanger is therefore limited to the use of clean fluids on the shellside. If, however, a satisfactory chemical cleaning process could be employed, the fixed tubesheet construction could be employed for fouling services in the shellside as well [1, 2].

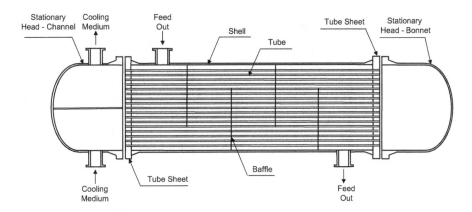

FIGURE 7.2 Fixed tubesheet heat exchanger.

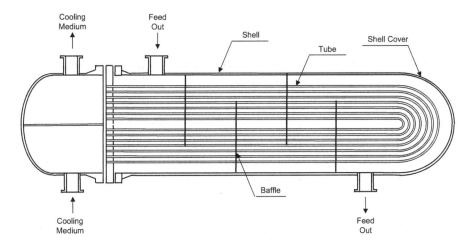

FIGURE 7.3 U-tube heat exchanger.

U-Tube: The tubes of a U-tube heat exchanger (Figure 7.3) are bent in the shape of a U, hence the name. In such exchangers, there is only one tubesheet. It may, however, be noted that while a single tubesheet will incur a lower cost, this is to some extent offset by the additional costs incurred in bending the tubes. Further, the somewhat larger shell diameter due to the minimum U-bend radius eliminates any advantage of having a single tubesheet.

However, a U-tube heat exchanger definitely enjoys certain advantages. Since one end is free, the bundle can expand or contract in response to stress differential. The other advantage is that the outsides of the tubes can be cleaned, since the tube bundle can be removed.

Until a few years ago, it was believed that the insides of the U-tubes cannot be cleaned effectively, and therefore cannot be employed for fouling services in the

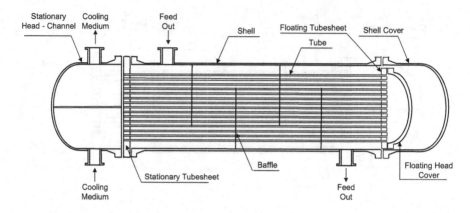

FIGURE 7.4   Floating head heat exchanger.

tubeside. This was counted as a drawback of these types of exchangers [1, 2]. However, in the meantime, new technologies have been invented, such as the Sentinel™ Technology by StoneAge [3] which uses the process of waterblasting inside of the tubes. Refer to the following video link for better understanding:

YouTube link: Sentinel™ Automated Cleaning Technology by StoneAge - YouTube

Floating Head: The most versatile type of shell-and-tube exchanger is the floating head construction (Figure 7.4). Incidentally it is also the most expensive of the three. In this type of construction, one tubesheet is fixed relative to the shell, and the other is free to float within the shell. This type of construction therefore allows free expansion of the tube bundle. In this type of construction, cleaning of both the inside and the outside of the tubes is possible. Thus, floating head exchangers can be suitably deployed for services where both the shellside and the tubeside fluids are dirty.

The advantage of this construction is that the tube bundle can be removed from the shell without removing either the shell or the floating head cover. This reduces maintenance time [1, 2].

Refer to the following video links on assembling floating head and U-tube heat exchangers:

YouTube links: FLOATING HEAD SHELL & TUBE HEAT EXCHANGER - YouTube

"U TUBE TYPE" SHELL & TUBES HEAT EXCHANGER - YouTube

7.2.2   SELECTING SHELLSIDE AND TUBESIDE FLUIDS

This is an area of heat exchanger design where experience comes in useful. An improper choice of shellside and tubeside fluids could create many problems during plant operation.

The selection could be relatively easy for certain services. For example, cooling or condensing hydrocarbons is normally carried out with cooling water and this is almost always placed inside the tubes since it is easier to clean the inside of tubes rather than the outside. However, if the fluid being cooled or condensed is corrosive and requires the use of expensive material of construction, then both the shell as well as the tubes would need to be made of the expensive material. In such cases, it is better to place cooling water in the shellside and the corrosive fluid on the tubeside.

However, not all cases are as simple. The following parameters need to be considered while making a selection [4]:

Corrosive Fluids: Corrosive fluids call for use of a superior metallurgy. If the more corrosive fluid is allocated to the tubeside, only the tubes, channel, channel cover, floating head cover and tubeside tubesheet face would need to be of superior metallurgy. However, if the same is allocated to the shellside, the shell, shell cover, floating head cover, shellside face of tubesheet, as well as the tubes would need to be of superior metallurgy. For an exchanger of identical geometry, the cost of the latter heat exchanger would be higher. Hence, it is general practice to place the more corrosive fluid on the tubeside.

Pressure: In addition, tubes of diameters normally used in the process industry, viz., ¾" and 1", can withstand high internal pressures. In comparison, shells would need to be of much higher thicknesses if they were to withstand similar pressures. In addition, there are fewer components on the tubeside than on the shellside and, hence, few thicker and expensive components would be required in the tubeside. To summarize, the cost of a heat exchanger will be lower if the stream with higher pressure is placed on the tubeside.

Viscosity: It is well known that turbulence increases heat transfer coefficients. Further, considerably higher heat transfer coefficients can be obtained on the shellside than on the tubeside due to higher turbulence. Therefore, it is always preferable to put viscous fluids on the shellside.

Fouling Tendency: There are more dead spaces on the shellside than on the tubeside; therefore, the shellside is more susceptible to fouling. Therefore, after several hours of operation, fouling builds up to a certain level and it becomes necessary to clean the heat transfer surface. Mechanical cleaning of the inside of the tubes is much easier than cleaning the outside of the tubes. Therefore, a fluid which is fouling should preferably flow through the tubeside. However, there is a contradiction here. The dirtier the fluid, normally, the more viscous it will be. We have discussed above that the more viscous fluid should be on the shellside. The designer's experience comes in useful here to make a proper judgement.

Flowrate: For a stream with low flowrates, if placed on the tubeside, its velocity could be increased by increasing the number of tube passes, subject to pressure drop limitations. Likewise, if the stream is placed on the shellside, for a given shell diameter, velocity could be increased by reducing the baffle spacing and baffle cut. However, very low baffle spacing or cut would not give a good performance. Thus, for a low flowrate stream flowing in the shellside, it is difficult to obtain a satisfactory heat transfer coefficient. One way to tackle this problem is to deploy two or more shells. However, this is an expensive option.

7.2.3   Choosing the TEMA Type

Shell and tube heat exchangers are described by the following (refer Figure 7.1):

- front end stationary head type
- shell type
- rear end head type

7.2.3.1   Front End Stationary Head

Type A: This configuration has ease of dismantling the cover and is therefore used when there is need for frequent tubeside cleaning. In doing so, the channel and the piping are undisturbed. Further, it is used for low pressure applications so that a flat head could be accommodated rather than an ellipsoidal one.

Type B: This is used for medium pressure applications. However, in this case, the dismantling of the head is more difficult. Therefore, it is used for cleaner tubeside fluids that do not require frequent cleaning.

Type C: Here the tubesheet end of the channel is welded to the tubesheet, and hence the channel is integral with the tubesheet. However, this type of head is rarely used.

Type N: Here the channel head is integral with the tubesheet as well. However, unlike the TEMA C head, the tubesheet is integral to both the channel and the head. Thus, for this type of head, there are only welded joints between channel and tubesheet and between tubesheet and shell, thus eliminating leakage between the shell and tubes. Such exchangers are useful when handling hazardous or toxic materials.

Type D: The TEMA D head has a special high-pressure closure and is used for applications of very high pressure on the tubeside. Most of these designs are patented.

7.2.3.2   Shell Type

The choice of TEMA type of exchanger is based upon how the fluid is supposed to flow inside the shell. The various TEMA shell types are E, F, G, H, J, K and X. The types which are more commonly used in the industry are described below.

Type E: This is the most common shell type and is inexpensive and simple. This is a single-pass shell where the shellside fluid enters at one end of the shell and leaves from the other end.

Type F: In this type of shell configuration, the shells are divided into two passes created by a longitudinal baffle. The shellside fluid enters the shell at one end, flows through the entire length of the shell, and then takes a turn and flows through the other half. It finally leaves from the other half. The longitudinal baffle does not traverse the entire length of the shell so as to allow sufficient space to the shellside flow to change directions and flow into the second pass. This configuration represents a true counter-current flow and thus avoids conditions where the cold fluid leaves at a temperature higher than that of the hot stream.

Type G, H: TEMA G shell is deployed in horizontal thermosiphon type reboilers and has a split flow arrangement. There are no baffles, but a central support plate.

In view of limitations of unsupported tube length, G type shells cannot be used in exchangers with tube lengths greater than 3000 mm. In cases where larger tube lengths are required, the TEMA H shell is used, which is basically two G shells placed side-by-side.

Type J: This is a divided flow shell which has either two inlets and one outlet, or one inlet and two outlets. Such shells are used when the allowable pressure drop is significantly low. In case of condensers, the inlet vapor line is expected to be much larger than the outlet liquid line, so in such cases, it is normal practice to have two inlets wherein the larger vapor line would be reduced to two smaller lines, and one outlet line.

7.2.3.3 Rear End Head Type

There are eight rear end head types: L, M, N, P, S, T and U. Practically speaking, they fall under three construction types. The head types L, M and N belong to the fixed tubesheet type heat exchangers. The rear head L is identical to front head A, and the rear head M is identical to front head B. The rear head N is identical to front head N.

The rear head type U applies to U-tube heat exchangers. Types P, S, T and W belong to the floating head type construction.

7.2.4 TUBESIDE DESIGN

Tube Layout Patterns: There are four tube layout patterns (refer to Figure 7.5), which are used in shell-and-tube heat exchangers, namely, triangular (30°), rotated triangular (60°), square (90°) and rotated square (45°).

A triangular or rotated triangular pattern will accommodate a larger number of tubes compared to that for a square or rotated square pattern. Furthermore, a triangular pattern will produce a higher turbulence and therefore a higher heat transfer coefficient. For identical tube pitch and flowrates, the tube layouts in decreasing order of shellside heat transfer coefficient and pressure drop are: 30, 45, 60, and 90°.

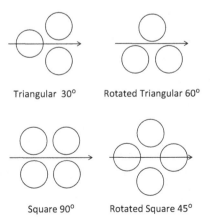

Triangular 30°   Rotated Triangular 60°

Square 90°   Rotated Square 45°

FIGURE 7.5 Tube pitch.

Thus, the 30° layout will have the highest heat transfer coefficient and the highest pressure drop [5].

In spite of its advantages, a typical pitch of 1.25 times the tube OD provides difficulties in mechanical cleaning of the outside of the tubes, since access is not available. Therefore, use of a triangular layout is limited to clean fluids flowing though the shellside. However, if chemical cleaning on the shellside is effective, then a triangular layout could be used since chemical cleaning does not require access lanes.

When a dirty fluid flows through the shellside and therefore mechanical cleaning is required, the square pattern should be selected. If a square pitch is selected for ease of cleaning, then fixed tubesheet design cannot be used because cleaning is not feasible. In such cases, normally U-tube or floating head exchanger are used where the tube bundle can be pulled out for cleaning.

It may, however, be pointed out that as explained earlier, the square pitch offers the lowest shellside heat transfer coefficient. However, when using a square pitch to facilitate shellside cleaning, if the Reynold's number is lower than 2000, it would be better to go for a rotate square pitch (i.e., 45° pattern) since this would give a better heat transfer coefficient while retaining the advantage of ease of shellside cleaning [1].

Tube Pitch: The tube pitch is the center-to-center distance between adjoining tubes. The selection of tube pitch is a tradeoff between a smaller pitch leading to a comparatively higher shellside heat transfer coefficient and a larger pitch leading to decreased shellside plugging and ease of shellside cleaning [5]. In addition, a larger pitch gives a lower shellside pressure drop. For a triangular pattern, TEMA recommends a minimum tube pitch of 1.25 times the tube OD. For square patterns, TEMA recommends a minimum cleaning lane of 6.4 mm between adjacent tubes. Thus, for square patterns the minimum tube pitch should be 1.25 times the tube OD or the tube OD plus 6.4 mm, whichever is greater [1].

7.2.5 Shellside Design

Shellside Stream Analysis: Although one of the major functions of the baffles is to generate crossflow for better heat transfer coefficients, this objective is not really met in conventional shell-and-tube heat exchangers. For the construction of the heat exchangers, a number of clearances are required. A part of the shellside fluid bypasses through these clearances instead of following a crossflow pattern. Flow across the bundle is divided into five separate streams (A, B, C, E and F) as illustrated in Figures 7.6(a) and 7.6(b). In order of decreasing effectiveness of heat transfer, the following is a description [5]:

A *Stream:* This is the leakage stream through the annular clearance between the tubes and the tube holes in the baffle. This stream is also considered effective for heat transfer since the heat transfer coefficients in the annular spaces are quite high.

B *Stream:* This is the actual crossflow stream flowing over the tubes and is fully effective for heat transfer.

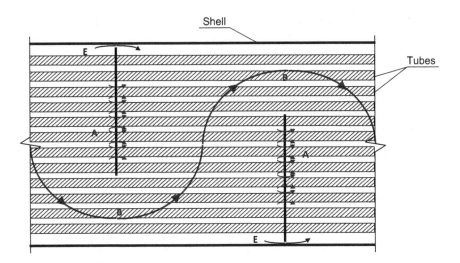

FIGURE 7.6(a)   Slip streams A, B and E.

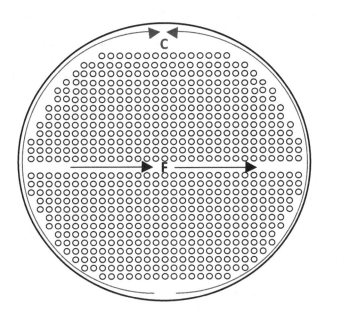

FIGURE 7.6(b)   Slip streams C and F.

C Stream: This is the stream that flows through the annular space between the
  tube bundle and the shell. This steam is partially effective for heat transfer.

E Stream: This is the leakage stream that passes through the clearance between
  the baffle and the shell. This stream is the least effective for heat transfer
  since it may not come in contact with any of the tubes.

F Stream: For exchangers having multiple tube passes, open passages exist between tube passes. The F stream flows between these passages. This stream is less effective compared to the A stream because it comes into contact with a lower heat transfer area per unit volume.

In order to have an optimum thermal design, the following guidelines may be followed [2, 5]:

Ideally, the design of a baffled shell-and tube heat exchanger should be such that about 80% represents the B stream; however, this is rarely the case. Sometimes, narrow baffle spacings and higher pressure drops result in substantial flows going as A, C or E streams. In the extreme case, the B stream should be at least 40% of the total flow; otherwise, the baffle geometry and the various clearances should be checked.

The fraction of A stream increases in case of narrow baffle spacings. However, this is not much of a concern since the A stream is effective for heat transfer.

The C and F streams are partially effective for heat transfer. The tube bundle should be so designed that the flow fractions of each of these should not normally exceed 10% of the total flow. Sealing strips for C streams and seal rods for F streams are normally used to minimize the flows of these streams.

The E stream is ineffective and also does not contact the heat transfer surface. Unfortunately, there is not much a designer can do, as the necessary clearances between baffle and shell are dictated by manufacturing restrictions. However, if the flow fraction of the E stream exceeds 15%, the design engineer should consider multi-segmental baffles instead of single-segmental baffles since in the latter case, the shellside pressure drop is lower, and this results in higher flow fractions going with the A, B and C streams.

Baffle and Baffle Spacing: Baffles provide a desirable velocity for the shellside fluid, are used for supporting the tubes, and also prevent failure of tubes due to tube-induced vibrations [1, 2].

The TEMA standard specifies minimum baffle spacing as one-fifth of the shell diameter or 50 mm, whichever is higher. A minimum spacing of 50 mm is required for cleaning the bundle. A very low baffle spacing increases the leakage streams due to the increase in the resistance of the main crossflow path. Thus, any advantage gained in terms of increase in heat transfer coefficients is eliminated [1, 2].

The maximum baffle spacing is the shell ID. Higher baffle spacings will create flow in the longitudinal direction rather than the preferred cross-flow. Moreover, large baffle spacings lead to large unsupported lengths in the exchanger which may cause tube failure due to flow induced vibrations. Optimum ratio of baffle spacing to shell diameter is between 0.3 and 0.6 [1].

In Table 7.1, a heat exchanger design with three different baffle spacings, i.e, 460, 510 and 575 mm is shown. As the baffle spacing increases, the cross-flow bypass (B stream), the bundle-to- shell bypass (C stream) and the tubefield partition bypass (F stream) increase, whereas the rest decrease. The shellside heat transfer coefficient decreases and the pressure drop decreases even faster.

Baffle Cut: Baffle cut can vary between 15 and 45% of the shell diameter. Very small and very large baffle cuts are counter-productive to efficient heat transfer. This is

TABLE 7.1
Slip Streams at Difference Baffle Spacings

Parameter		Case 1	Case 2	Case 3
TEMA Type		AES	AES	AES
Pressure Drop Allowable Shellside	kg/cm^2	0.3	0.3	0.3
Pressure Drop Calculated	kg/cm^2	0.1	0.082	0.066
Shell ID	mm	850	850	850
Tube OD	mm	20	20	20
Tube Count		676	676	676
Tube Length	mm	6000	6000	6000
Baffle Spacing	mm	460	510	575
Baffle Cut	%	30	30	30
Cross-flow Velocity	m/s	1.24	1.13	0.99
Windowflow Velocity	m/s	1.32	1.34	1.31
Shellside Heat Transfer Coefficient	kcal/m^2.hr.$^\circ$C	790	775	758
Overall Heat Transfer Coefficient	kcal/m^2.hr.$^\circ$C	360	356	353
Over-design	%	5.23	4.41	3.52
A Stream		0.072	0.067	0.062
B Stream		0.563	0.568	0.574
C Stream		0.169	0.171	0.173
E Stream		0.094	0.088	0.081
F Stream		0.103	0.106	0.110

illustrated in Figure 7.7 and is self-explanatory. Reducing baffle cut below 20% to increase the shellside heat transfer coefficient, or increasing the baffle cut beyond 35% to decrease the shellside pressure drop usually leads to poor designs [1, 2]. The recommended baffle cut is between 20 and 35%.

Refer to Table 7.2 which illustrates the effect of varying the baffle cut in a given heat exchanger. The baffle is varied from 20 to 40%. The cross-flow bypass (B stream) increases rapidly. The baffle-to-shell leakage (E stream) and tube-to-baffle leakage (A stream) decrease steadily. The bundle-to-shell bypass (C stream) and the tubefield partition bypass (F stream) remain somewhat steady. It can be seen that a baffle cut of 25% gives the higher heat transfer coefficient and over-design. As the window flow velocities decrease as a result of the increasing baffle cut, the shellside coefficient and the overall coefficient decrease and so do the over-design values.

For single-phase fluids on the shellside, perpendicular cut baffles are normally used. However, in case of two-phase flow, viz. condensers, parallel cut baffles are recommended.

Cross-flow and Window Velocities: Flow across tubes is referred to as "cross-flow," whereas flow parallel to the tubes (i.e., through the baffle cut area) is referred to as "window flow." The cross-flow velocity and the window velocity should be of the same order of magnitude, preferably within 20% of each other [1, 2].

Seal Strips and Seal Rods: Sealing devices are provided to stop/restrict bypass streams. Sealing strips are provided to stop the C stream from bypassing the bundle. Seal rods are provided to stop the F stream from bypassing the bundle.

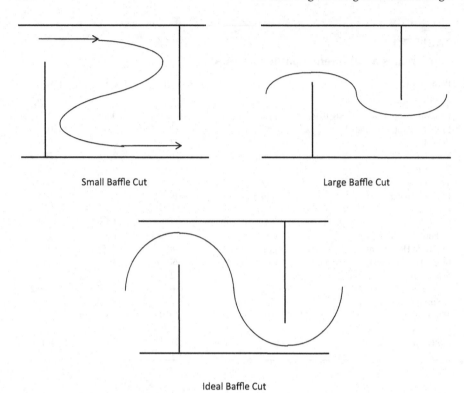

FIGURE 7.7   Effect of baffle cut on shellside flow.

TABLE 7.2
Slip Streams at Difference Baffle Cuts

Parameter		Case 1	Case 2	Case 3	Case 4	Case 5
TEMA Type		AES	AES	AES	AES	AES
Pressure Drop Allowable Shellside	kg/cm^2	0.3	0.3	0.3	0.3	0.3
Pressure Drop Calculated	kg/cm^2	0.114	0.108	0.100	0.091	0.071
Shell ID	mm	850	850	850	850	850
Tube OD	mm	20	20	20	20	20
Tube Count		676	676	676	676	676
Tube Length	mm	6000	6000	6000	6000	6000
Baffle Apacing	mm	459	459	459	459	459
Baffle Cut	%	20	25	30	35	40
Cross-flow Velocity	m/s	1.24	1.24	1.24	1.24	1.24
Windowflow Velocity	m/s	2.18	1.65	1.32	1.07	0.90
Shellside Heat Transfer Coefficient	kcal/m^2.hr.°C	807	810	790	737	667
Overall Heat Transfer Coefficient	kcal/m^2.hr.°C	363	364	360	348	332
Over-design	%	5.89	6.22	5.23	2.00	−2.59

(Continued)

A Stream	0.107	0.089	0.072	0.055	0.038
B Stream	0.515	0.540	0.563	0.594	0.616
C Stream	0.170	0.164	0.169	0.156	0.185
E Stream	0.115	0.105	0.094	0.083	0.067
F Stream	0.093	0.102	0.103	0.111	0.094

Refer to Figure 7.8(a) which shows a tube bundle having parallel segmental baffles. The F stream moves horizontally as shown. To restrict it, F stream seal rods are provided. The C stream is also shown and, to restrict the same, sealing strips are shown.

Refer to Figure 7.8(b) which shows a tube bundle having perpendicular segmental baffles. The F stream moves vertically as shown, and F stream seal rods are provided as shown. Similarly, the C stream is also shown and, to restrict the same, the sealing strips as shown are self-explanatory.

Impingement Plate: Impingement plates are generally provided in the shellside, just below the inlet nozzle to protect the row of tubes nearest to the inlet nozzle against direct impingement by the shellside fluid. According to TEMA, an impingement plate is required for non-corrosive, non-abrasive single-phase liquids if the ρv^2

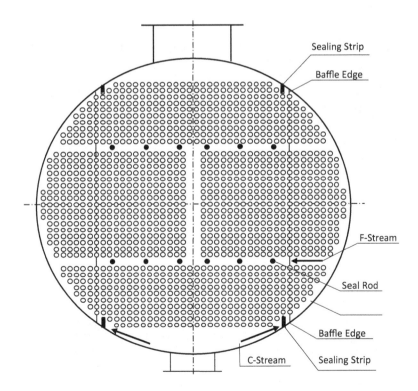

FIGURE 7.8(a)   F stream seal rods and sealing strips – parallel segmental baffles.

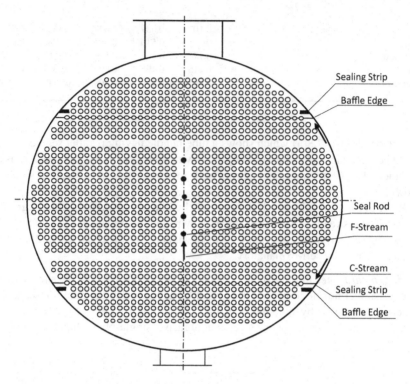

FIGURE 7.8(b) F stream seal rods and sealing strips – perpendicular segmental baffles.

exceeds 2232 kg/m.s². For all other liquids, including a liquid at its boiling point, the limit for ρv^2 is 744 kg/m.s².

Such devices should be placed sufficiently below the shell so that the area in between is enough to provide a flow without excessive velocity. As a result, a few rows of tubes would need to be sacrificed [2].

7.2.6 HEAT EXCHANGER CONFIGURATIONS

7.2.6.1 Multiple Shells in Series

Case 1: Consider the configuration shown in Figure 7.9(a). The tubeside fluid enters the exchanger at 30°C and leaves at 44°C. The shellside stream enters at 61°C and leaves at 37°C. For an exchanger with one shell pass and one tube pass, this is possible because the flow is close to counter-current (Figure 7.10(a)) and there is always a temperature difference between the hot fluid and the cold fluid. However, for an exchanger with one shell pass and two tube passes, one pass is counter-current, while the other is concurrent (Figure 7.10(b)). In such cases, the temperatures of the hot and cold fluids cross each other. In other words, the temperature of the cold fluid becomes hotter than the temperature of the hot fluid (Figure 7.9(b)).

With this condition, if one shell pass and two tube passes is the preferred design of the heat exchanger, then to achieve proper heat transfer with the given temperature

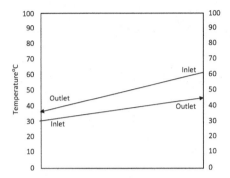

FIGURE 7.9(a)   Temperature cross in heat exchangers.

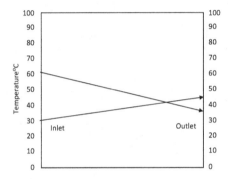

FIGURE 7.9(b)   Temperature cross in heat exchangers.

conditions, one has to go for multiple shells in series (Figure 7.10(c)). The explanation is illustrated in Figure 7.11(a) and 7.11(b) and is self-explanatory.

Case 2: Consider the case of a gas oil – cooling water heat exchanger having a single shell. The shell diameter is 1300 mm. The shellside coefficient is 304 kcal/m^2. hr.$^\circ$C and the tubeside coefficient is 4665 kcal/m^2.hr.$^\circ$C for a tube diameter of 20 mm. The overall coefficient is 185 kcal/m^2.hr.$^\circ$C. The over-design is 13.41% Table 7.3(a).

It may be seen that the shellside coefficient is too low, and thus the overall heat transfer coefficient is also low. This is, therefore, not a good design. Effort should be

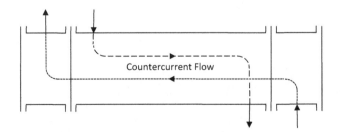

FIGURE 7.10(a)   1 - 1 Exchanger (counter-current flow).

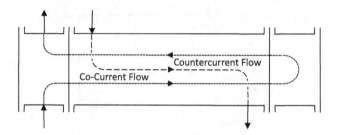

FIGURE 7.10(b)   1 - 2 Exchanger (co-current and counter-current flow).

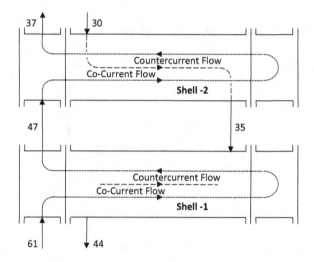

FIGURE 7.10(c)   Two exchanger shells in series.

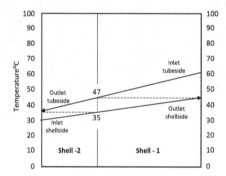

FIGURE 7.11(a)   Break-up into two shells.

made to improve the shellside coefficient. Consider now a two shell (series configuration with shell diameter of 585 mm (Table 7.3(b))). The same quantity of fluid now flows through a smaller shell diameter. The increased turbulence leads to a higher shellside coefficient, i.e., 695 kcal/m².hr.°C. The overall coefficient has also improved

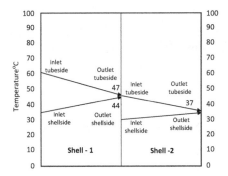

FIGURE 7.11(b)   Actual conditions (no temperature cross).

336 kcal/m^2.hr.^{o}C. However, the over-design is 5.43%, which could be improved. Further, the tubeside pressure drop has crossed the allowable limit.

We now go for some fine-tuning. We take a shell diameter of 735 mm and tube diameter of 25 mm. The over-design is now 12.61% and the pressure drops are within allowable limits (Table 7.3). This is a good design.

The above case illustrates another use of shells in series.

7.2.6.2   Multiple Shells in Parallel

Sometimes, exchangers may be too large to be accommodated in a single shell. In such cases, multiple shells in parallel are used. This is particularly critical for

TABLE 7.3(A)
Single Shell Configuration

Parameter	Units	Shellside		Tubeside
TEMA Type			AES	
Flowrate	1000 kg/hr	97		227
Inlet/Outlet Temperature	deg C	65/40		35/40
Heat Duty	MMKcal/hr		1.1364	
Pressure Drop Calculated	kg/cm^2	0.086		0.476
Pressure Drop Allowable	kg/cm^2	0.5		0.5
Fouling	m^2.hr.C/Kcal	0.0004		0.0006
Shell h	Kcal/m^2.hr.C	305		
Tube h	Kcal/m^2.hr.C			4665
Overall U	Kcal/m^2.hr.C		185	
EMTD	deg C		9.3	
Shell ID	mm	1300		
Tube OD	mm			20
Tube Count			1810	
Tube Length	mm			6000
Pitch Ratio				1.25
Tube Passes				6
Area	m^2		661	
Over-design	%		13.41	

TABLE 7.3(B)
Double Shell Configuration (Case 1)

Parameter	Units	Shellside		Tubeside
TEMA Type			AES	
Flowrate	1000 kg/hr	97		227
Inlet/Outlet Temperature	deg C	65 / 40		35 / 40
Heat Duty	MMKcal/hr		1.1364	
Pressure Drop Calculated	kg/cm^2	0.363		0.74
Pressure Drop Allowable	kg/cm^2	0.5		0.5
Fouling	m^2.hr.C/Kcal	0.0004		0.0006
Shell h	Kcal/m^2.hr.C	695		
Tube h	Kcal/m^2.hr.C			6698
Overall U	Kcal/m^2.hr.C		336	
EMTD	deg C	12		
Shell ID	mm	585		
Tube OD	mm			20
Tube Count			380	
Tube Length	mm			6000
Pitch Ratio				1.25
Tube Passes				6
Area	m^2		282	
Over-design	%		5.43	

TABLE 7.3(C)
Double Shell Configuration (Case 2)

Parameter	Units	Shellside		Tubeside
TEMA Type			AES	
Flowrate	1000 kg/hr	97		227
Inlet/Outlet Temperature	deg C	65/40		35/40
Heat Duty	MMKcal/hr		1.1364	
Pressure Drop Calculated	kg/cm^2	0.288		0.259
Pressure Drop Allowable	kg/cm^2	0.5		0.5
Fouling	m2.hr.C/Kcal	0.0004		0.0006
Shell h	Kcal/m^2.hr.C	540		
Tube h	Kcal/m^2.hr.C			4119
Overall U	Kcal/m^2.hr.C		269	
EMTD	deg C		12	
Shell ID	mm	735		
Tube OD	mm			25
Tube Count			380	
Tube Length	mm			6000
Pitch Ratio				1.25
Tube Passes				6
Area	m^2		352	
Over-design	%		12.61	

exchangers which have removable tube bundles and hence are limited by capacities of crane handling. In such cases, tube bundle weights are usually limited to 10,000 kg. Of course, fixed tubesheet heat exchangers do not have this limitation. But in these cases, the sizes are limited by the fabrication capabilities and transport limitations [2].

7.2.7 HEAT EXCHANGER OVER-DESIGN

Over-design on Surface: As the name suggests, the over-design on surface is the extra surface area provided beyond that required to achieve the given heat duty. Suppose the calculated surface area required for a given heat duty is 250 m^2, and the actual surface area provided is 265 m^2, the over-design on surface is 6%.

For small heat exchangers, a 10% or even a 20% over-design may be acceptable. For such cases, where going to the next lower pipe size may result in under-design, there is no option but to go for a slightly higher over-design than what is needed. However, for large exchangers, we normally do not face such limitations, and one can restrict the over-design close to 5%, keeping in mind the cost factor.

7.2.8 FOULING

Fouling in a heat exchanger occurs as a result of the deposition of undesirable material on the heat transfer surface. In case of shell-and-tube heat exchangers, it takes place on both the inside and outside of the tubes and leads to a reduction in the overall heat transfer coefficient, resulting in the requirement of a larger heat transfer area. It also leads to restriction of the flow area, particularly on the inside of the tubes, resulting in higher pressure drops.

It has been seen that higher pressure drops as a result of fouling are a bigger source of trouble compared to reduction in the overall heat transfer coefficients. It has also been seen that in severe cases, of the total heat transfer area provided, as high as 30% needs to be provided to account for fouling alone [2].

Parameters Affecting Fouling: There are a number of parameters that affect fouling, namely [2]:

Nature of fluid: Heat exchanges handling clean fluids such as steam or light hydrocarbons are not really affected by fouling. On the other hand, dirty fluids such a cooling water or heavy hydrocarbons cause a lot of nuisance in performance.

Flow velocity: At high velocities, fouling is minimized, because the shear forces on account of high velocities tend to remove the fouling deposits.

Temperature of the fluid: Temperature is a parameter in fouling if the flowing fluid consists of dissolved salts. For normal solubility salts, the solubility decreased on cooling and therefore deposition of salts occurs resulting in fouling. In case of inverse-solubility salts, deposition occurs on heating, resulting in fouling.

Surface finish: Smooth surfaces are less likely to retain dirt compared to rough surfaces. Rough surfaces foul significantly more than smooth ones.

Fouling allowance: Fouling allowance is the extra heat transfer area added to the theoretically calculated area to account for this extra resistance. One of the ways of doing this is to assign a fouling resistance to the shellside and the tubeside streams. It may, however, be emphasized that the application of the fouling resistance should be done to the extent necessary and not with the intention of adding an extra margin on the design. The use of unduly high fouling resistance and therefore higher surface areas may give rise to the following problems [2]:

- In the tubeside, a higher surface area would lead to a larger number of tubes. This may require an increase in the number of tube passes resulting in a higher pressure drop. If the pressure drop is higher than the maximum allowable value, it will again call for a reduction in the tube passes. The lower velocities in the tubes will again result in fouling.
- In the shellside, a larger number of tubes will result in a large shell diameter leading to lower shellside velocities. To take care of this, the baffle spacing would need to be reduced resulting in a higher pressure drop. If the pressure drop is higher than the maximum allowable value, the baffle spacing would need to be increased. The lower velocities in the shellside will again result in fouling.

Bottom line: The use of unduly high fouling resistances does more harm than good. Exercise the choice of fouling resistance with caution.

7.2.9 PRESSURE DROP

The pressure drop available to drive the fluids through the exchanger is normally set by the process conditions. The available pressure drop can range from a few millibars in vacuum service to several bars in pressurized systems.

In turbulent flow conditions, the tubeside heat transfer coefficient is proportional to the 0.8th power of the tubeside mass velocity, while the pressure drop increases in proportion to the square of the mass velocity. It may, therefore, be inferred that there is an optimum velocity beyond which there is no real benefit in increasing the velocity. Normally for liquids, a pressure drop of 0.7 kg/cm^2 is permitted per shell. Recommended liquid velocities in the tubeside range from 1.0 to 3.0 m/s [2]; however, when the designer has a free hand in selecting the pressure drop, an economic analysis can be made to determine an optimum exchanger design. The values suggested below can be used as a general guideline [6]:

For Liquids:	Viscosity (<1 cP)	0.35 kg/cm^2
	(1–10 cP)	0.5–0.7 kg/cm^2
For Gases and Vapors:	High vacuum	0.004–0.008 kg/cm^2
	Medium vacuum	10% of absolute pressure
	1–2 kg/cm^{2}g	50% of system gauge pressure
	Above 10 kg/cm^{2}g	10% of system gauge pressure

7.2.10 REBOILERS

A reboiler is a heat exchanger which fully or partially vaporizes the column bottom liquid from a distillation column, absorber or a stripper and returns the vapors upwards through the trays or packed sections.

Distillation columns can be equipped with mainly five types of reboilers:

- Vertical thermosiphon
- Horizontal thermosiphon
- Kettle
- Forced circulation
- Internal

Vertical thermosiphon reboilers: This is the most common type of reboiler in distillation operation. In such a reboiler, the stream from the column bottom enters at the bottom of the tubeside of the reboiler, and partially vaporizes as it moves up the tubes. The two-phase liquid finally discharges into the column. The driving force is the difference in densities between the liquid in the column and the froth (two-phase fluid) in the tubes. It is compact with little space requirement, simple piping configuration and low capital cost [7].

As an example, refer to the configuration shown in Figure 7.12. Consider the process data as follows:

Specific gravity of the liquid:	0.60
Reboiler inlet static height:	7.0 m
Specific gravity of the reboiler outlet froth:	0.09
Height of froth:	10.5 m
Driving force:	$[(7.0 \times 0.60) - (10.5 \times 0.09)] \times 1000$
	$= 3255 \text{ kg/m}^2$
	$= 0.33 \text{ kg/cm}^2$

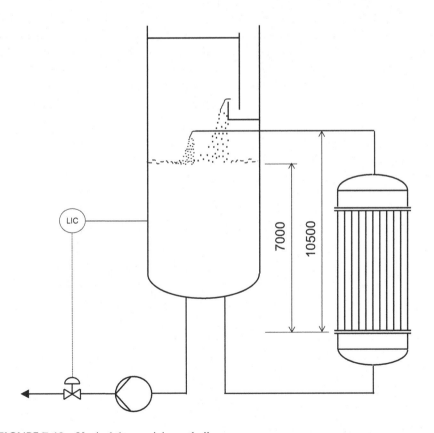

FIGURE 7.12 Vertical thermosiphon reboiler.

The differential pressure of 0.33 kg/cm² is used to overcome the frictional losses in the reboiler circuit.

There are two types of column reboiler configurations under the vertical thermosiphon category, namely, the once-through and the recirculating types. In the once-through type, the vapor generated in the reboiler does not mix with the column bottom product. In the recirculating type, the vapors mix with the column bottom product. Such a configuration is deployed as well in horizontal reboilers. For more details, refer to Chapter 9.

In vertical thermosiphon reboilers, boiling occurs inside the tubes and this phenomenon is well understood. Further, since boiling is inside the tubes, it is important to have a single-pass arrangement, as a two-phase flowing liquid cannot be subjected to movement downwards, which would be the case for a multiple-pass configuration [2].

In such configurations, the heating medium should be clean such that it can be kept on the shellside. However, in case of a dirty heating medium requiring flow through the tubes, such a reboiler is no longer suitable.

Horizontal thermosiphon reboilers: In a horizontal thermosiphon reboiler, liquid from the column bottom enters the shellside of the reboiler where it is vaporized. The heating medium flows through the tubes. Normally, the TEMA G shell or the TEMA H shell configuration is usually deployed for such reboilers (Figure 7.13). With the TEMA G or H type shell, the pressure drop also gets reduced substantially [2].

Such reboilers are preferred when a liquid head provided in a vertical thermosiphon reboiler or a forced circulation reboiler is not available [7]. A typical example is very tall columns where increase of skirt height to provide additional head for circulation would prove to be expensive.

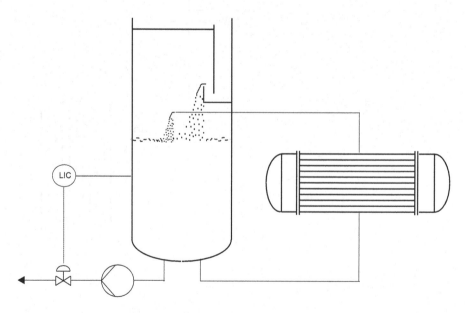

FIGURE 7.13 Horizontal thermosiphon reboiler.

Kettle type reboilers: In a kettle type reboiler, the liquid from the column bottom enters the shellside where it is vaporized. In such a reboiler, full vaporization of the column bottom takes place; therefore, such a reboiler serves as an additional distillation stage. A weir located beyond the tube bundle maintains the liquid level slightly above the tube bundle (Figure 7.14). Sometimes a level control is deployed for better control.

The kettle reboiler could be of a fixed tubesheet, U-tube or a floating head construction. If the heating medium flowing inside the tubes is clean (viz., steam), then the U-tube kettle is the preferred choice since it is cheaper; otherwise, the following options can be deployed:

- Fixed tubesheet construction if the tubeside fluid is dirty
- Floating head construction if both the shellside and tubeside fluids are dirty

For a U-tube or a floating head construction, the number of tube passes would be two or a higher even number. Needless to state, for a fixed tubesheet construction, the number of tube passes would be one.

The boiling liquid enters from the bottom of the exchanger, and the vapors leave from the top. For wide boiling mixtures, it is preferable that the liquid enters at multiple locations in the shell to ensure a better distribution. For reboilers with large tube lengths, two outlet nozzles are preferable since this reduces the vapor outlet velocity

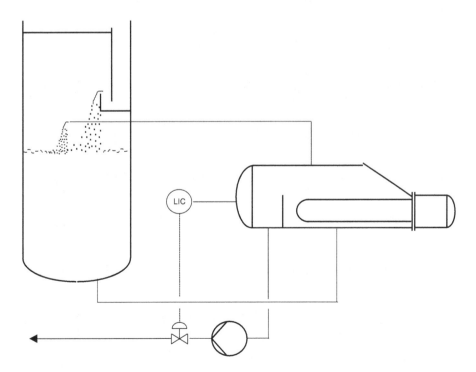

FIGURE 7.14 Kettle type reboiler.

and thus prevents entrainment [2]. Such reboilers are preferred where instabilities are expected, frequent cleaning of the shellside is anticipated, or where liquid in the reboiler outlet is not desirable [7].

Forced circulation type reboilers: Under normal circumstances, such reboilers are normally avoided to eliminate the cost of pumping and also the addition of another rotating equipment and are therefore used only in special situations. However, these types of reboilers are used in cases where fouling liquids having rather high viscosities are expected. In such cases, vaporization is better avoided and this is achieved using a pump which circulates the liquid at high velocities at a pressure above the saturation. The liquid receives sensible heat and is then flashed across a valve or an orifice before entry into the column. Since only sensible heat is received, large flowrates have to be achieved (Figure 7.15).

Such reboilers have the flexibility that the column bottom liquid can flow either through the shellside or the tubeside [2]. Such reboilers are also deployed where (1) the reboiler is located at a distance from the column, (2) a controlled flow is needed, and (3) in vacuum systems (<0.3 kg/cm^{2}a) where thermosiphon reboilers are not reliable [6].

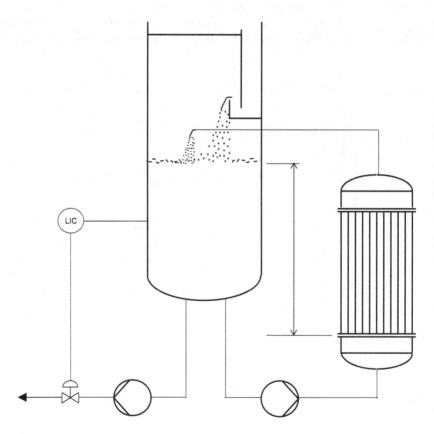

FIGURE 7.15   Forced circulation reboiler.

TABLE 7.4
Comparison of Various Types of Reboilers

Parameter	Vertical Thermosiphon	Horizontal Thermosiphon	Kettle	Forced Circulation	Internal
Boiling side	Tube	Shell	Shell	Tube	Shell
State of exit stream	Vapor fraction typically 20%	Vapor fraction typically 20–30%	Vapor fraction 100%	Normally it is 100% liquid	–
Sensitivity to fluctuations in operating conditions	Sensitive	Sensitive	Less sensitive	Less sensitive	Less sensitive
Turndown operation	Turndown limited. A rough approximation could be 3:1	A rough approximation could be 4:1	No limit (an oversized reboiler does not normally present operational problems)	Not a problem	–
Differential expansion	Not accommodated	Can be accommodated	Easily accommodated	Can be accommodated	Easily accommodated
Space requirement	Small	Large	Large	Large (if horizontal)	None
Cost	Low	Expensive in case of U-tube or floating head configuration	Shell expensive	Extra cost of pump	Lowest
Cleaning of process side	Difficult	Can be cleaned after removing bundle. Possible only if square pitch is used	Can be cleaned after removing bundle. Possible only if square pitch is used	Assuming that there is no vaporization, the liquid could be allocated to the tubeside and cleaning would not be a problem with a horizontal configuration	Can be cleaned after removing bundle. Possible only if square pitch is used and only during shutdown.

(Continued)

TABLE 7.4 (Continued)

Parameter	Vertical Thermosiphon	Horizontal Thermosiphon	Kettle	Forced Circulation	Internal
Additional column skirt requirement	Required	Required but less compared to that in vertical thermosiphon	Required but less compared to that in vertical thermosiphon	As such, not required. But may be required for the pump in case of low NPSH	Not required
Process side fouling tendency	Fouling if any is easy, to clean by virtue of liquid in the tubeside	Reduced by virtue of higher velocity and lower exit vapor fraction	High, in view of very little turbulence	Very low	Moderate
Vapor disengagement area	None	None	Available	None	—
Piping requirement	Minimum piping of simple configuration required	Additional piping required	Additional piping required	Additional piping required	None
ΔT requirement	High	Moderate	Operate efficiently even at low ΔT	High	Moderate to high
Reliability	Good, but comparatively less reliable than kettle type	Medium	Excellent. There is no concern of two-phase piping from the reboiler to the column	Very good	Medium

Source: Adapted from Mukherjee [2], Kister [7] and GBH [8].

Internal reboilers: In such reboilers, a tube bundle is inserted into the column. Such reboilers by far offer the lowest cost since there is no shell or shellside piping. In such applications, however, the tube length is limited by the column diameter and the number of tubes is also limited, since the maximum opening in the column shell is normally limited to one-half the column diameter from the mechanical point of view. Another drawback of such a reboiler is that for any maintenance work in the column, the reboiler would need to be removed [2].

Applications of such reboilers are therefore limited to:

- Clean services which require little maintenance
- Low duty services where the reboiler bundle is small and the column diameter is large enough to accommodate it
- Dirty services which deploy batch distillation, so that periodic cleaning could be taken care of [6]

In view of small heat transfer areas and other limitations, in most refinery and petrochemical applications, such reboilers are rarely deployed since they seldom match the criteria stated above.

Table 7.4 lists a comparison of various types of reboilers [2, 7, 8].

7.3   ENHANCED HEAT TRANSFER

Conventional shell-and-tube heat exchangers have limitations from the point of view of efficient heat transfer. On the tubeside, a boundary layer exists which poses a limitation to heat transfer. Similarly, on the shellside, the formation of eddies and dead zones lead to inefficient heat transfer. By introducing modifications in the bare tubes and (or) in the baffles, we can achieve enhanced heat transfer wherein, for a given pressure drop, an improved heat transfer is achieved [2].

By using such modifications, we can have a lower cost by achieving the given heat transfer with an exchanger having a lower heat transfer area. Further, by increasing turbulence using enhancement devices, we can achieve reduced fouling. In addition, such modifications come in very handy during revamp applications, wherein, instead of replacing the exchangers, one can replace only the tube bundles by modified tube bundles equipped with the enhancement devices.

7.3.1   MODIFICATIONS ON TUBESIDE

Wire-tube Inserts: In many cases, the fluid passing through the tubeside has a high viscosity. When such a fluid flows through a bare tube, the movement of the layer of fluid nearest to the tube wall is subjected to frictional drag. This slows down the movement of the fluid at the wall. The laminar boundary layer formed as a result can significantly reduce the tubeside heat transfer coefficient, resulting in loss of performance of the heat exchanger. Inserting such tube insert elements disrupts the laminar boundary layer by removing stagnant fluid from the tube wall and replenishing it with fluid from the center of the tube.

TABLE 7.5
Bare Tube Design vs. Hitran® Enhanced Tubes

Parameter		Plain Tubes	Hitran® Enhanced Tubes
TEMA designation		BEM	BEM
Shell diameter	mm	1524	689
Number of tubes		1829	371
Number of tube passes		8	1
Tube length	mm	6096	6096
Tube diameter	mm	25.4	25.4
Effective surface area	m²	874	178.5
Overall service coefficient	kCal/h.m².K	34.4	156.5
Tubeside coefficient		43.9	253.7
Tubeside pressure drop	kg/cm²	0.71	0.71

Source: Copyright and courtesy of CALGAVIN Ltd.

Although the heat transfer increase is greatest in the laminar flow region (up to 16 times), significant benefits can be obtained in the transitional flow regime (up to 12 times) and turbulent flow regime (up to 3 times) [9]. Table 7.5 illustrates comparative figures of performance using a conventional shell-and-tube exchanger and an exchanger with wire inserts.

7.3.2 MODIFICATIONS ON SHELLSIDE

Helical Baffles: Conventional shell-and-tube heat exchangers with perpendicular baffles could give rise to flow maldistribution, excessive leakage of E stream, dead zones and vapor blanketing on the shellside. In the dead zones, fouling gets pronounced.

Heat exchangers with helical baffles were developed with the intention of improving the flow pattern in the shellside. The helical baffles attain a "near-plug flow" behavior. They reduce the shellside turning losses and give a higher shellside coefficient compared to the segmental baffle heat exchangers for a given pressure drop. For applications where the shellside resistance governs, such baffles provide appreciably higher heat transfer coefficients.

7.3.3 OTHER ENHANCEMENTS

There are several other types of enhancements, a detailed discussion of which is beyond the scope of this book. The use of fins in tubes (both inside and outside) is one of the oldest methods. The fins increase both the heat transfer coefficient as well as the heat transfer area. Tube deformations, such as corrugated or twisted tubes, have been developed which result in enhancement of heat transfer. The reader may refer to reference [10] for further details.

7.4   PLATE AND FRAME HEAT EXCHANGERS

Plate and frame heat exchangers are assemblies of pressed corrugated plates on a frame. Fluids flow into and out of the spaces between the plates. Hot and cold fluids flow on opposite sides of the plates. Close spacing and the presence of the corrugations result in high coefficients on both sides several times those of shell-and-tube exchangers. Thicknesses of the plates are normally in the range of 0.5–3.0 mm and are placed at a gap of 1.5–5 mm between them. Surface areas of plates range from 0.03 to 1.5 m^2. Surface areas of plate heat exchangers can vary from 0.03 m^2 to as large as 1500 m^2. Typically, such exchangers have surface areas of the order of 1200 m^2/m3 and therefore offer heat transfer rates per volume, which are four times those achieved in shell-and-tube heat exchangers. These exchangers can handle flowrates as high as 2500 m^3/h [6]. Fouling coefficients are lower. The accessibility of the heat exchange surface for cleaning makes them particularly suitable for fouling services [11, 12]. Figure 7.16 illustrates a schematic diagram of a plate heat exchanger.

The following are the advantages and drawbacks of plate and frame heat exchangers [2, 6, 12]:

Advantages:

- Attractive when using expensive materials of construction
- Higher heat transfer coefficients
- Fouling is significantly lower
- Easy to clean and maintain
- Capacity can be enhanced by adding more plates
- Closer approach temperatures can be used (as low as 1°C)
- Lower space requirements
- Lower costs

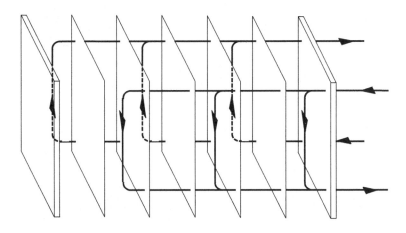

FIGURE 7.16   Plate heat exchanger.

Drawbacks:

- Lower operating pressures (not suited for pressures above 30 kg/cm²g)
- Maximum operating temperature is limited to about 250°C due to the gasket materials

Recent developments, however, include semi-welded and fully welded types which permit pressures of up to 40 kg/cm²g and temperatures of up to 350°C.

7.5   DOUBLE PIPE HEAT EXCHANGERS

Double pipe heat exchangers normally consist of two concentric pipes (Figure 7.17). One of the fluids flows through the inner pipe. The other fluid flows through the annulus between the inner and the outer pipes in counter-current flow. These types of exchangers have a true counter-current flow pattern, unlike in certain multi-pass shell-and-tube exchangers where such a flow may not be achieved. The straight length is limited to approximately 6 m to avoid sagging of the inner pipe, thus causing poor distribution in the annulus. For applications where gases or viscous liquids are used in the annular space, external longitudinal fins can be used to enhance the heat transfer [4, 8].

This is by far the simplest type of heat exchanger. Cleaning is carried out easily by dismantling. Maintenance and repair are also easily carried out in such exchangers. Further, there are larger number of suppliers for such exchangers and the delivery periods are also shorter compared to conventional shell-and-tube heat exchangers. However, such exchangers require a larger space for a given heat transfer area.

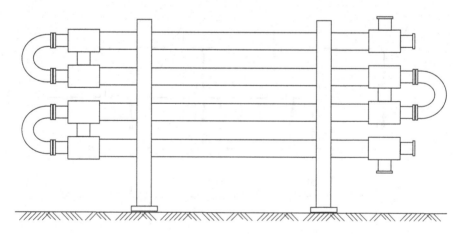

FIGURE 7.17   Double pipe heat exchanger.

TABLE 7.6
Shell-and-Tube vs. Double Pipe Heat Exchangers

Parameter		Shell-and-Tube Heat Exchanger	Double Pipe Heat Exchanger
Heat duty	kCal/hr	77736	77738
Shell diameter	mm	250	77.93
Tube OD	mm	19	60.32
Tube count		42	1
Tube passes		1	1
Tube length	mm	6000	10×6000
Fouling factor – Shellside	m².K/W	0.0034	0.0034
Fouling factor – Tubeside	m².K/W	0.0034	0.0034
EMTD	°C	14	14
Over-heat transfer coefficient	kCal/hr.m².K	433.4	643.97
Effective area	m²	14.922	10.028

Double pipe can specifically be used for the following applications [4, 8]:

1. For small heat duty applications where the total heat transfer surface required is less than 20 m². In such cases, double pipe exchangers are the cheaper option. In cases where greater surfaces are needed, a stack of double-pipe exchangers are also used in some process applications.
2. In shell-and-tube exchangers, where the shellside coefficient is less than half the tubeside coefficient, the use of double pipe exchangers can make the annular side coefficient comparable to that of the tubeside.
3. For shell-and-tube exchangers where issues of temperature crosses are faced, double pipe exchangers make the job simpler since true counter-current flow takes place here.
4. Such exchangers are also suitable when both fluids are at high pressures. For a given wall thickness, pipes can withstand higher pressures than shells and are therefore less expensive.

Table 7.6 illustrates a comparison between shell-and-tube heat exchangers and double pipe exchanges with areas in the range of 10–15 m².

REFERENCES

1. Mukherjee, R., "Effectively Design Shell-and-Tube Heat Exchangers", *Chemical Engineering Progress*, February 1998.
2. Mukherjee, R., *Practical Thermal Design of Shell-and-Tube Heat Exchangers*, Begell House, Inc., New York, 2004.
3. http://www.stoneagetools.com/rotary-waterblast-tools/banshee-tube-cleaners.html)
4. Mukherjee, R., "Broaden Your Heat Exchanger Design Skills", *Chemical Engineering Progress*, March 1998, pp. 35–43.

5. Shah, R. K. and Sekulic, D. P., *Fundamentals of Heat Exchanger Design*, John Wiley & Sons Inc., New Jersey, 2003.
6. Towler, G. and Sinnott, R., *Chemical Engineering Design – Principles, Practice and Economics of Plant and process Design*, Butterworth-Heinemann, Burlington, MA, 2008.
7. Kister, H. Z., *Distillation Operations*, McGraw-Hill, Inc., New York, 2016.
8. www.GBHEntrprises.com, "Process Engineering Guide: Selection of Reboilers for Distillation Columns".
9. CALGAVIN Limited, www.calgavin.com
10. Lunsford, K. M., "*Increasing Heat Exchanger Performance*", Bryan Research & Engineering Inc., Bryan, Texas, 2006. (http://www.bre.com/technicalpapers/technical-paper.asp?articlenu)
11. Walas, S. M., *Chemical Process Equipment – Selection and Design*, Butterworth-Heinemann, Newton, MA, 1990.
12. www.ipieca.org/resources/energy-efficiency-solutions/efficient-use-of-heat/heat-exchangers/

8 Air Coolers and Fired Heaters

8.1 INTRODUCTION

In the chemical process industry, the importance of cooling of process streams need not be over-emphasized. Whether it is cooling of a column overhead stream, a reactor outlet stream following an exothermic reaction, or a process stream routed to the plant battery limit, cooling is widely used. Historically, cooling water has been the most widely used medium for cooling of process streams in spite of the fact that air is more abundantly available at practically no cost. Air cooling offers a number of advantages compared to water cooling and is discussed later in this chapter.

Fired heaters are major consumers of energy in the chemical process industry, especially in refineries and petrochemical units. Of the total energy consumed in a typical refining unit, as high as 70–75% of the total consumption can be accounted for in the form of hydrocarbon fuel, burnt in fired heaters [1, 2], in which the heat liberated by combustion of the fuel is transferred to the fluid contained in coils.

This chapter presents a broad practical picture of the design aspects as well as various practical aspects of air coolers and fired heaters.

8.2 AIR COOLERS

Air cooled heat exchangers consist of banks of finned tubes across which air flows. The air is driven by fans located above or below the unit. The tube side performance is similar to that of shell-and-tube heat exchangers.

8.2.1 REASONS FOR USING AIR COOLERS

CAPEX (capital expenditure) point of view: The thermal conductivity of air is considerably lower than that of water. This results in a much lower heat transfer coefficient. Ambient air temperatures are of the order of 40–45°C. Cooling water temperatures are in the range of 30–33°C. Thus, for air coolers, for the process fluid at a given temperature, the mean temperature difference is always lower.

Summary: Air coolers are considerably larger than water cooled heat exchangers for the same duty. The larger area requires an elaborate structural support system, which increases the cost further. Based on a bare tube heat transfer area, air cooled heat exchangers cost substantially more than water cooled heat exchangers for the same heat duty [3].

OPEX (operating expenditure) point of view: The operating costs for water cooling are much higher than those for air cooling. These include:

DOI: 10.1201/9780429284656-8

TABLE 8.1
Air Cooler vis-à-vis Shell-and-Tube Heat Exchanger

Parameter	Air Cooled Exchanger	Shell-and-Tube Exchanger
Flowrate of air, kg/hr	38,42,000	–
Flowrate of hot water, m³/hr	840	840
Shell diameter, mm	–	1200
Tube OD × thickness, mm	25.4 × 2.77	19.05 × 2.11
Number of tubes	938	1712
Tube length, mm	10360	6096
Number of tube passes	2	1
Pitch, mm	67	25.4
Bare tube area, m²	1550	616
Extended area, m²	38,600	–
Power consumption, kW	21.5	–
Cooling water consumption, m³/hr	–	1680

Source: Adapted from Mukherjee [4].

- Cost of initial raw water itself
- Make-up water
- Treatment chemicals
- Cost of plant cooling tower
- Pumping cost

Illustrated example: Consider 840 m³/hr of hot water to be cooled from 79 to 61°C. Design was done both with cooling water available at 34°C and with air. The comparative details for both these types of heat exchange methods are summarized in Table 8.1. Refer to [4] for further details.

For the given configurations in Table 8.1, the costs of the air cooler and the heat exchanger have been worked out. The calculations are as follows:

Estimated cost of the air cooler: $269,000
Estimated cost of the heat exchanger: $117,000
Additional cost of installing the air cooler: $269,000 – 117,000 = $152,000
Let us take the cost of electric power as $0.04/kWh
Power consumption in air cooler: 21.5 kW
In 1 year, cost incurred in electric power consumption works out to: $21.5 × 0.04 × 8000 × 1
= $6880
Let us take the cost of cooling water as $0.014/m³
Cooling water consumption in heat exchanger: 1680 m³/hr
In 1 year, cost incurred in cooling water consumption works out to: $1680 × 0.014 × 8000 × 1 = $188,160

Savings in utility cost by using air cooler works out to: $188160 – $6880 = $181,280 per year

In summary, the additional cost of installing the air cooler instead of the heat exchanger (i.e., $152,000) would be recovered in approximately 10 months.

8.2.2 CONSTRUCTION FEATURES

Before discussing these principal components, let us first consider some important terms with reference to air-cooled heat exchangers. Refer to Figures 8.1 and 8.2 for details.

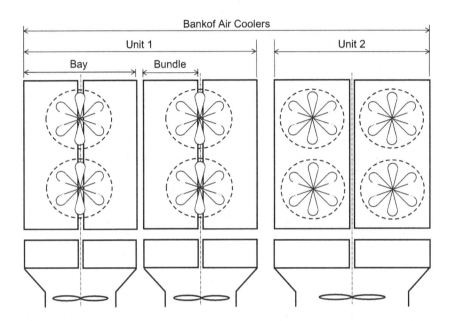

FIGURE 8.1   Air cooler – bundle, bay, unit, bank.

Source: Adapted from [3], [4].

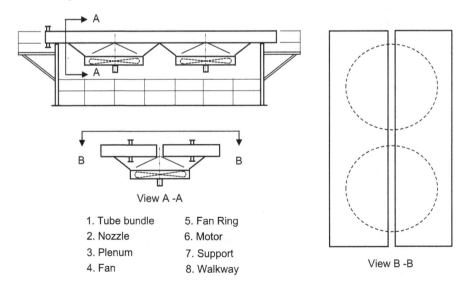

1. Tube bundle	5. Fan Ring
2. Nozzle	6. Motor
3. Plenum	7. Support
4. Fan	8. Walkway

FIGURE 8.2   Air cooler – construction features.

Tube bundle: An assembly of headers, tubes, tube supports, and frames constitute a tube bundle. Although standard bundle lengths are available, maximum bundle dimensions are dictated by logistics.

Bay: A bay is a setup in which one or more tube bundles may be installed. A bay is equipped with two or more fans along with the structure, plenum, and other associated equipment. An air cooler may consist of a single bay for small duties and multiple bays for higher duties.

Unit: An air cooler may consist of a single bundle equipped with fans and associated equipment, a single bay consisting of one or more bundles, or several bays, depending upon the size of the unit. A unit caters to an individual heat exchange service and has a specific equipment tag number.

Bank: A bank of air-cooled heat exchangers comprises one or more units of air-cooled exchangers arranged side-by-side on a structure.
Refer to the video links below on assembling air coolers:

YouTube links: Kelvion Air Fin Cooler - Petrochemical Application - YouTube
AIR COOLER HEAT EXCHANGER - ANIMATED ASSEMBLY - YouTube

8.3   FIRED HEATERS

In fired heaters, the thermal energy liberated by the combustion of fuel is transferred to fluids flowing through tubular coils within an internally insulated enclosure. A typical fired heater consists of three major components: radiant section, convection section and the stack. Figure 8.3 shows a typical view of a vertical cylindrical fired heater.

The fired heater is by oil or gaseous fuel. The process fluid, passing through the tubes in the heater, absorbs the heat mostly by radiant heat transfer, and by convective heat transfer from the flue gases. The flue gases are vented through the stack. Burners are located on the floor or on the sidewalls of the heaters (Figure 8.3).

8.3.1   APPLICATIONS OF FIRED HEATERS

In the chemical process indu*stry*, fired heaters are deployed in several services, namely the following [1]:

Column reboilers: Here, the feed is taken from the bottoms of a distillation column and is partially vaporized in the heater. The mixed vapor stream re-enters the column from where the two phases separate. The vapor stream moves up the column and the liquid stream drops to the column bottom. In such applications, the feed is at its boiling point and the purpose is vaporization. Hence, there is little temperature difference between inlet and outlet fluid. This application is the least critical of fired heater applications.

Fractionation column feed preheater: Fired heaters in this application are amongst the most common in the process industry. The feed, normally in liquid state, is fed to the heater where the fluid temperature is raised to achieve partial vaporization.

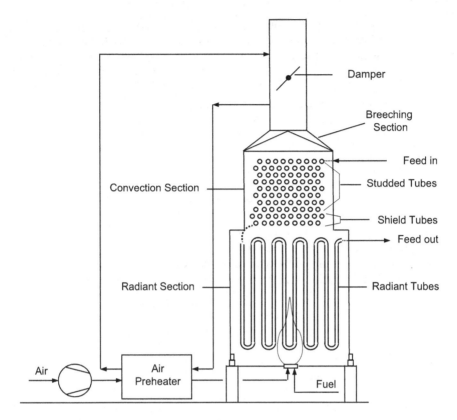

FIGURE 8.3   Schematic diagram of a fired heater.

Typical example is an atmospheric distillation column. The crude entering the heater at 230°C gets further heated in the convection section. It finally leaves the radian at section 370°C with about 60% of the feedstock vaporized.

Reactor feed preheater: In this application, the feed is heated to a temperature necessary for carrying out a reactor in a downstream reactor. A typical example is heating of a mixture of liquid hydrocarbons and recycled hydrogen in a hydroprocessing unit. Fluid temperatures could range from 370°C at the inlet to 450°C at the outlet.

Cracking furnace: A typical example is the cracker which produces olefins from feedstocks such as naphtha or natural gas. In such heaters, a cracking reaction occurs in the coils of the heater at temperatures in the range of 815 to 900°C.

Hot oil systems: In some applications, heat is transferred to individual consumers through an intermediate medium. A typical example is a hot oil system. The heater transfers heat to the hot oil which circulates through the system and in turn transfers heat to the individual users. The return oil at a lower temperature is again heated to the supply temperature by the heater and recirculated.

Catalytic reactors: A typical example of such an application is the steam-methane reformer. In this application, the feed, typically natural gas, is fed to tubes filled with nickel bearing catalyst. The endothermic heat is provided by firing the outside of the tubes. The reactor outlet stream consists of hydrogen and carbon monoxide, in addition to some carbon dioxide and methane. This is further processed downstream to get hydrogen as the product. Reactor outlet temperatures are in the range of 780 to 900°C.

8.3.2 COMBUSTION

Combustion air is drawn from the atmosphere. The exothermic reaction resulting in rapid combination of fuel with oxygen produces heat and flue gases. The fuel and air must be mixed thoroughly for complete combustion. However, if only the theoretical amount of combustion air were provided, then some fuel would not burn completely. So excess air is needed for complete combustion [2].

Burners start and maintain the combustion in the firebox. They introduce fuel and air in the correct proportion, mix them, provide a source of ignition, and stabilize the flame.

8.3.3 TYPES OF HEATER CONSTRUCTION

While there are various configurations of the types of fired heaters, the most common ones are the cylindrical shell type and the box type.

Radiant section: The radiant section of the heater contains radiant tubes located along the walls and receives radiant heat directly from the burners. Combustion takes place in the open space. The radiant zone with its refractory lining is the costliest part of the heater where approximately 75% of the heat of combustion is absorbed. The section consists of either a rectangular or cylindrical chamber refractory lining. The fluid, which is to be heated, passes through the tubes which are placed along the wall in either horizontal or vertical arrangement.

Convection section: Nowadays, in most designs, the radiant section is followed by the convection section through which the flue gas from the radiant section flows. The feed charge enters the coil inlet in the convection section where it is preheated before entering the radiant tubes. The convection section removes heat from the flue gas to preheat the contents of the tubes.

Shield section: Just below the convection section is the shield (or shock tube) section, containing rows of tubing which shield the convection tubes from the direct radiant heat.

Breeching sections and stack: The transition from the convection section to the stack is called the "breeching." By the time the flue gas exits to the stack, most of the heat should be recovered and the temperature is much lower.

For low temperature duties, carbon steel is sufficient. For higher temperatures stainless steel or special alloy steels are used. The burners are located at the base or at the sides of the radiant section. The heat from the flue gas leaving the heater is sometimes utilized to preheat the combustion air before it enters the radiant section. This serves to increase the heater efficiency by utilizing the heat content of the flue gas leaving the system. This portion of the heater is called the "air preheater."

Convection zones are normally located at the top. The process fluid first flows through the convection section and then enters the radiant section. In some designs, some of the tube rows in the convection may also be used for steam generation [5].

8.3.4 DRAFT CONFIGURATIONS

Draft is the pressure differential between the air or flue gas in the heater and the ambient air. It materializes because the hot flue gases inside the firebox and stack are lighter (and thus at a lower pressure) than the ambient air outside. Combustion air is drawn into the burners from the atmosphere, and the hot gases rise due to the buoyancy and flows out of the stack to the atmosphere. While passing through the heater's convection section and the stack, the flue gases encounter friction resistance, known collectively as "draft losses." Sufficient stack height is provided to ensure the buoyancy effect needed to overcome these losses and to ensure that the pressure is always negative inside the firebox [2].

Natural draft: This is the most common type. Air is drawn into the burners by means of the draft created by the radiant section. The taller this section, the greater will be the available draft. This is the most simple and reliable type since the air supply does not fail (Figure 8.4(a)).

Forced draft: In this type of system, the air is supplied by means of a centrifugal fan, commonly known as a "forced draft" fan. An forced draft fan provides air at a relatively high pressure, leading to better air-fuel mixing and smaller burners (Figure 8.4(b)).

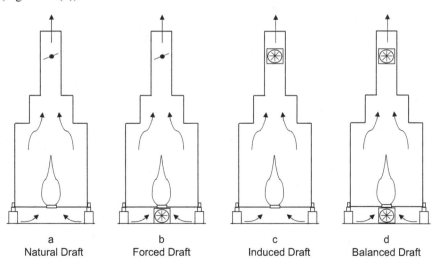

a	b	c	d
Natural Draft	Forced Draft	Induced Draft	Balanced Draft

FIGURE 8.4   Draft configurations in fired heaters.

Induced draft: When the height of the stack is inadequate, to compensate for the draft-loss requirements, an induced draft fan is provided on top of the convection section, to draw the flue gases out of the heater (Figure 8.4(c)). The resulting negative pressure inside the heater ensures adequate draft for the burners from the atmosphere. Such a system is employed in heaters where the convection section is very large.

Balanced draft: When both forced draft and induced draft fans are used in a fired heater, the combination is known as a "balanced draft" system. Most air preheating installations are equipped with this type of system. In a typical air preheating system, the draft loss across the preheater could be fairly high. The stack by itself cannot compensate for a loss of this magnitude. Instead, the forced draft fan supplies the combustion air, and the induced draft fan takes care of flue gas disposal (Figure 8.4d).

8.3.5 DESIGN GUIDELINES AND FORMULAE

The rating of a fired heater is a rather complicated exercise. Lobo and Evans [6] had initially developed an elaborate rating method which was later simplified by Wimpress [7] by eliminating minor variables. The example presented here uses many of these correlations. However, the method gives approximate equipment sizes. Even the method of Lobo and Evans yields equipment sizes with accuracies which could go up to 16% [8].

Thermal efficiency: Thermal efficiency of a fired heater is given by the formula (refer to Figure 8.5):

$$Thermal\ efficiency = \frac{Heat\ input - Stack\ loss - Radiation\ heat\ loss}{Heat\ input} \times 100$$

Which can also be detailed as,

$$\eta_{th} = \frac{Q_{fc} + Q_a + Q_f - Q_s - Q_{rl}}{Q_{fc} + Q_a + Q_f} \times 100 \tag{8.1}$$

Radiant zone calculations: Heat transfer in the radiant section is given by Equation (8.6). However, before we get to that, certain other terms need to be discussed.

Approximately 75% of the total heat can be considered to be transferred in the radiant section. A part of the radiation from the hot gases is absorbed in the tubes, and the remainder passes [7]. If the tubes are in front of a refractory wall, the heat that passes through is radiated back into the furnace, wherein a part is absorbed by the tubes and the remainder passes through. This complication is handled by defining a cold plane area A_{cp} wherein the tube area is expressed as an equivalent plane surface. A_{cp} is expressed as:

$$A_{cp} = (exposed\ tube\ length) \times (centre-to-centre\ spacing)$$
$$\times (number\ of\ tubes\ excluding\ shield\ tubes) \tag{8.2}$$

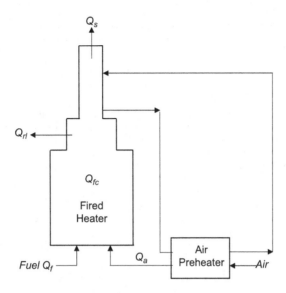

FIGURE 8.5 Thermal efficiency of fired heater.

Further, since the tube bank does not absorb the entire energy radiated to the cold plane, an absorption efficiency factor needs to be applied. Values of α are provided by Wimpress [7] and simplified by Walas in the form of the following Equations [8]:

$$\alpha = 1 - \left[0.0277 + 0.0927(\varkappa - 1)\right](\varkappa - 1) \tag{8.3}$$

where,

$$\varkappa = (\text{center} - \text{to} - \text{center spacing}) / \text{outside tube diameter} \tag{8.4}$$

Moreover, all the components of the gas in the firebox are not good radiators. The only components in the flue gas that contribute to significant radiation are carbon dioxide, water and sulfur dioxide. This issue is addressed by applying the radiant exchange factor expressed as F. Values of F are furnished in Table 8.2 as a function of the emissivity (φ) and $A_w / \alpha A_{cp}$ [6, 7, 9].

Values of emissivity are given in literature as a function of PL_b and flue gas temperature in the form of graphs [7]. These are illustrated in the form of Table 8.3 in this book.

There are two more terms which need to be mentioned, i.e., A_r and A_w [10].
A_r = total area of furnace surfaces, m^2
A_w = effective refractory surface, m^2
The relation between these two terms is given by

$$A_w = A_r - \alpha A_{cp} \tag{8.5}$$

The total heat transfer in the radiant section is the sum of the radiant and the convective heat transfer and is expressed as:

TABLE 8.2
Exchange Factor, F

Emissivity, φ	Aw/αAcp							
	0.5	1	2	3	4	5	6	7
0.20	0.265	0.326	0.430	0.510	0.555	0.600	0.630	0.658
0.25	0.330	0.390	0.502	0.578	0.640	0.673	0.695	0.718
0.30	0.370	0.450	0.563	0.642	0.679	0.720	0.745	0.758
0.35	0.425	0.495	0.610	0.671	0.727	0.755	0.775	0.788
0.40	0.465	0.550	0.655	0.724	0.765	0.783	0.802	0.813
0.45	0.520	0.585	0.690	0.760	0.791	0.814	0.825	0.834
0.50	0.560	0.630	0.728	0.783	0.820	0.831	0.839	0.850
0.55	0.605	0.668	0.765	0.818	0.835	0.848	0.860	0.865
0.60	0.640	0.705	0.785	0.838	0.855	0.865	0.874	0.876
0.65	0.677	0.735	0.819	0.854	0.868	0.874	0.878	0.880
0.70	0.720	0.767	0.835	0.870	0.879	0.885	0.885	0.886

Source: Adapted from Lobo and Evans [6], Wimpress [7] and Mekler and Fairall [9].

TABLE 8.3
Flue Gas Emissivity

P × L_b, atm-m	Flue Gas Temperature, °C					
	600	800	1000	1200	1400	1600
0.2	0.285	0.240	0.208	0.174	0.155	0.124
0.4	0.375	0.341	0.307	0.258	0.224	0.204
0.6	0.430	0.410	0.375	0.326	0.295	0.270
0.8	0.486	0.453	0.425	0.378	0.345	0.308
1.0	0.530	0.495	0.470	0.423	0.385	0.352
1.2	0.560	0.530	0.500	0.454	0.420	0.383
1.4	0.590	0.575	0.532	0.490	0.450	0.420
1.6	0.615	0.590	0.554	0.518	0.480	0.445
1.8	0.635	0.620	0.590	0.549	0.502	0.475

Source: Adapted from Wimpress [7].

$$\frac{Q_R}{\alpha A_{cp} F} = 4.92 \times 10^{-8} \left(T_{gr}^4 - T_{wr}^4 \right) + 35 \left(T_{gr} - T_{wr} \right) \tag{8.6}$$

The radiant heat flux to the tubes normally lies in the range of 0.02–0.05 MKCal/hr.m^2.

The mean length of radiant beam is provided by Wimpress [11] and furnished in Table 8.4.

The clearance between tube bundle and refractory is normally taken as 0.50 m.

Convection zone calculations: Overall heat transfer coefficient in the convection section U_c is given by Walas [8]:

TABLE 8.4
Mean Radiant Beam Length

Dimensional Ratio	Mean Beam Length
Rectangular Furnaces, Length-Width-Height, in any order	
1-1-1 to 1-1-3	
1-2-1 to 1-2-4	2/3 × (Furnace Volume)$^{1/3}$
1-1-4 to 1-1-∞	1.0 × Smallest Dimension
1-2-5- to 1-2-∞	1.3 × Smallest Dimension
1-3-3- to 1-∞-∞	1.8 × Smallest Dimension
Cylindrical Furnaces, Diameter-Height	
1-1	2/3 × Diameter
1-2 to 1-∞	1 × Diameter

Source: Reprinted from "Rating Fired Heaters – A Special Report", October 1963, with permission from *Hydrocarbon Processing and Petroleum Refiner.*

$$U_c = \frac{\left(16a + 3.27bG + 0.67cG^2\right)}{d^{0.25}} \qquad (8.7)$$

where,

$$a = 2.461 - 0.759z + 1.625z^2$$

$$b = 0.7655 + 21.373z - 9.6625z^2$$

$$c = 9.7938 - 30.809z + 14.333z^2$$

T_f = average outside film temperature of convection section, °C

$z = (T_f + 17.8) / (556)$

Recommended flue gas mass velocities through the convection section are in the range of 1.5–2.0 kg/m².sec. This is useful while calculating the dimensions of the flue gas duct in the convection section.

The exposed length of the tubes is 0.5 m shorter than the actual length of the convection section.

Tube side liquid: Acceptable flow velocities of liquid in tubes range between 1.5 and 2.0 m/sec.

8.3.6 THERMAL DESIGN OF FIRED HEATERS

The method used here would provide an approximation of the equipment sizes and is similar to that used by Walas [8]. It would give a fairly good idea to the readers on how a thermal rating of a fire heater is carried out.

Example: Consider a fired heater of a hydrocarbon feedstock to be designed for the following duty:

Flowrate of feedstock   : 136000 kg/hr
Density of feed   : 889.6 kg/m³
Inlet temperature of feed   : 258°C
Outlet temperature of feed   : 370°C
Specific heat of feed   : 0.675 Kcal/kg.°C
Heat duty   : 13.20 MKcal/hr

(The reader may think that product of mass, specific heat and temperature difference gives a value of 10.28 MKcal/hr. The reason for the higher heat duty is that partial vaporization of the feed also takes place.)

Combustion calculations:

Let us target a heat duty with a 25% margin, i.e., 16.50 MKcal/hr.
The fuel gas composition is provided in Table 8.5.
The lower heating value of the fuel gas is 9931 Kcal/kg.

TABLE 8.5
Fuel Gas Composition

Component	Mole, %
H_2	26.77
H_2S	0.20
CH_4	16.77
C_2H_6	14.55
C_2H_4	5.65
C_3H_4	1.22
C_3H_8	12.79
C_4H_{10}	8.19
C_4H_8	0.11
C_5H_{12}	0.30
O_2	1.60
CO	0.36
CO_2	0.05
N_2	11.44
Total	**100**

Fuel gas firing rate   = $16.50 \times 10^6/9931$
                                         = 1661.4 kg/hr
                                         = 68.61 kmol/hr (refer to Table 8.6 for a summary)

The element-wise break-up of   H   145.62 kmol/hr (58.95%)
this is (refer to Table 8.6)
                                  O   1.26 kmol/hr (0.51%)
                                  C   92.15 kmol/hr (37.31%)
                                  S   0.14 kmol/hr (0.06%)
                                  N   7.85 kmol/hr (3.17%)

TABLE 8.6
Fuel Gas Characteristics

	Molecule-wise Break-up			Element-wise Break-up					
	Composition Mole, %	Firing Rate, kg/hr	Firing Rate, Mole/hr	H	O	C	S	N	Total
H_2	26.77	36.73	18.37	18.37					
H_2S	0.20	4.67	0.14	0.14			0.14		
CH_4	16.77	184.09	11.51	23.01		11.51			
C_2H_6	14.55	299.48	9.98	29.95		19.97			
C_2H_4	5.65	108.54	3.88	7.75		7.75			
C_3H_4	1.22	33.48	0.84	1.67		2.51			
C_3H_8	12.79	386.11	8.78	35.10		26.33			
C_4H_{10}	8.19	325.91	5.62	28.10		22.48			
C_4H_8	0.11	4.23	0.08	0.30		0.30			
C_5H_{12}	0.30	14.82	0.21	1.23		1.03			
O_2	1.60	35.13	1.10		1.10				
CO	0.36	6.92	0.25		0.12	0.25			
CO_2	0.05	1.51	0.03		0.03	0.03			
N_2	11.44	219.77	7.85					7.85	
Molal Flow, Mole/hr			68.61	145.62	1.26	92.15	0.14	7.85	247.0
Mole, %	100.0			58.95	0.51	37.31	0.06	3.17	100.0
Mass Flow, kg/hr		1661.4		291.2	40.2	1105.8	4.4	219.8	1661.4

Stoichiometric quantity of oxygen $145.62/2 + 92.15 + 0.14 - 1.26$ kmol/hr
required for complete combustion of
hydrogen, carbon and sulfur is given by

                                                         $= 163.84$ kmol/hr
Assume excess air required $= 10\%$
Hence, actual oxygen required $= 163.84 \times 1.1$
                                                         $= 180.22$ kmol/hr
Hence, total air required $= 180.22/0.21$
                                                         $= 858.2$ kmol/hr $= 24750$ kg/hr
Total accompanying nitrogen $= 858.2 \times 0.79$
                                                         $= 678.0$ kmol/hr

The flue gas composition is calculated as
  follows:
Carbon dioxide generated $= 92.15$ kmol/hr
Moisture generated $= 145.62$ kmol/hr
Nitrogen in flue gas $= 7.85 + 678.0 = 685.85$ kmol/hr
Oxygen in flue gas (excess oxygen) $= 163.84 \times 0.1 = 16.38$ kmol/hr
Sulphur dioxide generated $= 0.14$ kmol/hr
Average molecular weight of flue gas $= 28.09$ (refer to Table 8.7)

Specific heat of flue gas (800°C)   : 8.09 Kcal/kgmol.°C
Specific heat of flue gas (500°C)   : 7.48 Kcal/kgmol.°C

Total flue gas flowrate (Table 8.7) $= 940.1$ kmol/hr
                                                         $= 26413$ kg/hr

---

TABLE 8.7
Flue Gas Characteristics

Component	Molal Flow, kgmol/hr	Mole, %	Molecular Weight, kg/kgmol	Mass Flow, kg/hr
Carbon Dioxide	92.15	0.0980	44	4054.6
Moisture	145.62	0.1549	18	2621.2
Nitrogen	685.85	0.7295	28	19203.8
Oxygen	16.38	0.0174	32	524.2
Sulphur Dioxide	0.14	0.0001	64	9.0
Total	**940.14**	**1.00**	**28.09**	**26413**

Flue Gases from Air Preheater (APH):
Let us consider air temperatures as follows (refer to Figure 8.6):

Average sp. heat of air $= 0.205$ Kcal/kg.°C

Inlet temperature to (APH) $= 25$°C

Outlet temperature from (APH) $= 273$°C

APH heat duty $= 24750 \times 0.205 \times (273-25)$

                                 $= 1.26$ MKcal/hr

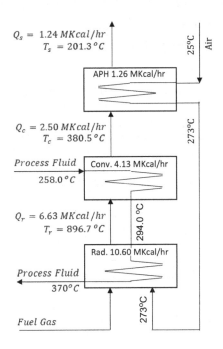

FIGURE 8.6 Temperature distribution.

Heat from fuel combustion Q_{fc}	= 16.50 MKcal/hr
Heat entering with air	= APH heat duty = 1.26 MKcal/hr
Heat entering with fuel	= nil (since fuel is not being preheated)

Therefore, total heat into the fired heater is given by:

Heat input
$$= Q_{fc} + Q_a + Q_f$$
$$= 16.50 + 1.26 + 0$$
$$= 17.76 \text{ MKcal/hr}$$

Assume percent radiant heat loss = 3%

Assume thermal efficiency = 90%

Rearranging Equation (8.1), we have

$\eta_{th} \times (heat\ input) =$
$(heat\ input) - Q_s - Q_{rl}$

$$0.90 \times 17.76 = 17.76 - (0.03 \times 17.76) - Q_s$$
$$Q_s = 17.76 \times (1 - 0.03 - 0.90)$$
$$= 1.24 \text{ MKcal/hr}$$

Hence, heat content of flue gas outlet from APH (Q_s) is 1.24 MKcal/hr

Now,

Therefore, the temperature of flue gas leaving from the stack T_s

$$1.24 = 940.1 \times 7.48 \times (T_s - 25)/1000000$$
$$= 25 + (1.24 \times 1000000)/(940.1 \times 7.48)$$
$$= 201.3°C$$

Flue Gases from Convection Section:

Assume 25% of the heat is transferred
in the convection section.
Therefore,

Heat transferred in convection section   = 0.25×16.50
     = 4.13 MKcal/hr

Heat content of flue gas leaving
convection section, Q_c

     = Q_s + Duty of APH
     = 1.24 + 1.26
     = 2.50 MKcal/hr

Now, $2.50 = 940.1 \times 7.48 \times (T_c - 25)/1000000$
Therefore, temperature of flue gas
leaving convection section T_c

     = $25 + (2.50 \times 1000000)/(940.1 \times 7.48)$
     = 380.5°C

Flue Gases from Radiant Section:

Heat content of flue gas leaving
radiant section, Q_r

     = Q_c + Duty of Convection Section
     = 2.50 + 4.13
     = 6.63 MKcal/hr

Now, $6.63 = 940.1 \times 7.48 \times (T_c - 25)/1000000$
Temperature of flue gas leaving
radiant section, T_r

     = $25 + (6.63 \times 1000000)/(940.1 \times 8.09)$
     = 896.7 °C

Calculations for Process Side Fluid (convection section):

Flowrate of feedstock	136000 kg/hr
Flowrate of feedstock (incl. 25% margin)	$136000 \times 1.25 = 170000$ kg/hr
Avg. specific heat of feedstock	0.675 Kcal/kg.°C
Feedstock inlet temperature T_{1c}	258°C
Heat duty of convection section	4.13 MKcal/hr

Temperature rise of feedstock Δt in
convection section is given by
$$4.13 \times 1000000 = 170000 \times 0.675 \times \Delta t$$
$$\Delta t = 36°C$$
Hence, feedstock outlet temperature T_{2c} = $258 + 36.0 = 294.0°C$
The log men temperature difference in
the convection section is given by

LMTD in convection section   = $[(896.7\text{-}294.0)\text{–}(380.5\text{-}258.0)]/$
     $\ln[(896.4\text{-}294.0)/(380.5\text{-}258.0)]$
     = 301.5

Flowrate of feedstock (incl 25% margin)	= 170000 kg/hr
Density of feedstock	= 889.6 kg/m^3
Volumetric flowrate of feedstock	= 191.1 m^3/hr
Assume a velocity in the tube of 2 m/sec	
Area of cross-section required for flow	= 191.1/(3600 × 2) = 0.027 m^2
For a 5" tube, area of cross-section	= 3.142 × (5.0 × 2.54/100)^2 / 4 = 0.013 m^2
Hence, there will be two tube passes in the convection section.	
Suppose thickness of the tubes is 0.3".	
Therefore, outside diameter of tubes	= [5 + (2 × 0.3)] × 25.4
	= 142.24 mm
Tube pitch	= 1.70 × 142.24
	= 241.81 mm

Let us assume the width of convection section (W) as 2300 mm.

Number of tubes in a row
$$= \frac{2300 - (0.5 \times 241.81)}{241.81} - 1$$

$$= 8.01$$

Let us take 8 (refer to Figure 8.7).
Let the length of the convection section be L.

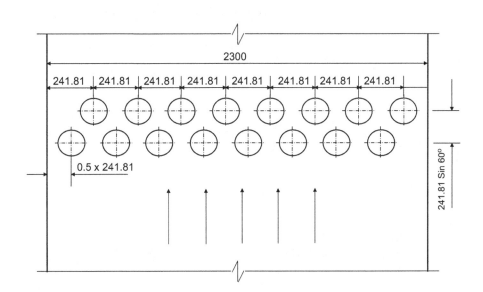

FIGURE 8.7 Tube configuration in convection section.

Free flow area for flue gas	$= \dfrac{L}{1000} \times \dfrac{2300-(8 \times 142.24)}{1000}$

Flue gas flowrate = 26413 kg/hr = 7.34 kg/sec
Consider flue gas mass velocity G as = 1.5 kg/m².sec
Hence, free flow area for flue gas = 7.34/1.5
 = 4.89 m²

Now, $\dfrac{L}{1000} \times \dfrac{2300-(8 \times 142.24)}{1000} = 4.89$

Or, L = 4208 mm
 L/W = 4208 / 2300
 = 1.83 (which is acceptable)

Overall heat transfer coefficient in convection section U_c is given by Equation (8.7).

Temperature of flue gases at = 896.7°C
  inlet to convection section

Temperature of flue gases at = 380.5°C
  outlet from convection section

Average flue gas temperature = (896.7 + 380.5)/2 = 638.6°C
For film temperature assume a T_f = 638.6 + 100 = 738.6°C
  100°C difference.
Hence,
From Equation (8.7), z $= (T_f + 17.8) / (556)$
 = 1.360
 a = 2.461 – (0.759 × 1.360) + (1.625 × 1.36²)
 = 4.434
 b = 0.7655 + (21.373 × 1.36) - (9.6625 × 1.36²)
 c = 11.961
 = 9.7938 – (30.809 × 1.36) + (14.333 1.36²)
 = -5.596

and, $U_c = \dfrac{(16 \times 4.434)+(3.27 \times 11.961 \times 1.5)-(0.67 \times 5.596 \times 1.5^2)}{(142.24)^{0.25}}$

 = 35.09 Kcal/m².°C
Bare tube surface required for = (4.13 × 10⁺⁶) / (35.09 × LMTD)
  convection section

 = 4.13 × 10⁺⁶) / (35.09 × 301.5)
 = 390.4 m²
For a convection section length
  (L) of 4208 mm,
Effective tube length = 4208 – 500
 = 3708 mm

Heat transfer area available in
  each bare tube

$$= \pi \times 142.24 \times 3708 \times 10^{-6}$$
$$= 1.66 \text{ m}^2$$

Number of bare tubes

$$= 390.4 \ / \ 1.66$$
$$= 235.2 \text{ (say, 236)}$$

Tubes in the convection section
  normally have studs to
  increase their surface area.
Assume stud surface area/bare $= 3.6$
  tube surface area

Assume a stud tube efficiency $= 85\%$
Number of studded tubes in $= 236/(3.6 \times 0.85)$
  convection section

                                       $= 77.1 \text{ (say, 78)}$
Number of rows in convection $= 78/8 \text{ (8 tubes in a row)}$
  section

                                       $= 9.75 \text{ (say, 10)}$
Hence, actual number of stud $= 10 \times 8 = 80$
  tubes in convection section

Center-to-center spacing $= 142.24 \times 1.70$
  between tubes

                                       $= 241.81 \text{ mm}$

Assume a clearance between
  shock tubes and convection
  section tube rows as 500 mm.
Assume a space for three
  additional rows of tubes for
  the future.
Therefore, height of convection $= 500 + (241.81 \, \mathrm{Sin}\, 60^\circ \times 9) + (241.81 \, \mathrm{Sin}\, 60^\circ \times 2)$
  section

                                       $= 3013 \text{ mm}$
                                       $\simeq 3020 \text{ mm}$

Cross-checking of Radiant Zone Assumptions:
1ˢᵗ Iteration

From the previous sections, we have the following duties (refer to Figure 8.6):

Heat duty of fired heater: 16.50 MKcal/hr
Heat loss with stack: (-) 1.24 MKcal/hr
Heat removed in convection section: (-) 4.13 MKcal/hr
Heat loss from radiant section: (-) 0.533 MKcal/hr
Heat duty of radiant section: 10.60 MKcal/hr

Now, assume a heat flux in the radiant section as 0.0305 MKcal/hr.m².
  Hence,

| Radiant heat transfer surface area | = 10.60/0.0305 |
| | = 347.54 m^2 |

Select a tube length of 12 m.
Select 5" tubes.

Outside diameter is 5.6"	= 0.142 m
Number of tubes	= 347.54 / (π × 0.142 × 12)
	= 64.90 (say, 65)

Consider a tube pitch ratio of 1.70.

Tube bundle perimeter	= 1.70 × 0.142 × 65
	= 15.69 m
Hence, tube bundle diameter	= 15.69 / π
	= 4.99 m

Take clearance between tube bundle and refractory as 0.50 m.

Firebox diameter	= 4.99 + (2 × 0.50)
	= 5.99 m
Height of firebox	12 + 0.6 = 12.6 m
Firebox height/diameter	≥ 2
Hence beam length, L_b	= firebox diameter = 5.99 m

Cold plane area is given by:

A_{cp} = exposed tube length × center-to-center spacing × number of tubes (excluding shield tubes)

Hence,

$$A_{cp} = 12 \times 1.70 \times 0.142 \times 65$$
$$= 188.29 \text{ m}^2$$

Absorptivity α of the tube surface with a single row of tubes is given by:

	α	= 1 − [0.0277 + 0.0927(x − 1)] (x − 1)
	x	= (center-to-center spacing)/outside tube diameter
		= 1.70
Hence,	α	= 1 − [0.0277 + 0.0927(1.70 − 1)] (1.70 − 1)
		= 0.935

Partial pressure P, of $CO_2 + H_2O + SO_2$	= (92.15+145.62+0.14)/940.14 (refer to Table 8.7)
	= 0.253 atm
We have	PL_b = 1.515 atm-m
Gas temperature	T_{gr} = 896.7°C
From Table 8.3, Emissivity φ	= 0.564

Refer to Equation (8.5). We calculate:

Total refractory area	A_r	= (π × 5.99 × 12) + [2π × (5.99)^2/ 4] − (2.3 × 4.2)
		= 272.55 m^2
Effective refractory area	A_w	= A_r - αA_{cp}
		= 272.55 − (0.935 × 188.29)
		= 96.50 m^2
	$A_w / \alpha A_{cp}$	= 96.50/(0.935 × 188.29) = 0.548
From Table 8.2 we have,	F	= 0.621

Firebox heat duty is given by:

$$\frac{Q_r}{\alpha A_{cp} F} = 4.92 \times 10^{-8} \left(T_{gr}^4 - T_{wr}^4 \right) + 35 \left(T_{gr} - T_{wr} \right)$$

TABLE 8.8
Summary of Iterations

Parameter	Unit	1	2	3
Assumed heat transfer in radiation section	MKcal/hr	10.60	10.60	10.60
Assumed heat flux	MKcal/hr.m²	0.0305	0.030	0.031
Number of tubes		65	66	64
Firebox diameter	m	5.99	6.07	5.92
Beam length, L_b	m	5.99	6.07	5.92
Cold plane area, A_{cp}	m²	188.29	191.19	185.40
PL_b	atm-m	1.515	1.536	1.498
Emissivity, φ		0.564	0.566	0.563
$A_w / \alpha A_{cp}$		0.548	0.550	0.546
Exchange factor, F		0.621	0.623	0.621
Calculated heat transfer in radiation section	MKcal/hr	10.99	11.19	10.48

Temperature of flue gas leaving T_{gr} = 896.7 + 273
  the radiant section,   = 1169.7 K
Mean tube wall temperature T_{wr} = 56 + 0.5 (T_{1r} + T_{2r})
(radiant section)   = 56 + 0.5 (294.0 + 370)
              = 388°C
              = 661 K

> T_{1r} = Feedstock inlet temperature to radiant section
>   (Feedstock outlet temperature from convection
>   section)
>   = 294°C
>
> T_{2r} = Feedstock outlet temperature from radiant
>   section
>   = 370°C

$$\frac{Q_R}{\alpha A_{cp} F} = 4.92 \times 10^{-8}(1.681 \text{ x } 10^{12}) + 35(896.7 - 388)$$

$$= 82705 + 17805 = 100510$$
$$Q_R = (100510) \times (0.935 \times 188.29 \times 0.621)$$
$$= 10.99 \text{ MKcal/hr}$$

Other Iterations

Two other iterations have been performed with heat fluxes of 0.030 and 0.031 MKcal/hr.m². The summary is provided in Table 8.8. It can be seen that for Iteration No. 3, for an assumed heat flux of 0.031 MKcal/hr.m², the calculated value of heat transfer nearly matches with the assumed value.

Hence, we can consider this for our final design with respect to the number of tubes and the firebox dimensions.

Symbols

A_r Total area of furnace surface, m²
A_w Effective area of furnace surface, m²

d Tube outside diameter, mm

G   Flue gas mass velocity, kg/m².sec

L   Length of convection section, mm

L_b Length of beam, m

Q_{fc} Heat release from fuel combustion, MKcal/hr

Q_a Heat input from air, MKcal/hr

Q_c Heat content of gas leaving the convection section, MKcal/hr

Q_f Heat input from fuel, MKcal/hr

Q_r Heat content of gas leaving the radiant section, MKcal/hr

Q_{rl} Radiant heat loss, MKcal/hr

Q_R Heat transfer in radiant section, MKcal/hr

Q_s Heat lost with stack gases, MKcal/hr

T_{1c} Feed inlet temperature to convection section, ℃

T_{2c} Feed outlet temperature from convection section, ℃

T_{1r} Feed inlet temperature to radiant section, ℃

T_{2r} Feed outlet temperature from radiant section, ℃

T_f Average outside film temperature of convection section, ℃

T_c Temperature of flue gas leaving the convection section, ℃

T_r Temperature of flue gas leaving the radiant section, ℃

T_s Temperature of flue gas leaving the stack, ℃

T_{gr} Temperature of flue gas leaving the radiant section, K

T_{wr} Tube wall temperature – radiation section, K

T_{wc} Tube wall temperature – convection section, ℃

U_c Overall heat transfer coefficient in convection section, Kcal/m².hr.℃

W   Width of convection section, mm

η_{th} Thermal efficiency

REFERENCES

1. Berman, H., "Fired Heaters-I: Finding the Basic Design for your Application", *Chemical Engineering*, June 19, 1978.
2. Garg, A., "Get the Most from your Fired Heaters", *Chemical Engineering*, March 2004.
3. Mukherjee, R., "Effectively Design Air-Cooled Heat Exchangers", *Chemical Engineering Progress*, February 1997.
4. Mukherjee, R., *Practical Thermal Design of Air Cooled Heat Exchangers*, Begell House Inc., 2004.
5. Towler, G. and Sinnott, R., *Chemical Engineering Design – Principles, Practice and Economics of Plant and Process Design*, Butterworth-Heinemann, 2008.
6. Lobo, W.E. and Evans, J. E., "Heat Transfer in the Radiant Section of Petroleum Heaters", *Trans. of AIChE*, Vol. 34, 1939, pp. 743–778.
7. Wimpress, N., "Generalised Method Predicts Fired-Heater Performance", *Chemical Engineering*, May 22, 1978.
8. Walas, S., *Chemical Process Equipment – Selection and Design*, Butterworth-Heinemann, 1990.
9. Mekler and Fairall, Petroleum Refiner, June 1952.
10. Kern, D. Q., *Process Heat Transfer* Tata McGraw-Hill, 1997.
11. Wimpress, N., "Rating Fired Heaters", *Hydrocarbon Processing*, 42(10), 1963.

9 Mass Transfer Equipment

9.1 INTRODUCTION

All separation processes involving gas-liquid contacting, e.g., distillation, flashing, stripping, absorption, etc. require column and internals for mass transfer operation. The various column internals include trays, grid and random packing, demisters, weirs, downcomers, spray nozzles, etc. To enable the vessels department to make detailed specifications for the column internals to be issued to the fabricator, a process engineer is required to furnish the necessary information to them in the form of a process data sheet for column trays or column packings. This tray data sheet should be accompanied with the process data sheets for columns showing the relevant nozzles, draw-offs, etc. The following guidelines should help the process engineer understand the important features of mass transfer equipment and prepare the relevant process documents [1].

Before we delve deeper into this chapter, it would be a good idea to look at the following video link to get a more practical perspective on what happens inside columns (both tray and packed columns):

YouTube link:

Distillation column working guide details of packing and tray columns - YouTube.

9.2 TRAY COLUMNS

9.2.1 TYPES OF TRAYS

There is an extensive literature available on different types of trays. However, before going to the engineering aspects, it is worth having a recap of the common types of trays [2].

Trays can be classified under the following heads:

- Bubble-cap Trays
- Sieve or Perforated Trays
- Valve Trays

The important features of each of these trays are described below. These are useful guidelines to select the right type of trays for a given service.

Bubble Cap Trays:
General Notes: Bubble cap trays used to be the workhorse of distillation operations until the 1960s. Now they are used only for special applications, while sieve and valve trays are more popularly used.

Capacity: Moderately high
Efficiency: Moderately high
Entrainment: About three times that of sieve trays
Turndown: Excellent
Pressure Drop: High
Cost: High, about 2–3 times that of sieve trays

Perforated Trays with Downcomers:

General Notes: In these trays, the vapor rises through small holes. The vapor velocity prevents liquid from weeping through the holes. Hole sizes are in the range of 5–6 mm. Normal practice is to have a percentage hole area of 5–15% of the bubbling area. The optimum range is 11–12% [3]. As the percentage hole area increases, the turndown gets compromised. Liquid flows across the tray deck over the weir and through the downcomer to the tray below. Such trays are used in systems where capacities close to rated loads are to be maintained. They are not suitable for variable loads. In view of the small holes required to minimize weeping, these trays are not very suitable when solids and corrosive materials are to be handled.

Capacity: High
Efficiency: As good as other tray types, but falls when capacity reduces below 60%
Ent *Entrainment:* Moderate
Turndown: About 50%
Pressure Drop: Moderate
Cost: Low

Perforated Trays without Downcomers:

General Notes: These are sieve trays without downcomers (also known as "dual-flow trays"). The vapor-liquid traffic is through the same opening and does not require a separate downcomer for liquid. Since liquid continuously weeps through the holes, the efficiencies are low. Such trays are often used while increasing the capacity of existing columns.

Capacity: Very high. Performance falls off at lower rates.
Efficiency: Lower than other types of trays
Entrainment: Low to moderate
Turndown: Lower than sieve trays
Pressure Drop: Low to moderate
Cost: Low

Valve Trays:

General Notes: These are similar to sieve trays except for one important feature. In such trays, the perforations are covered with valves which rise and fall depending upon the vapor load. The valve lids thus act as check valves to limit liquid weeping through the perforations at low vapor rates. Maximum hole areas can go up to 13-14% [3]. In some design, two types of valves provided i.e., the light and the heavy ones. At low loads, only the light valves will open and at high loads, the heavy ones will open as well. This option provides a high flexibility at differing loads. Thus, high efficiencies can be maintained over a wide range of operating throughout.

Capacity: High to very high
Efficiency: High
Ent *Entrainment:* Moderate
Turndown: About 25%
Pressure Drop: Moderate
Cost: About 20% higher compared to that of sieve trays

9.2.2 WHEN TO USE TRAY COLUMNS

Tray columns are recommended for use under the following conditions:

- If the operation involves liquid that contains dispersed solids or cause fouling, use of a tray column is preferred because the trays are more accessible and convenient for cleaning.
- Tray towers can be designed to handle wide ranges of liquid rates without flooding.
- Tray towers are preferred if interstage cooling is required to remove heats of reaction, because the heat exchange required in the liquid delivery line from tray to tray can be achieved by passing through an external cooler.
- Tray columns are preferred if side streams are to be withdrawn like in refinery crude distillation columns.
- Design information for tray towers is generally more readily available and more reliable than that for packed towers.

9.2.3 DESIGN GUIDELINES FOR TRAY COLUMNS

Tray Spacing: A major factor in deciding the tray spacing is the economic trade-off between column height and column diameter. Enlarging tray spacing adds to the column height requirement but permits a smaller column diameter. Usually, the cost of lengthening the column matches the savings achieved from the corresponding reduction in the column diameter. In some cases, there are constraints that govern tray spacing other than the economic trade-off between column height and column diameter. Tray spacings in industrial columns vary between 450 and 900 mm. Lower spacings have also been used but are not normally recommended. Guidelines for tray spacings are illustrated in Table 9.1. For further details, refer to Kister [4].

Tray Passes: In cases where the liquid load is relatively high for a single pass, multiple passes are used. This enhances tray and downcomer capacity and lowers tray pressure drop, but at the expense of a shorter path length. Shorter path lengths reduce tray efficiency, and if extremely short, may be inadequate for accommodating tray manways. Trays containing more than two liquid passes are prone to liquid and

TABLE 9.1
Guidelines for Selection of Tray Spacing

Description	Tray Spacing (mm)	Comments
Column diameters larger than 3000 mm	>600	Tray support beams restrict crawling space available, hence the large tray spacing
Column diameters between 1200 and 3000 mm	600	This spacing is sufficiently wide to allow a worker to freely crawl between trays
Column diameters between 750 and 1200 mm	450	Crawling between the trays is seldom required, because the worker can reach the column wall from the tray manways
Fouling and corrosive service	>600	Frequent maintenance is expected
Systems with a high foaming tendency	At least 450 mm but preferably 600 mm or higher	Required to avoid premature flooding
Columns operating in spray regime	At least 450 mm but preferably 600 mm or higher	Required to avoid excessive entrainment
Columns operating in froth regime	450	Lower tray spacing restricts allowable vapor velocity, thereby promoting froth regime operation

Source: Reprinted by special permission from *Chemical Engineering*, Copyright© September 2005, by Access Intelligence, Rockville, MD 20850.

vapor maldistribution because of non-symmetrical panels. Preferably, odd pass trays should be avoided. The following are certain guidelines for selection of tray passes. For further details, refer to Kister [4]:

1. Begin with an assumption of the number of passes such that weir load does not exceed 70 m³/hr.m of weir length.
2. Having chosen the number of passes, check the path length. Installation of internal manways is not feasible if the path length works out to be lower than 400 mm. In case multiple passes become mandatory, the issue is even more complicated and the column diameter may need to be adjusted to arrive at the minimum path length of 400 mm.
3. To avoid short liquid paths, the following is recommended:
   - for column diameters of 1200 to 2100 mm, maximum 2 pass trays
   - for column diameters of 2100 to 3000 mm, maximum 3 pass trays
   - for column diameters above 3000, maximum 4 pass trays

Downcomers and Types: Downcomers are passages which have circular, segmental or rectangular cross-sections. These passages convey liquid from an upper tray to a lower tray in a column. Different types of downcomers are shown in Figure 9.1.

The straight segmental vertical downcomer is the type most commonly used in distillation columns. These have a cost and simplicity advantage over the sloped downcomers. The sloped downcomers represent the best utilization of column area for downflow. They provide sufficient volume for vapor liquid disengagement at the top of the downcomer while maximizing the active area on the tray below.

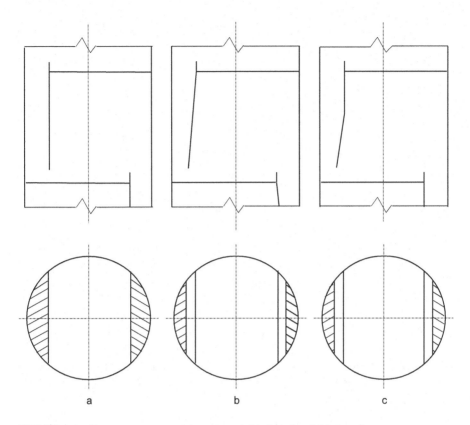

FIGURE 9.1   Downcomer types: (a) segmental (b) sloped and (c) sloped.

Source: Reprinted by special permission from *CHEMICAL ENGINEERING*, Copyright© September 2005, by Access Intelligence, Rockville, MD 20850.

Downcomer Velocity and Area: The velocity of clear liquid in the downcomer is selected based on two criteria, i.e., the prevention of choking and the achievement of satisfactory disengagement of vapor bubbles from the downcomer liquid. This is particularly important when liquids with high foaming tendencies are used. Recommended downcomer velocities are given by [4]:

Low Foaming Liquids:	0.12–0.21 m/s
Medium Foaming Liquids:	0.09–0.18 m/s
High Foaming Liquids:	0.06– 0.09 m/s

Downcomers smaller than 5–8% of the column diameters should be avoided. Narrow downcomers distort the liquid flow pattern as it approaches the weir. This increases the pressure drop and leads to the formation of stagnant regions near the tray periphery.

Downcomer Clearance: It is important to have the downcomer clearance less than the outlet weir height. If the clearance is larger than the outlet weir, there is a high probability of the vapor flowing up the downcomer rather than through the tray deck above. The vapor flowing up pushes the liquid up onto tray # 5 above, which is a cause of flooding (Figure 9.2(a)). If, however, the downcomer clearance is too small,

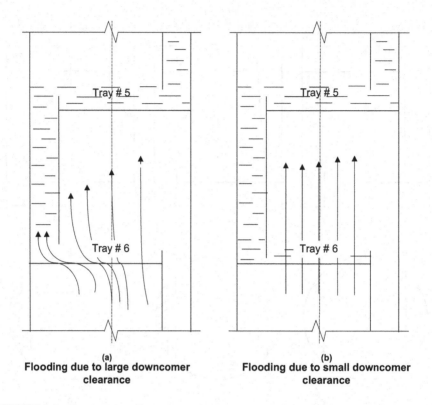

(a)
**Flooding due to large downcomer
clearance**

(b)
**Flooding due to small downcomer
clearance**

FIGURE 9.2   Downcomer clearance.

the high pressure drop needed for the liquid to flow from the downcomer onto tray #
6 causes the liquid level in the downcomer to back up onto tray # 5 (Figure 9.2(b)).
This causes flooding of tray # 5. Once tray # 5 floods the downcomer above, that will
also back up and flood, and subsequently, all the trays and downcomers above get
flooded. For further details, refer to Liebermann [5].

To achieve a proper downcomer seal, the bottom edge of the downcomer should
be about 10 mm below the top edge of the outlet weir. However, the reader is cautioned that this is more of a guideline. There are cases when at high liquid loads there
is enough backup liquid in the downcomer, which acts as a seal even if the downcomer clearance is greater than the weir height. The exact downcomer clearance can
only be specified by the tray vendor.

Outlet Weirs: The liquid enters the downcomer by flowing over the outlet weir. The
weir height directly sets the liquid level and hold-up on the tray, if the tray operates
in the froth regime. Consequently, in this regime, the higher the liquid level, the
higher the tray pressure drop, the downcomer backup, entrainment rate and weeping
tendency. Weir heights in this regime are therefore restricted to 50–80 mm tall.

However, at low liquid rates, there is a chance that instead of liquid being in continuous phase on the tray, the vapor exists in continuous phase. This is the spray

regime operation. In this regime, the liquid no longer flows over the weir but is carried as a spray over the weir into the downcomer. When trays operate in the spray regime, liquid enters the downcomer as a shower of droplets. Under these conditions, liquid hold-up on the tray is independent of weir height. However, even with columns designed to operate normally in the spray regime, it is good policy to provide outlet weirs because at low vapor rates, the column may operate in the froth regime, and also the presence of weir limits spray regime entrainment. At still lower loads, the phenomenon called "blowing" occurs [6]. The preferred weir height for columns operating in the spray regime is 20–25 mm.

The liquid level on a tray is a function of two factors: weir height and crest height. The weir height has been discussed above. Crest height is the height of liquid over the weir. It can be expressed as:

$$C_h = 239.5 \left(\frac{Q_L}{W_L} \right)^{0.67} \tag{9.1}$$

The sum of the crest height and the weir height equals the depth of liquid on the tray.

Weir Loading: As a general guideline, weir loadings should fall within the range of 15–70 m³/hr.m of weir length. Although, in some cases, loadings as low as 5.5 m³/hr.m and as high as 90 m³/hr.m have also been used, the probability of blowing increases below weir loads of 9 m³/hr.m, and is highly probable at loads of 4.5 m³/hr.m.

Tray Pressure Drop: The pressure drop across a tray consists of two components. The first component is the dry tray pressure drop. This is the pressure drop of the vapor as it flows through the tray holes. Dry tray pressure drop is given by [5]:

$$\Delta P_{dry} = 273.4K \left(\frac{\rho_V}{\rho_L} \right) V_g^2 \tag{9.2}$$

The orifice coefficient K can be as low as 0.3 for a smooth hole in a thick plate, and 0.6–0.95 for various valve tray caps.

The second component is the hydraulic tray pressure drop. This is the head of liquid which the vapor has to overcome to rise up the tray. Two points are important here: (1) The actual height of liquid over the weir is greater than the weir height and is given by the so-called crest height, (2) the liquid above the tray in reality does not exist as a clear liquid but as a foam. Therefore, the head of the liquid is somewhat reduced as a result of the effect of foam. Hydraulic tray pressure drop is thus expressed as:

$$\Delta P_{hyd} = \left(A_r \times W_h \right) + C_h \tag{9.3}$$

The aeration factor is the ratio of the density of the foam to the density of the clear liquid. It is typically 0.5.

The total tray pressure drop is the sum of the dry tray pressure drop and the hydraulic tray pressure drop:

$$\Delta P_{total} = \Delta P_{dry} + \Delta P_{hyd} \tag{9.4}$$

Seal Pans: The downcomer descending from the bottom tray of a column must be sealed to prevent vapor flow through it. The following are some guidelines for seal pans (Figure 9.3):

1. The bottom seal pan is one of the most sensitive areas for solids accumulation. Therefore, the clearance between the bottom downcomer and the floor of the seal pan should be greater than the normal clearance under the tray downcomers, and it should be as a minimum, 50 mm.
2. The submergence of the downcomer within the seal pan should be approximately the same as the clearance between the bottom of the downcomer and the seal pan floor (refer to Figure 9.3 for clarity).
3. The distance between the floor of the bottom tray and the seal pan should be 150 mm larger than the normal tray spacing [1].

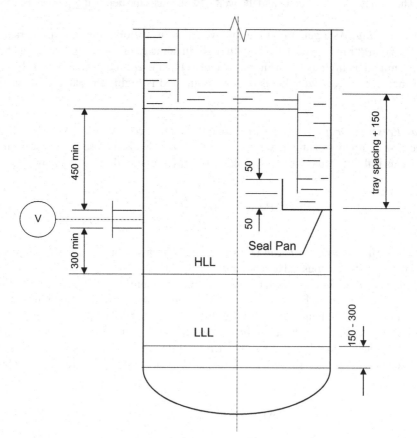

FIGURE 9.3 Column bottoms arrangement and seal pan.

9.2.4 PROCESS DATA SHEETS FOR TRAY SECTIONS

It is the job of the process engineer to fill in the process data sheet for column trays and furnish all the pertinent information asked in the data sheet. This data sheet is then sent to tray suppliers. Liquid and vapor rates should be specified for critical trays. In general, for pressure services, step load changes occur at feed and draw-off points. Rapid change in the liquid and vapor loads will occur on the top tray if cold reflux is being fed. In cases where the variation in the vapor-liquid traffic across the column varies widely, the column is split into a number of sections and the vapor liquid data are provided for the decisive tray of each section. In case of any doubt, complete vapor and liquid profile along the tower should be supplied.

System frothiness characteristics and the properties like viscosity and surface tension should always be indicated. The design of downcomers is greatly affected by the foam-forming properties of the system. Refer to Table 9.2 for system factors.

Column Simulation: The basic input for filling up a tray data sheet comes from column simulation results. These are nowadays carried out in process simulation software, viz., Aspen Hysis, PRO II, Promax, etc. while carrying out the mass and energy balance.

Table 9.3 illustrates a typical column simulation output for a stripper column. Each tray is a theoretical stage. The overhead condenser is also considered as a theoretical stage. Feed is introduced into the top tray. Reflux is also introduced into the top tray. The stripped product is drawn from the bottom of the column.

Preparing the Tray Data Sheet: The simulation output provides vapor liquid traffic at every tray. In general, step load changes occur at feed and draw-off points. Rapid change in the liquid and vapor loads will occur on the top tray if cold reflux is being fed.

It is not practical to reproduce data for every tray in the data sheet. In cases where the variation in the vapor-liquid traffic across the column varies widely, the column

---

TABLE 9.2
Suggested System Factors

System	Factor
Non-foaming	1
Mild or Slight Foaming	
Depropanizers	0.9
Hot Carbonate Regenerators	0.9
Moderate Foaming	
Deethanizers	0.85
Demethanizers	0.85
Amine Regenerators	0.85
Crude Towers	0.85
High Foaming	
Amine Absorbers	0.73–0.80
Sour Water Strippers	0.5–0.7

Source: Adapted from Kister [1].

TABLE 9.3

Typical Simulation Output

Stage	Pressure kPa	Heat duty MW	Liquid from Tray					Vapor to Tray			
			Flow, kg/hr	Temperature, °C	Viscosity, Pa-sec	Density, kg/m³	Surf. Tension, mN/m	Flow, kg/hr	Temperature, °C	Viscosity, Pa-sec	Density, kg/m³
1	116.5	−5.95	2241	77.0	0.000407	973.5	63.74	9561	114.4	1.27E-05	0.694
2	117.3	0	10987	114.4	0.000269	950.5	57.23	10322	114.5	1.27E-05	0.682
3	118.3	0	11008	114.5	0.000269	950.5	57.15	10343	114.5	1.27E-05	0.685
4	119.2	0	11043	114.6	0.000269	950.5	57.05	10378	114.6	1.27E-05	0.688
5	120.1	0	11098	114.6	0.000269	950.4	56.92	10433	114.6	1.27E-05	0.692
6	120.9	0	11175	114.6	0.000269	950.4	56.75	10510	114.7	1.27E-05	0.697
7	121.8	0	11295	114.7	0.000270	950.4	56.51	10629	114.8	1.27E-05	0.705
8	122.7	0	11446	114.8	0.000271	950.5	56.19	10788	115.0	1.27E-05	0.710
9	123.6	0	11670	115.0	0.000272	950.5	55.74	11004	115.1	1.27E-05	0.731
10	124.5	0	11915	115.1	0.000273	950.5	55.11	11268	115.3	1.28E-05	0.753
11	125.3	0	12201	115.3	0.000276	950.6	54.22	11584	115.5	1.28E-05	0.786
12	126.3	0	12531	115.5	0.000280	950.6	52.91	11977	115.9	1.28E-05	0.812
13	127.2	0	12919	115.9	0.000285	950.6	50.96	12445	116.6	1.28E-05	0.878
14	128.0	0	13372	116.6	0.000294	950.3	48.05	13017	117.7	1.28E-05	0.986
15	128.9	0	13893	117.7	0.000306	949.5	43.93	13759	119.7	1.27E-05	1.271
16	129.8	0	14560	119.7	0.000320	947.9	38.76	14667	122.7	1.27E-05	1.435
17	130.7	0	15375	119.7	0.000320	947.9	38.76	15767	122.7	1.27E-05	1.435
18	131.6	0	16436	122.7	0.000335	945.3	33.30	17107	126.3	1.23E-05	2.029
19	132.5	0	17751	124.8	0.000341	943.8	30.89	18647	129.4	1.22E-05	2.204
20	133.4	0	19526	126.3	0.000348	942.3	28.48	20512	132.5	1.20E-05	2.379

is split into a number of sections. For each such tray section, the vapor liquid data provided should correspond to that of the decisive tray. This means that for each tray section, the tray geometries for all the trays will be identical and will be designed corresponding to data of the decisive tray.

Table 9.4 shows how the simulation output illustrated in Table 9.3 is broken into different tray sections. A tray efficiency of 65% is assumed. Based on the tray efficiency, the number of actual stages is calculated, and the same are entered adjacent to the theoretical trays.

It is seen that from the column top the vapor-liquid traffic increases progressively and becomes two times at the bottom. The column is therefore divided into two sections in such a way that the vapor-liquid traffic in each section remains within an acceptable range. The decisive tray in each section is that tray which would require the largest diameter for that particular section. Thus, the top section consists of trays 1–20 and the bottom section of trays 21–31. Based on the fact that total pressure drop is 0.16 kg/cm² and there are 31 trays, the allowable pressure drop across each tray is specified at 0.005 kg/cm².

Table 9.5 illustrates the tray data sheet generated using data from Table 9.4 and is self-explanatory. Table 9.6 illustrates the tray vendor's calculations with details of tray geometry. This will be further discussed in Chapter 15.

9.2.5 Miscellaneous Details of Tray Columns

Column Data Sheet: The column data sheet consists of data on the column shell, viz., column height, diameter, operating and design temperatures and pressures, material of construction, corrosion allowance, nozzle table and column sketch. Figure 9.4 illustrates a typical column sketch. It should be noted that while the tray data sheet is sent to the tray vendor, the column data sheet is sent to the fabricator of the column shell.

Top Tray Feed Nozzle/Reflux Nozzle: Feed to a stripping column or reflux to a distillation column is fed to the top tray. In such cases, the normal practice is to try and orient the feed nozzle perpendicular to the downcomer of the top tray (Figure 9.5). While there could be exceptions to this arrangement, this is the most preferred orientation. If the feed contains some vapor in addition to liquid, only arrangements A and B are suitable. However, when the feed is fully liquid, any of the three arrangements A, B or C may be used. However, arrangement C would most likely be the cheaper one.

Intermediate Tray Feed: Nozzles feeding liquid at intermediate trays can be oriented with greater flexibility. In Figure 9.6, we have a case for liquid feed on tray #10. In this case, the feed nozzle can be oriented anywhere except in the downcomer segment. Two such feed arrangements are shown in Figure 9.6 (one at a nozzle orientation of 270° and the other at 180°). The liquid is injected onto the downcomer in such a way that it flows down the wall and then along the tray. At locations where such feed arrangements are to be installed, the space between the two trays should be at least 800 mm to facilitate their installation.

Column Manways: It is normal practice to provide column manways after every 12–15 trays. This is useful during tray installation and plant shutdown. At locations where manways are installed, the tray spacings need to be at least 800 mm. It is a

TABLE 9.4
Simulation Output Divided into Tray Sections

Section	Tray Decisive	Actual Stage	Theoretical Stage	Pressure kg/cm²(g)	Liquid from Tray Flow, kg/hr	Temp, °C	Viscosity, Pa-sec	Density, kg/m³	Surf. Tension, mN/m	Vapor to Tray Flow, kg/hr	Temp, °C	Viscosity, Pa-sec	Density, kg/m³
			1	1.17	10987	114.4	0.000269	950.5	57.23	10322	114.5	1.27E-05	0.682
			2	1.18	11008	114.5	0.000269	950.5	57.15	10343	114.5	1.27E-05	0.685
		5	3	1.19	11043	114.6	0.000269	950.5	57.05	10378	114.6	1.27E-05	0.688
			4	1.20	11098	114.6	0.000269	950.4	56.92	10433	114.6	1.27E-05	0.692
			5	1.21	11175	114.6	0.000269	950.4	56.75	10510	114.7	1.27E-05	0.697
1-20	12		6	1.22	11295	114.7	0.000270	950.4	56.51	10629	114.8	1.27E-05	0.705
		11	7	1.23	11446	114.8	0.000271	950.5	56.19	10788	115.0	1.27E-05	0.710
			8	1.24	11670	115.0	0.000272	950.5	55.74	11004	115.1	1.27E-05	0.731
			9	1.25	11915	115.1	0.000273	950.5	55.11	11268	115.3	1.28E-05	0.753
		16	10	1.25	12201	115.3	0.000276	950.6	54.22	11584	115.5	1.28E-05	0.786
			11	1.26	12531	115.5	0.000280	950.6	52.91	11977	115.9	1.28E-05	0.812
			12	1.27	12919	115.9	0.000285	950.6	50.96	12445	116.6	1.28E-05	0.878
		21	13	1.28	13372	116.6	0.000294	950.3	48.05	13017	117.7	1.28E-05	0.986
			14	1.29	13893	117.7	0.000306	949.5	43.93	13759	119.7	1.27E-05	1.271
			15	1.30	14560	118.6	0.000313	948.7	41.35	14667	121.2	1.27E-05	1.353
21-31	26	26	16	1.31	15375	119.7	0.000320	947.9	38.76	15767	122.7	1.27E-05	1.435
			17	1.32	16436	122.7	0.000335	945.3	33.30	17107	126.3	1.23E-05	2.029
			18	1.33	17751	124.8	0.000341	943.8	30.89	18647	129.4	1.22E-05	2.204
	31	31	19	1.33	19526	126.3	0.000348	942.3	28.48	20512	132.5	1.20E-05	2.379

TABLE 9.5
Tray Data Sheet

Item			C-03	
Designation			Stripping Column	
Column Section Number			1	2
Tray Section			1–20	21–31
Decisive Tray No.			20	31
Vapor to tray	Operating pressure	kPa(a)	122.7	128
	Operating temperature	°C	116.6	132.5
	Normal load	kg/h	12445	20512
	Required load min/max	%	50/110	50/110
	Density	kg/m³	0.878	2.379
	Dynamic viscosity	cP	0.0128	0.012
Liquid from tray	Operating temperature	°C	115.9	126.3
	Normal load	kg/h	12919	19526
	Required load min/max	%	50/110	50/110
	Density	kg/m³	950.3	942.3
	Dynamic viscosity	cP	0.285	0.348
	Surface tension	N/m	50.96	28.48
Allowable Pressure Drop per Tray		kg/cm²	0.005	0.006
Allowable Flooding Factor		%	80	80
Foaming Factor (system factor)			1	1
Type of Tray			Valve	Valve
Number of Passes			1	1
Tray Spacing		mm	450	450
Estimated Column Diameter		mm	1250	1850

Notes: 1. Flooding factor of 0.80 at 110% load
2. Pressure drop per tray of 0.005 kg/cm² at 100% load.

good idea to install manways above feed trays, where the spacing is anyway kept higher to ensure sufficient space for the distributor pipes.

Instrument Nozzles: Nozzles for pressure instruments are located on the vapor space. For example, for pressure indication for Tray #20, the best location for the pressure instrument nozzle is 100 mm below Tray #21. For temperature measurement, the nozzle should be placed on the downcomer of the tray for which the temperature needs to be measured. Refer to Figure 9.4, Nozzle P and Nozzle T, respectively).

Bottom Feed and Reboiler Return: It has been reported that 50% of the problems in the lower part of a column are initiated in the space between the bottom tray or packing support plate and the liquid level [4]. It is therefore important to follow these guidelines:

- The bottom feed and reboiler return should be above the high liquid level (HLL). The space between the "bottom of pipe" for the bottom feed or reboiler return should be at least 300 mm from the high liquid level (refer to Figure 9.3).
- The bottom feed and reboiler return should not impinge on the bottom seal pan, seal pan overflow, or the bottom downcomer. Vapor impingement on the seal pan overflow may result in liquid entrainment to the bottom tray. Impingement

TABLE 9.6
Vendor Tray Hydraulic Calculations

Item							C-03
Designation					Stripping Column		
Column Section							
Number			1			1	
Tray Section			1–20			21–31	
Decisive Tray No.			20			31	
Type of Tray			Sieve			Sieve	
Percentage Load	%	50	100	110	50	100	110
Vapor to Tray	kg/hr	6223	12445	13690	10256	20512	22563
Density	kg/m³	0.878	0.878	0.878	2.379	2.379	2.379
Vapor Volume Rate	m³/s	1.97	3.94	4.33	1.99	3.97	4.37
Vapor Viscosity	mPa-s	0.0128	0.0128	0.0128	0.0120	0.0120	0.0120
Liquid from Tray	kg/hr	6460	12919	14211	9763	19526	21479
Density	kg/m³	950.5	950.5	950.5	942.3	942.3	942.3
Liquid Volume Rate	m³/hr	6.80	13.59	14.95	10.4	20.7	22.8
Surface Tension	N/m	50.96	50.96	50.96	28.48	28.48	28.48
Liquid Viscosity	mPa-s	0.285	0.285	0.285	0.348	0.348	0.348
Column Diameter	mm	1300			1650		
Tray Spacing	mm	450			450		
Tray Passes		1			1		
Column Area	m²	1.33			2.01		
Downcomer Chord Height	top/bottom mm	120/120			150/150		
Downcomer Area	top/bottom m²	0.061			0.099/0.099		
Flow Path Length	mm	1060			1350		
Active Area	m²	1.2			1.94		
Exit Weir Length	mm	752.6			948.68		
Downcomer Clearance	mm	35			35		
Exit Weir Height	mm	40			40		
Weir Load	m³/hr.m	9.03	18.06	19.87	10.86	21.71	23.88
Downcomer Backup Liquid	mm	46.23	81.70	89.79	54.02	81.97	89.4
Jet Flood Percent	%	43.48	77.50	84.82	39.50	76.00	82.94
Dry Tray Pressure Drop	kg/cm²	0.0008	0.0039	0.0048	0.0007	0.0027	0.0033
Total Tray Pressure Drop	kg/cm²	0.0018	0.0050	0.0059	0.0022	0.0042	0.0049

Notes: 1. Weeping at minimum loads (for tray sections 21–31).

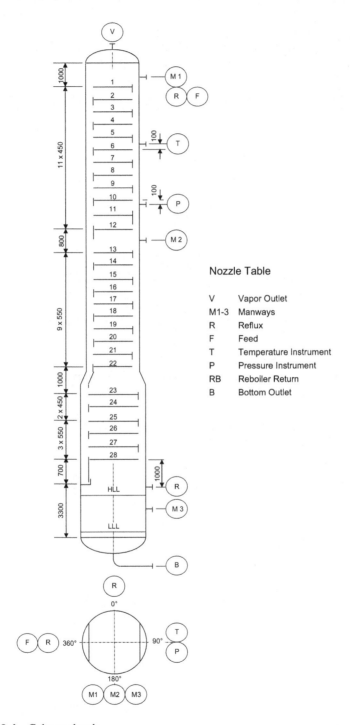

FIGURE 9.4 Column sketch.

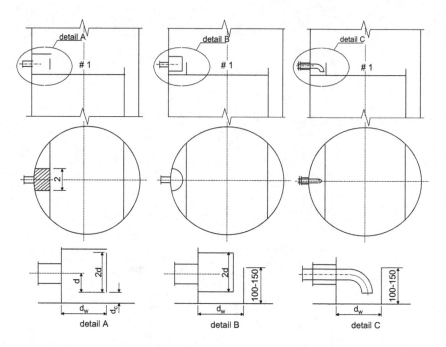

FIGURE 9.5   Top tray feed/reflux nozzle arrangement.

Source: Reprinted by special permission from *CHEMICAL ENGINEERING*, Copyright© September 2005, by Access Intelligence, Rockville, MD 20850.

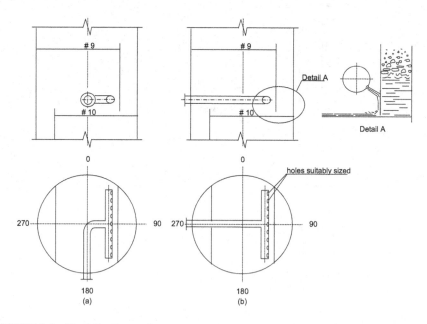

FIGURE 9.6   Feed distributor pipes.

Source: Reprinted by special permission from *CHEMICAL ENGINEERING*, Copyright© September 2005, by Access Intelligence, Rockville, MD 20850.

on the downcomer or seal pan may cause vaporization and lead to premature flooding, particularly if the feed is superheated. The preferable arrangement is to introduce these nozzles parallel to the edge of the seal pan.
- The "top of pipe" of the bottom feed and reboiler return nozzle should be located at least 400–450 mm below the bottom tray (Figure 9.3).
- Tangential bottom feed and reboiler return nozzles should be avoided, since they impart a swirl on the sump and promote vortexing.

Column Bottom Sump: The depth of the column sump is decided based on the residence time of the column bottom's product. In normal practice, this is taken at 4–7 minutes. This sets the HLL in the column sump. The low liquid level (LLL) is normally between 150–300 mm (Refer to Figure 9.3 for typical details).

9.2.6 HIGH PERFORMANCE TRAYS

For a Greenfield plant, a well-designed conventional tray column usually provides the best solution for vapor-liquid contacting. However, with the passage of time and increasing demand for throughput, the operators tend to run the plant to a point where conventional trays are no longer able to perform.

The 1990s saw the emergence of the so-called "high performance trays" as a major development. They offer a practical solution to achieve higher capacities while retaining the same column shell, i.e., keeping the same column diameter. High performance trays can increase throughput by 10–25% over a good conventional tray design [7].

Limits of Conventional Trays: A conventional tray has three functional zones [7]:

1. Active area for mixing vapor and liquid
2. Vapor space above the active area
3. Downcomer between trays

The job of a good designer is to balance these three zones to achieve the desired objectives. The trick here is in striking the best balance between the active and downcomer areas.

High performance trays function by improving performance in one or more of the three above functional zones. There are three possible configurations of high performance trays.

Cross-Flow Trays with Truncated Downcomers
Increasing the Active Area: A conventional tray has an area under the entering downcomer and over the exiting downcomer where liquid-vapor contact cannot take place. Typically, the downcomer top area accounts for 15–25% of the column cross-section area, leaving as little as 50% for active contact of the vapor and the liquid. This reduces the tray's capacity to handle vapor-liquid traffic significantly [4, 8, 9]. High performance trays precisely attempt to address this inactive area. One way to address the issue is to slope the downcomers from the top and truncate

them before they approach the tray deck. The truncated downcomers still provide a sufficiently large downcomer top area allowing entrained vapor along with the liquid to disengage. The downcomer area requirement at the bottom is therefore reduced, since the liquid is clear. The area under the downcomer is thus increased and can be used as a "true active area" for fractionation. The resultant increase in capacity is of the same order as the percentage gain in the active area (refer to Figure 9.7). The VGPlus™ trays by Sulzer Chemtech and SUPERFRAC® trays by Koch-Glitsch work on this principle.

In addition, such trays also combine the following facilities to achieve the best possible results.

Decreasing Weir Load: Increasing the length of the outlet weir reduces the weir load and weir crest and, as a result, lowers tray pressure drop (refer to Figure 9.8). This leads to an increase in the tray capacity.

Low Pressure Drop Valve: Low pressure drop valves reduce the liquid height in the downcomer. So, more liquid can be fed into the downcomer, increasing the liquid handling capacity [9].

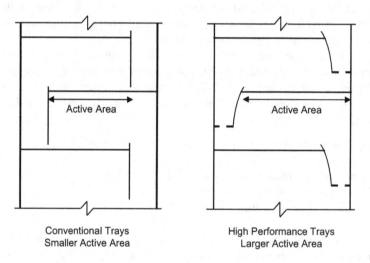

FIGURE 9.7   Conventional trays vs. high performance trays.

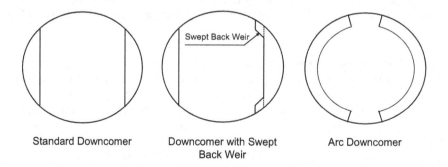

FIGURE 9.8   Increasing lengths of downcomers.

Because of the lower liquid height in the downcomer, the effective height of the downcomer on a high-capacity tray is less than that of a standard tray. Thus, tray spacing can be reduced and the number of distillation stages increase in the same shell.

Such trays have the added advantage of higher efficiencies compared to conventional trays on account of a better tray deck design.

Multi-downcomer Trays: Increasing the weir length through which a liquid flows decreases the pressure drop for a given liquid flowrate. In a conventional tray, a swept-back weir achieves this. Multi-downcomer tray also uses this approach. However, its downcomer length is much longer than that of a conventional tray [7].

In multi-downcomer trays, the flow path from inlet to outlet is comparatively short, resulting in reduced tray efficiencies. However, recent developments claim higher efficiencies for such trays. Such trays have large total weir length. This, along with large downcomer areas, provides high liquid handling capability. These trays can therefore handle large liquid loads, particularly when the volumetric ratio between vapor and liquid rates is low [8]. Multi-downcomer trays can have tray spacings as low as 300 mm. This helps to accommodate a larger number of trays for a given column height. The MD™ trays offered by UOP and the HiFi™ trays offered by Shell/Sulzer Chemtech work on this principle [8].

Trays without Downcomers: Downcomers, including truncated ones, reduce the active area available [7]. Further reduction in the active area can be achieved by using trays without downcomers, i.e., the dual flow tray. Without downcomer, the liquid goes down the same hole through which the vapor rises.

It is stated that in dual flow trays, the liquid and vapor should pass through the same hole at the same time. However, this is unlikely the case. Either vapor and liquid pass through separate holes at the same time, or the liquid and vapor flows in pulses at different times through most of the holes. In the latter case, vapor will pass through most of the holes while the liquid level builds up. Thereafter, when the liquid reaches sufficient static height to overcome the vapor flow, liquid will flow through most of the holes.

Dual flow trays have open areas as high as 25%. High hole areas, however, reduce the stable operating range of the tray. In addition, dual flow trays have another drawback: lower efficiencies. The path length of dual trays is essentially zero. This reduces efficiencies by 30–40%. On the other hand, dual flow trays have higher capacities than conventional trays [7].

Over the years, various modifications have been made to improve the dual flow design. Prime among them are Stone and Webster's Ripple Tray® and Shell Turbo grid tray. These designs provide greater operating flexibility compared to the standard dual flow design [7].

Applications of High Performance Trays: High performance trays have been successfully used in many applications including ethylene plants. Important ones are the Primary Fractionator, Caustic Scrubber, Demethanizer, HP Depropanizer, LP Depropanizer, Deethanizer, C2 Splitter, and C3 Splitter [8].

9.3 PACKED COLUMNS

In packed columns, random packings made of ceramic, steel or polypropylene are used. The latter is increasingly preferred because of its low weight, easy handling and resistance to corrosion. The use of packing grids of a variety of proprietary designs is gaining importance these days. A few typical proprietary designs include Glitsch Grids, Sulzer packings, among others. The flow could be counter-current or co-current, gas being the continuous phase. Typical applications include absorption, stripping and distillation operations. Typical column packings are well documented in the literature and are not reproduced in this book [2].

9.3.1 WHEN TO USE PACKED COLUMNS

Packed columns are recommended for use under the following circumstances:

- Packed columns are always considered for columns with diameters smaller than 900 mm in which case trays are difficult to install.
- Pressure drop through packed columns is usually less than the pressure drop through a tray column for the same duty. This advantage makes packed towers particularly desirable for vacuum operation.
- The advantage of lower pressure drop can be effectively used when carrying out capacity expansion activities for existing plants. The high vapor-liquid traffic to be handled for future loads is effectively addressed by replacing the existing trays with packing while keeping the column shell unchanged.
- Packed towers are usually preferred if the liquids have a large tendency to foam.
- Corrosive materials can be more easily handled in packed columns because construction could be of ceramic, polymer or other corrosion resistant materials. Further, for such cases, a packed column would normally be cheaper compared to a tray column.

9.3.2 DESIGN GUIDELINES FOR PACKED COLUMNS

A typical packed column consists of essentially the following components:

- Shell
- Packing
- Distributor
- Feed Pipe
- Flash Feed Gallery Tray
- Redistributor
- Collector Tray
- Support Plate
- Bed Limiter

Figure 9.9 illustrates a bird's eye view of a packed column. While the shell needs no description, there is enough description of various types of packings elsewhere in literature. Hence, they are not covered here. The remaining components are briefly detailed below:

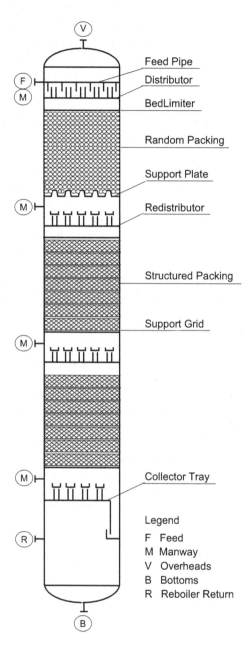

FIGURE 9.9 Packed column.

Selecting the Packing Size: The size of the packing should be so selected that the ratio of the column diameter to the packing size is greater than 30 for Raschig rings, 15 for ceramic saddles, and 10 for slotted rings or plastic saddles [10]. The following are general guidelines for selecting the size of packings [11]:

- for column diameters below 300 mm, select packing size between 15–25 mm
- for column diameters of 300–900 mm, select packing size between 25–40 mm
- for column diameters above 900 mm, select packing size between 50–80 mm

Height of Packed Section: The height of a packed section is usually calculated using the concept of height equivalent to a theoretical plate (HETP). It is the height of packing that will give the same separation as a theoretical stage. While the values of HETP can be predicted from mass transfer models, rules of thumb and data interpolation [12], it has been reported that such values are remarkably constant for a large number of organic and inorganic systems [10]. According to another study, in distillation operations, HETP values for a given type and size of packing are essentially constant and, independent of the system physical properties provided, the liquid is properly distributed and the pressure drop across the packed section is at least 0.0017 kg/cm^2 per meter of packing height [13]. An example on the use of HETPs is illustrated later in this chapter.

Distributors: Proper distribution of liquid across the packing is achieved by means of liquid distributors. Spray nozzles/pipes can be used for small diameter columns. For large diameter columns, orifice type distributors are normally used. Vapor flows up through the large chimneys, and liquid drains through the smaller distribution holes in the tray deck. Distributors are normally located 100–150 mm above the packing to permit vapor disengagement from the bed before passing through the distributor. Some designers may prefer a 200 mm distance. However, a larger distance above the bed may lead to liquid streams moving away from their desired trajectory and falling at an undesirable location, leading to uneven wetting [14]. Orifice sizes are in the range of 6–12 mm. Orifice diameters smaller than this are generally avoided to prevent plugging.

As per industry practice, 60–100 liquid drip points per square meter of column cross-section are recommended. Excessive liquid depth in the risers may lead to spillage of the liquid into the risers. Therefore, the design of the distributor should be such that normal liquid head above the deck is well below 70% of the riser height.

Orifice or pan type distributors are not suitable for very low liquid loads since all holes may not be effective, leading to poor distribution. Trough type distributors are more expensive but more versatile. Liquid is fed proportionately to one or more parting boxes and thence to troughs. The exact type and geometry of the distributors is usually specified by manufacturers of column internals. Figure 9.10 illustrates typical distributor types.

Feed Pipes: In large diameter columns, if the feed is introduced through a nozzle directly into the distributor, it is possible that the velocity through the nozzle creates ripples on the distributor tray affecting the quality of liquid distribution. In such

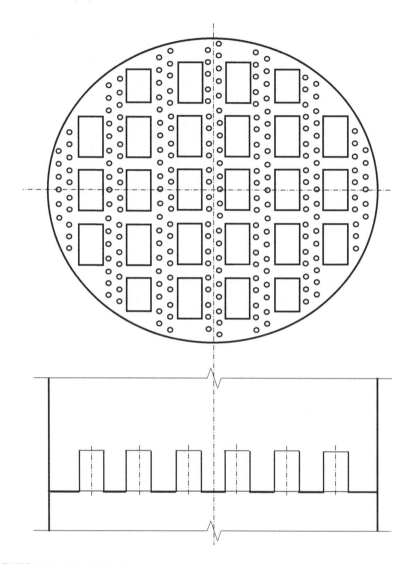

FIGURE 9.10 Liquid distributor (deck type).

cases, a feed pipe is used to feed the liquid into the distributor such that minimum turbulence is caused to the liquid surface (Figure 9.11).

Flashing Feed Distributors: When a feed has a vapor component, a liquid distributor will not be suitable. Flashing feed distributors are used for such two-phase feeds. There are several flashing feed distributors, viz., galley type, vapor-liquid separator type, baffle type, and tangential type. For details the reader is referred to [4].

Redistributors: Liquid coming down through the packing and on the wall of the column should be redistributed to prevent any degradation in mass transfer efficiency.

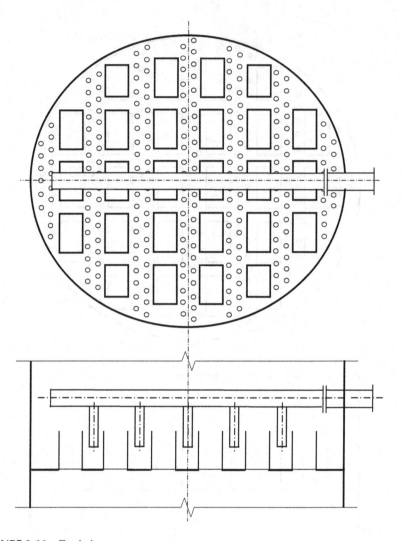

FIGURE 9.11   Feed pipe.

Redistribution brings the liquid off the wall and outer portions of the column and distributes it uniformly across the column. Rules of thumb say that liquid flow in packed columns should be redistributed after every 10 theoretical stages. For modern packings, as high as 20 theoretical stages are sometimes permitted without redistribution. For smaller column diameters there is another thumb rule which calls for redistribution once the ratio between the packing height and the column diameter reaches 5–10. Having said this, there are industrial columns which are designed to redistribute after packed heights of 3–5 m [15]. The exact type and geometry of the redistributor is specified by the supplier of the internals. Figure 9.12 is a redistributor that achieves total collection of the liquid prior to redistribution, typically used in columns of large diameter.

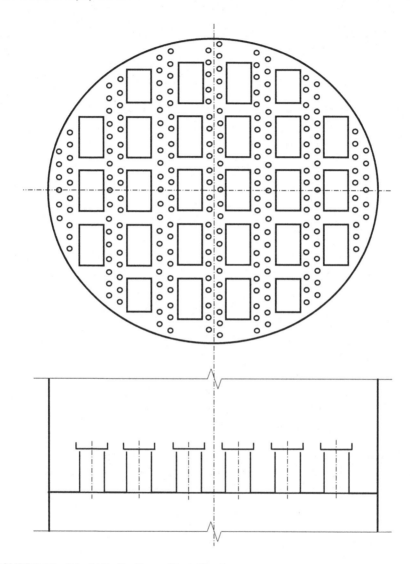

FIGURE 9.12 Liquid Redistributor (Deck Type).

Collector Trays: It may be desired to divert the entire liquid from the bottom of a certain packed section to a designated compartment. In such cases, a collector tray is used. Chimney tray, or a collector tray as it is often called is a device specifically designed to separate vapor from the and liquid. The liquid flows along the surface of the collector tray and through a downcomer to the designated section viz., as a side-stream draw-off, or is sent to a reboiler or to a pumparound. The vapor passes through risers of the collector tray. Figure 9.13 illustrates some typical chimney tray applications.

Support Plates: Packing support may be anything from cross-grid bars spaced to prevent free-fall of packing to more refined specialty units designed to direct the flow

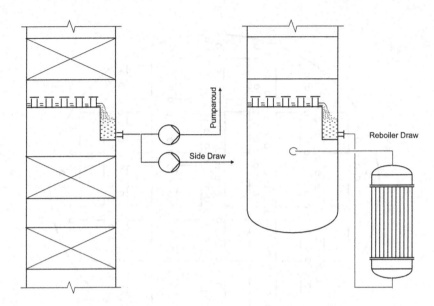

FIGURE 9.13   Chimney tray.

of gas and liquid. The net free flow cross-sectional area of the support is normally of the order of 70–100 of the column area. If the fractional open area at the support plate is significantly lower than that of the packing, premature flooding may occur at the support and may propagate upwards [4].

Bed Limiters: Bed limiters are placed at the top of the packing to prevent shifting, breaking and expanding of the bed during high pressure drops or surges. They are used primarily with ceramic packing, which is subject to breakage, and plastic packing, which may float out of the bed.

9.3.3   PROCESS DATA SHEET OF A PACKED SECTION

Column Simulation: Similar to what was discussed in Section 9.2.4 on tray sections, the design of a packed column starts with column simulation. Here we will take the example similar to that illustrated for tray columns earlier. We will take the case of small diameter columns (typically below 700 mm), where it is not possible to go for trays, and packed columns are the preferred option. We therefore consider a column loading data significantly lower than that discussed in Section 9.2.4, so that it calls for a column of diameter less than 700 mm. Consider the tray hydraulic calculations in Table 9.7, illustrating a tray section consisting of 12 theoretical and 20 actual trays.

It can be seen that for the top section a diameter of 900 mm has been calculated. Let us see how the dynamics work for packed columns for the same vapor-liquid traffic.

TABLE 9.7
Tray Hydraulic Calculations

Item			C-03
Designation			Stripping Column
Column Section Number			1
Tray Section			1–20
Decisive Tray No.			20
Type of Tray			Valve
Percentage Load		%	100
Vapor to Tray		kg/hr	4468
Density		kg/m³	0.878
Vapor Volume Rate		m³/s	1.41
Vapor Viscosity		mPa-s	0.0128
Liquid from Tray		kg/hr	4638
Density		kg/m³	950.3
Liquid Volume Rate		m³/hr	4.702
Surface Tension		N/m	50.96
Liquid Viscosity		mPa-s	0.285
Column Diameter		mm	900
Tray Spacing		mm	550
Tray Passes			1
Column Area		m²	0.636
Downcomer Chord Height	top/bottom	mm	
Downcomer Area	top/bottom	m²	0.028/0.028
Flow Path Length		mm	740
Active Area		m²	0.58
Exit Weir Length		mm	512.25
Downcomer Clearance		mm	35
Exit Weir Height		mm	40
Weir Load		m³/hr.m	9.53
Downcomer Backup Liquid		mm	57.44
Downcomer Flood		%	27.57
Jet Flood Percent		%	61.9
Dry Tray Pressure Drop		kg/cm²	0.0025
Total Tray Pressure Drop		kg/cm²	0.0034
Total Pressure Drop in Tray Section		kg/cm²	0.0680

For the packed section calculations, we proceed as follows:
We define a packing factor F_s as

$$F_s = v\sqrt{\rho_G}$$

Figure 9.14 illustrates the variation of HETP with F_s for different sizes of random packings. Figures 9.15(a) and 9.15(b) illustrate typical pressure drops across IMTP® packings of size 25 and 40 mm, respectively, and at different liquid loadings.

***

EXAMPLE 9.1   LET US CARRY OUT A TYPICAL PACKED COLUMN CALCULATION BASED ON DATA IDENTICAL TO THE TRAY SECTION ILLUSTRATED IN TABLE 9.7

1st trial:

Column diameter assumed	= 900 mm = 0.90 m

For such a column diameter, we choose an IMTP packing size of 40 mm

ρ_G	= 0.878 kg/m^3
Mass flow of vapor	= 4468 kg/hr
Volumetric flow rate of vapor	= 4468 / 0.878 = 5089 m^3/hr
	= 1.414 m^3/s
Column cross-sectional area	= π (0.90)^2/ 4 = 0.636 m^2
v = vapor superficial velocity	= 1.414 / 0.636 = 2.22 m/s
F_s	= 2.22 x $\sqrt{0.878}$ = 2.08 m/s.(kg/m^3)$^{0.5}$

With an F_s of 2.08 m/s.(kg/m^3)$^{0.5}$ and for an IMTP packing size of 40 mm, Figure 9.14 gives:

HETP	= 420 mm
Number of theoretical stages	=12

Hence, the total height of packing works out to 420 x 12 = 5040 mm or 5.04 m.

The liquid loading is calculated as follows:

Liquid mass flow rate	= 4638 kg/hr
Liquid density	= 950.3 kg/m^3
Liquid volumetric flow rate	= 4.88 m^3/hr
Cross-sectional area of column	= 0.636 m^2
Liquid loading	= 4.88/0.636 = 7.67 m^3/hr.m^2

From Figure 9.15(b), for an IMTP packing size of 40 mm, and F_s of 2.08 m/s.(kg/m^3)$^{0.5}$

the pressure drop is	= 2.0 mbar/m
	= 0.002 kg/cm^2/m

For a packing height of 5.04 m, the pressure drop across the packed section works out

to	= 0.010 kg/cm^2.

The allowable pressure drop across the column section is 0.1 kg/cm^2. We therefore still have enough margin on the pressure drop. Let us try with a smaller column diameter, say 800 mm.

2nd trial:

Volumetric flow rate of vapor	= 1.414 m^3/s
Column cross-sectional area	= π (0.80)^2/ 4 = 0.503 m^2
v = vapor superficial velocity	= 1.414 / 0.503 = 2.81 m/s
F_s	= 2.81 x $\sqrt{0.878}$ = 2.63 m/s.(kg/m^3)$^{0.5}$

With an F_s of 2.63 m/s.(kg/m^3)$^{0.5}$ and for an IMTP packing size of 40 mm, Figure 9.14 gives:

HETP   = 450 mm
Number of theoretical stages   = 12
Hence, the total height of packing works out to 450 x 12 = 5400 mm or 5.4 m.

The liquid loading is calculated as follows:
Liquid volumetric flow rate   = 4.88 m³/hr
Cross-sectional area of column   = 0.503 m²
Liquid loading   = 4.88/0.503 = 9.70 m³/hr.m²

From Figure 9.15(b), for an IMT packing size of 40 mm, and F_s of 2.63 m/s.
$(kg/m^3)^{0.5}$ the pressure drop is

$$= 3.6 \text{ mbar/m}$$
$$= 0.0036 \text{ kg/cm}^2/\text{m}$$

For a packing height of 5.4 m, the pressure drop across the packed section
works out to 0.019 kg/cm², which is still much below the allowable figure of
0.1 kg/cm².

***

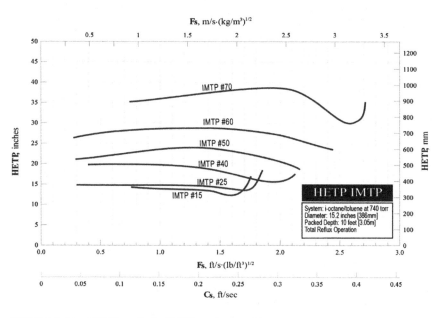

FIGURE 9.14   HETP vs. Fs for IMTP random packings.

Source: Courtesy of Koch-Glitsch LP, Wichita, Kansas, USA.

Column Data Sheet: Table 9.8 illustrates the complete packed section calculations.
The diameter which was calculated in Example 9.1 for a tray column with the same
vapor-liquid traffic was 900 mm (Table 9.7). The take-away from this exercise is that
packed columns require smaller diameters for a given vapor-liquid traffic. Further,

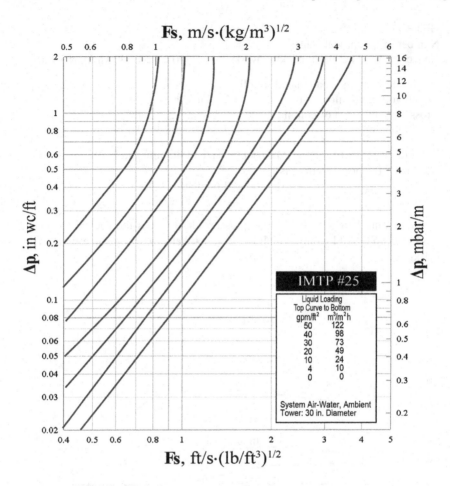

FIGURE 9.15(a)   Pressure drop vs. Fs for IMTP #25 random packings.

Source: Courtesy of Koch-Glitsch LP, Wichita, Kansas, USA.

according to Kister [16], pressure drop across packings is typically 3–5 times lower than that across trays for a given duty. In Table 9.7 the pressure drop across the tray section is 0.068 kg/cm^2. The pressure drop across the packed section as worked out in Example 9.1 above is 0.02 kg/cm^2. The difference is a factor of 3.4, which is exactly in line with what Kister says.

According to Lieberman [17], a properly designed packed column can have 20–40% more capacity than a tray column with an equal number of stages. Packed columns are, therefore, particularly useful in revamp activities when the existing trays do not meet the pressure drop criteria as a result of the increased revamp loads.

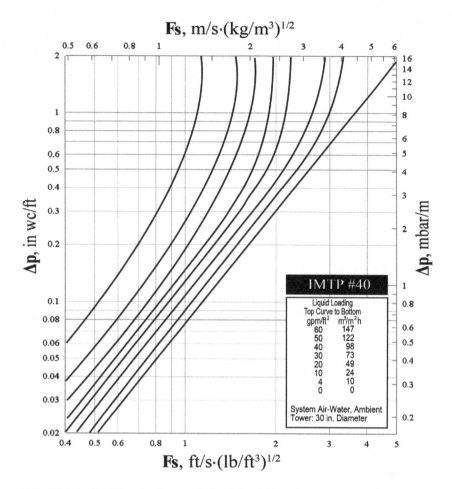

FIGURE 9.15(b) Pressure drop vs. Fs for IMTP #40 random packings.

Source: Courtesy of Koch-Glitsch LP, Wichita, Kansas, USA.

9.4 OTHER ASPECTS OF COLUMN DESIGN

9.4.1 COLUMN BOTTOM ARRANGEMENTS

Four different arrangements are commonly used in column bottoms:

1. *Circulating Thermosiphon Reboilers (Unbaffled Arrangement):* Here, both the bottom product and the reboiler liquid are withdrawn from a common sump (Figure 9.16a). This arrangement is preferred in the following cases [4]:
   - Due to simplicity and low cost, in columns having diameters less than 1000 mm where baffles are difficult to install.

TABLE 9.8
Packing Hydraulic Calculations

Item		C-03
Designation		Stripping Column
Column Section Number		1
Type of Packing		IMTP #40
Percentage load	%	100
Vapor to Packing Section	kg/hr	4468
Density	kg/m³	0.878
Vapor Volume Rate	m³/s	1.41
Vapor Viscosity	mPa-s	0.0128
Liquid from Packing Section	kg/hr	4638
Density	kg/m³	950.3
Liquid Volume Rate	m³/hr	4.702
Surface Tension	N/m	50.96
Liquid Viscosity	mPa-s	0.285
Column Diameter	mm	800
Fs	m/s.(kg/m³)^0.5	2.63
HETP		450
No. of Theoretical Stages		12
Height of Packing	mm	5400
Liquid Loading	m³/hr.m	9.7
Pressure Drop per Meter of Packing	kg/cm²/m	0.0036
Total Pressure Drop in Packed Section	kg/cm²	0.02

- In columns with kettle type reboilers, because the reboiler return is 100% vapor which moves up the trays of the column. Furthermore, the column bottom product is not withdrawn from the column bottom but from the reboiler surge compartment. Hence, a baffle does not serve any purpose.
- With forced circulation reboilers, because the large reboiler circulation rate (compared to the bottom product rate) demands the availability of a common bottom sump rather than a baffled portion.
- In a packed column, where the presence of a baffle would call for diversion of the liquid from the bottom of the packing to the reboiler side of the baffle. This arrangement is complicated and also more expensive.

2. *Circulating Thermosiphon Reboilers (Baffled Arrangement):* Here, the space at the bottom of the column is divided into a bottom draw-off sump and a reboiler feed sump by a preferential baffle (Figure 9.16b). The arrangement has the advantage that it supplies a constant head to the reboiler. In such cases, the reboiler design does not need to be checked for operation at low column bottom level. Thus, even if the bottom draw-off sump is at a low level, the reboiler is always supplied with a constant elevated head. However, such arrangements are usually preferred in columns having diameters larger than 1000 mm [4].

3. ***Once-Through Reboilers (Unbaffled Arrangement):*** Here, reboiler liquid is withdrawn from the bottom downcomer seal pan (Figure 9.17a). This arrangement offers the following advantages [4]:

   - They provide an additional theoretical stage, or a fraction of a theoretical stage.
   - They lower the boiling point of liquid supplied to the reboiler. This maximizes reboiler ΔT and its capacity and minimizes its fouling tendency. This is a major consideration when the bottom liquid is a wide-boiling mixture. For a better explanation, refer to Figure 9.16(a), which is a case of circulating (unbaffled arrangement). The column bottom temperature has been calculated from process simulation as 166°C. A part of this stream gets heated to 184°C in the reboiler using MP steam. The liquid portion of this reboiler return mixes with the liquid at 149°C coming from the bottom tray of the column, and the mixed steam at 166°C forms the column bottoms liquid.

   Now refer Figure 9.17(a) which is a case of once-through (unbaffled arrangement). The liquid at 142°C coming from the bottom tray directly enters the reboiler, where it is heated to 166°C, and reenters the column. The liquid portion of the reboiler return at 166°C forms the column bottoms liquid.

   In both cases, it can be seen that the column bottoms liquid is at 166°C since this is determined from the process simulation. MP steam at 15 kg/cm²g and at 201°C is the heating medium. It can be clearly seen that for the once-through case, the ΔT for the reboiler is higher compared to that in the circulating case.

4. ***Once-Through Reboiler (Baffled Arrangement):*** In this case, the column bottom is equipped with a baffled arrangement, and the reboiler liquid is drawn from the reboiler feed sump (Figure 9.17(b)). The difference between this and

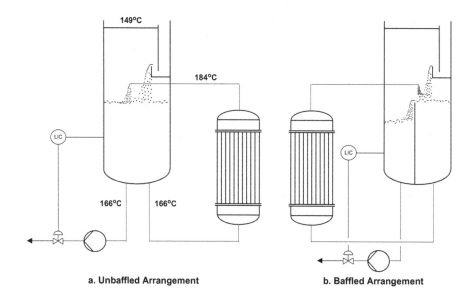

a. Unbaffled Arrangement   **b. Baffled Arrangement**

FIGURE 9.16   Vertical thermosiphon reboiler – recirculating type.

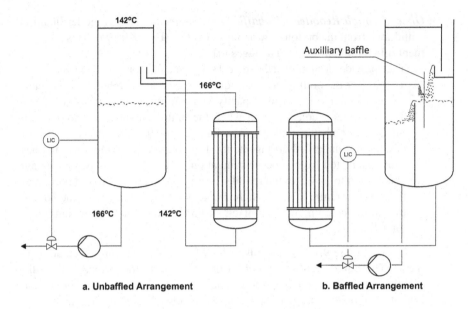

a. Unbaffled Arrangement   b. Baffled Arrangement

FIGURE 9.17   Vertical thermosiphon reboiler – once through.

the once-through thermosiphon reboiler (unbaffled arrangement discussed above) is that in the baffled arrangement, a constant head to the reboiler is maintained. However, the other advantages of an extra theoretical stage and maximization of the reboiler ΔT as described in case of the unbaffled arrangement are only partially met in the baffled arrangement. The reason for this is that in the latter case, a part of the liquid from the bottom tray or the packed section may bypass the reboiler and find its way to the bottom sump. Such arrangements are usually preferred in columns having diameters larger than 1000 mm [4].

9.4.2 MISCELLANEOUS DETAILS

Pressure Drop in Columns: Table 9.9 can be referred to while specifying pressure drops across various tower internals. The figures are indicative. The diameter and the distribution of pressure drops in various sections are recalculated by the tray vendor during the detailed design.

Other Internals: The foregoing sections have concentrated on the internals which are commonly used for vapor-liquid contacting. Additionally, one can consider the following internals which are of vital importance in the performance of a column:

Mist Eliminators: These are used to prevent carryover of entrained liquid in the vapors leaving a particular section of the column. Thus, they are provided above the wash zone packing in a vacuum column to prevent carryover entrained asphaltenes into the vacuum gas oil. Provision of the demisters in the top section of many columns prevents the carryover of liquid hydrocarbons to an overhead system connected to an ejector, compressor, etc. Such services may include vacuum towers, absorbers

TABLE 9.9
Pressure Drop in Columns

Sl. No.	Service	Pressure Drop
1	Trays in Atmospheric Distillation Columns, Absorbers, Strippers	0.01–0.005 kg/cm² per tray
2	Trays in Vacuum Towers	0.005 kg/cm² per tray
3	Distributors/Redistributors (across the risers)	0.0007–0.001 kg/cm² per tray
4	Demisters	0.0007 kg/cm²
5	Random Packing	0.001–0.003 kg/cm² per meter of bed
6	Structured Packing	0.0015–0.002 kg/cm² per meter of bed

Note: Pressure drops are indicative only. Tray/Packing vendor to be contacted for actual values.

and columns in foaming services. While giving the specifications for the demisters, the operating conditions and requirements (droplet size to be retained), allowable pressure drop, and material of construction should be clearly spelt out. The pertinent data for demisters should be filled in the standard data sheet.

Spray Nozzles: These are very vital to ensure the performance small diameter packed columns. Gravity flow through a downcomer will not irrigate the packing uniformly. It is, therefore, essential that the liquid be sprayed over the packing in a pattern which ensures uniform distribution over the cross-section. The liquid being sprayed through the nozzles should be clear of particulate matter to avoid erosion and choking of the spray nozzles. For this, appropriate strainers are generally used. A pressure drop of 2–3 kg/cm² through the spray nozzles should normally be considered when pumping requirements are calculated.

Nozzle Orientations and Locations: Nozzle orientation is very important especially in tray columns where vapor and liquid pass through different sections. Nozzles for vapor outlet, pressure tapping and vapor samples should be located in the vapor space. Nozzles for liquid draw-off and liquid temperature should be in the liquid space. Nozzles for liquid feed and reflux should be properly oriented. Figure 9.4 illustrates typical pressure and temperature tapping locations in a tray column. Column bottom configurations are very important when reboilers are connected. Especially for column bottoms, the orientation of the vapor return nozzle should be carefully checked. Chapter 15 discusses such issues in greater detail. Figure 15.9 shows typical configurations for column bottoms having reboilers.

Symbols

A_r	Aeration factor
C_h	Crest height of liquid, mm
F_s	Packing factor, m/s.(kg/m³)$^{0.5}$
ΔP_{dry}	Tray pressure drop, mm liquid
ΔP_{hyd}	Hydraulic tray pressure drop, mm liquid
ΔP_{total}	Total tray pressure drop, mm liquid

Q_L	Liquid volumetric flowrate, m^3/hr
ΔT	Temperature difference
W_L	Weir length, mm
W_h	Weir height, mm
v	Superficial velocity of vapor, m/s
V_g	Velocity of vapor or gas flowing through the hole in the tray, m/s
ρ_L	Density of liquid, kg/m^3
ρ_V	Density of vapor, kg/m^3

REFERENCES

1. Mukherjee, S., "Tray Column Design – Keep Control of the Details", *Chemical Engineering*, September 2005.
2. Kister, H. Z., *Distillation Design*, McGraw Hill Inc., 1992.
3. Kleiber, M., *Process Engineering – Addressing the Gap between Study and Chemical Industry*, Walter de Gruyter, 2016.
4. Kister, H. Z., *Distillation Operation*, McGraw Hill Inc., 1990.
5. Lieberman, N.P., Lieberman, E.T., *A Working Guide to Process Equipment*, McGraw Hill Inc., 2008.
6. Sloley, A. W., "Improve Tray Operations", *Hydrocarbon Processing*, June 2001, pp. 85–86.
7. Sloley, A. W., "Should you Switch to High Capacity Trays?", *Chemical Engineering Progress*, January 1999.
8. Gondolfe, J., High Capacity Tower Internals for Ethylene Plant Expansions, *European Ethylene Seminar*, October 2000, The Netherlands.
9. Khalil, A.M., Goswamy, V., Busaleh, H. A., Khaldi, Al., and Hood, T., "High-Capacity Trays Hike Demeth Capacity, Efficiency at Two Middle East Gas Plants", *Oil and Gas Journal*, December 13, 2004.
10. Chopey, N. P., *Handbook of Chemical Engineering Calculations,* McGraw Hill Inc., 2nd Edition, 1993.
11. Towler G. and Sinnott, R., *Chemical Engineering Design – Principles, Practice and Economics of Plant and Process Design*, Elsevier, 2008.
12. Schweitzer, P. A., *Handbook of Separation Techniques for Chemical Engineers*, 3rd edition, McGraw Hill, 1997.
13. Eckert, J. S., "How Tower Packings Behave", *Chemical Engineering*, April 1975.
14. https://www.engstack.com/kb/orifice-distributors-packed-towers/
15. Schultes, M., "Influence of Liquid Redistributors on the Mass-Transfer Efficiency of Packed Columns", *Industial Engineering Chemistry Research*, (39), pp. 1381–1389, 2000.
16. Kister, H. Z., Larson, K. F., and Yanagi, T., "How Do Trays and Packing Stack Up?" *Chemical Engineering Progress*, February 1994.
17. Lieberman, N. P., *Process Design for Reliable Operations*, 2nd Ed., Gulf Publishing Co., Houston, Texas, 1988.

10 Hazardous Area Classification

10.1 INTRODUCTION

When electrical equipment exists in the vicinity of an atmosphere that has flammable liquids, flammable gases or vapors, combustible dusts, ignitable fibers or flyings, there is always a chance of fire or explosion. This is because during normal operation, or during malfunction, such equipment produces sparks, which are likely to ignite the flammable liquids, gases or vapors. It is therefore mandatory that suitable precautions are taken to reduce the possibility of such an ignition. This requirement led to the development of the concept of hazardous area classification.

A "hazardous area" is defined as an area in which explosive gas-air mixtures are present in quantities sufficient to create a risk of fire or explosion. A process engineer has a major role to play in developing a hazardous area classification drawing. The activity involves classifying a plant into zones within which the chance of the existence of an explosive gas-air mixture is likely to be high, medium, low or negligible.

The classification drawing so developed provides a basis for the selection of electrical equipment that are protected suitably, depending upon the risk involved. The type of protection of the apparatus selected should be such that the possibility of it being a source of ignition, and the surrounding atmosphere being explosive, is eliminated.

10.2 DEFINITIONS

Before we delve deeper into this subject, it is important for the reader to be acquainted with some definitions. Here are certain important terms which should be understood [1–3]:

Flash point: The minimum temperature at which the liquid gives sufficient vapors which when mixed with air forms an ignitable mixture and gives a momentary flash on application of a pilot flame.

Flammable gas or vapor: A gas or vapor which when mixed with air in a certain proportion will form an explosive gas atmosphere.

Flammable liquid: A liquid that has a closed-cup flash point below 37.8°C, as determined by specified test procedures.

Combustible liquid: A liquid that has a closed-cup flash point at or above 37.8°C, as determined by specified test procedures.

Explosive gas mixture: A mixture of a flammable gas or vapor with air, under atmospheric conditions, in which after ignition, the combustion spreads throughout the unconsumed mixture.

Hazardous area: An area is deemed hazardous when it has:

-   Petroleum having a flash point below 65°C, or any flammable gas or vapor in a concentration capable of ignition is likely to be present.
-   Petroleum or any flammable liquid having a flash point above 65°C is likely to be refined, blended, handled or stored at or above its flash point

Lower explosive limit (LEL): The concentration of flammable gas, vapor or mist in air below which the gas atmosphere will not be explosive.

Upper explosive limit (UEL): The concentration of flammable gas, vapor or mist in air above which the gas atmosphere will not be explosive.

Adequately ventilated: An area is considered adequately ventilated when it prevents accumulation of significant quantities of vapor-air concentration from exceeding 25% of the lower flammable limit.

10.3 SYSTEMS FOR CLASSIFICATION OF HAZARDOUS AREAS

Broadly, there are two systems which are widely used to classify hazardous areas: class/division system and zone system. The class/division system is used predominantly in the United States and Canada. The rest of the world normally uses the zone system.

10.3.1 CLASS/DIVISION SYSTEM

In this system, the hazardous locations are classified according to the class, division and group.

Class: Class defines the general nature or type of the hazardous material in the surrounding atmosphere which may or may not be in sufficient quantities [4].

  Class I locations are those in which flammable vapors and gases may be present
      in the atmosphere in quantities sufficient to produce an explosive mixture.
  Class II locations are those in which combustible dust may be found.
  Class III locations are those in which easily ignitable fibers or flyings are present and, as a consequence, are hazardous.

Division: Each of the three classes I, II and III is further divided into divisions. The division defines the likelihood of the hazardous material present in a flammable concentration and therefore capable of producing an explosive or ignitable mixture based upon its presence.

TABLE 10.1

Gas Groups Based on Class/Division System

Gas Group	Flammable Gas	Gas Group	Combustible Dust
	Acetylene		Aluminum flake
A		E	Tin
			Iron
	Acrolein		Charcoal
	Hydrogen		Pitch, Coal Tar
B	1,3 Butadiene	F	Coal
	Ethylene Oxide		Cocoa Bean Shell
	Propylene Oxide		Coke
	Acetaldehyde		Benzoic Acid
C	Ethylene		Citrus Peel
	Butyl Mercaptan	G	Coconut Shell
	Dimethyamine		Cottonseed Meal
	Propane		
	Acetic Acid		
	Ammonia		
D	Benzene		
	n-Butane		
	Methanol		
	n-Pentane		
	Methane		

Source: Adapted from NFPA 497 (2012 Edition), and NFPA 497 (2013 Edition).

Division 1 indicates that the ignitable concentrations of hazards exist under normal operation conditions and/or the hazard may exist because of frequent maintenance, repair work or breakdown of fault operation of equipment [4].

Division 2 indicates that the ignitable concentrations of hazards are handled, processed or used, but they are present in closed containers or systems from which they can only escape through accidental rupture or breakdown of such containers or systems. It also indicates locations where ignitable concentrations of hazards are normally prevented by positive mechanical ventilation and therefore might become hazardous on account of failure or abnormal operation of the ventilation system [4].

Group: Class I and Class II divisions are further divided into groups of hazardous materials. Materials are allocated to groups depending upon their ignition temperatures and explosion pressures. Groups A, B, C and D are for gases (Class I only), while groups E, F and G are for dusts and flyings (Class II or III).

The table of groups as illustrated in the National Fire Protection Association (NFPA) is very elaborate. Table 10.1 provides an example of the materials that make up each of these groups [5].

10.3.2 ZONE SYSTEM

Zone: In this system, zone designations replace the divisions designations followed in the class/division system. The zone system divides hazardous materials into three levels of probability of the presence of a flammable gas to produce explosive or ignitable mixtures [3, 4].

  Zone 0: An area where ignitable concentrations of flammable gases or vapors are present continuously or for long periods.

  Zone 1: An area where ignitable concentrations of flammable gases or vapors are likely to occur under normal operating conditions, or may exist frequently because of repairs, maintenance operations or through leakage.

  Zone 2: An area where ignitable concentrations of flammable gases or vapors are not likely to occur under normal operating conditions, or occur only for a short period of time.

Flammable gases falling under Zones 0, 1 or 2 are further covered under one of the following gas groups:

  Group I: Equipment used in mines with atmospheres containing methane or gases and vapors of similar hazard will fall in this group.

  Group II: This gas group is for equipment intended for use in explosive atmospheres other than mines. Gas group II is further divided into three subgroups depending upon the nature of the explosive gas.

  Group IIA: This group includes propane, or gases and vapors of similar hazard.

  Group IIB: This group includes ethylene, or gases and vapors of similar hazard.

  Group IIC: Acetylene and hydrogen and gases, or vapors of similar hazard.

  Gas group IIC is the most severe of the above gas groups. Gases in this group could be easily ignited. Equipment suitable for this group is therefore also suitable for gas groups IIA and IIB. Likewise, equipment suitable for gas group IIB is also suitable for gas group IIA, but not for IIC.

Zone (for dust): This zone covers atmospheres where an explosive dust atmosphere is present. Combustible or non-combustible dust can fall under one of the following zones [4]:

  Zone 20: An area where ignitable concentrations of dusts or fibers and flyings are present continuously or for long periods of time.

  Zone 21: An area where ignitable concentrations of dusts or fibers and flyings are likely to occur under normal operating conditions.

  Zone 22: An area where ignitable concentrations of dusts or fibers and flyings are not likely to be present under normal operating conditions and, at worst, are present only for short periods of time.

Flammable gases falling under Zones 20, 21 or 22 are further covered under one of the following gas groups:

  Group IIIA: Combustible flyings (viz., cotton, rayon)
  Group IIIB: Non-conductive dust (viz., wood, plastic)
  Group IIIC: Conductive dusts (viz., magnesium).

10.4   AREA CLASSIFICATION METHODOLOGY

Based on the discussions above, it must now be clear that a hazardous area is an area in which explosive gas-air mixtures are present in quantities that may pose a potential risk of fire or explosion. Having understood this, one can proceed with classifying the plant into hazardous areas. The practice of hazardous area classification involves a knowledge of the behavior of flammable gases when they are released to the atmosphere accompanied by engineering judgement based on experience.

Each plant section or area of a process plant is considered in determining its classification. The first step is to assess the likelihood of occurrence of explosive gas atmospheres based on the definitions of Zones 0, 1 and 2. Thus, by assigning different degrees of probability with which flammable concentrations of combustible gases or vapors may arise in a process plant, both in terms of the frequency of occurrence and the probable duration of occurrence, Zones 0, 1 and 2 are assigned. Allocation of Zone 0 and Zone 1 should be minimized in number and extent. Plants should be mainly allocated to Zone 2 or non-hazardous. Design, operation and location of process equipment should be such that even in case of abnormal operation, the quantity of flammable material released into the atmosphere is minimized in order to limit the extent of hazardous area [2].

10.5   TEMPERATURE CLASS

Besides the danger of explosion caused by an electric spark, there is also a danger of a flammable gas getting ignited by coming into contact with a hot surface. The likelihood of ignition depends upon the area of the surface, temperature and concentration of the gas. It is therefore necessary to know the ignition temperature of the flammable gas or vapor that constitutes the hazard. A temperature classification system has been arrived at and each electrical equipment is assigned a temperature class based on the maximum temperature its surface can reach. Whenever an electrical equipment is selected, care should be taken to ensure that its maximum surface temperature is lower than the ignition temperature of the flammable gas or vapor.

Temperature classes exit in both the class/division system as well as the zone system. Both have classes from T1 through T6. In the class/division system, there are further subdivisions for T2, T3 an T4 temperature classes.

Figure 10.1 illustrates temperature classification according to the zone system [6]. The figure shows equipment maximum surface temperature as well as the ignition temperatures of the gas or vapor. Let us consider Class 6. The equipment maximum surface temperature is 85°C. The ignition temperatures of the surrounding gas or vapor could therefore be in the range of 85–100°C. One could argue that we can have a greater flexibility of ignition temperature, i.e., it could go beyond 100°C. Theoretically this is correct; but if the ignition temperature is above 100°C, then we would go for an electrical equipment with a temperature Class 5 which is cheaper.

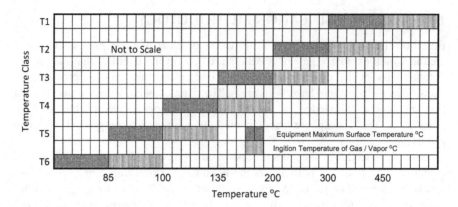

FIGURE 10.1 Temperature class.

10.6   EXTENT OF HAZARDOUS AREA

In the following sections, a short discussion is provided on the guidelines regarding the extent of hazardous areas for various cases. In arriving at the extent of hazardous areas, various parameters are considered such as inventory of the material, shut-off time, dispersion time, pressure and temperature. IEC 60079-10 [3] presents several illustrations of the extent of hazardous areas in the form of examples. In the following section, the extent of hazardous area is provided for two cases; however, the reader is informed that these are generic descriptions. Local or project specific guidelines should be adhered to in all cases.

10.6.1   Gases and Vapors Lighter Than Air

Zone 2: When the source of hazard is located in open air, the cylindrical span of hazardous area surrounding the point of hazard would be 4.5 m. The height of the cylinder would be 8 m above the source and 4.5 m below the source. If the source of hazard is at a height less than 4.5 m above the ground, the extent of hazardous area below the source shall extend up to the ground level (Figure 10.2). The entire area would come under Zone 2.

Zone 1: In cases when the source of hazard is inside enclosed premises with restricted ventilation, the entire area within the enclosed premises would be a Zone 1 area. In case the source of hazard is located inside enclosed premises with openings on the sides, the entire area within the enclosed premises would come under Zone 1 and the extremities of the openings would be considered as sources of hazard. The extent of hazardous area is shown in Figure 10.3.

10.6.2   Gases and Vapors Heavier Than Air

Zone 2: When a source of hazard gives rise to a hazardous area falling under Zone 2, the hazardous area shall extend horizontally 16 m in all directions from the source of hazard, and vertically 8 m above the source of hazard. Above 8 m from the ground,

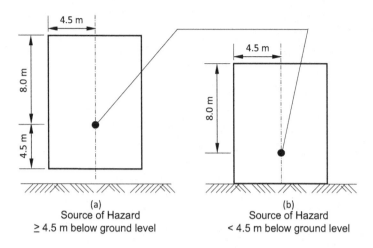

FIGURE 10.2   Gases lighter than air – source of hazard in open air.

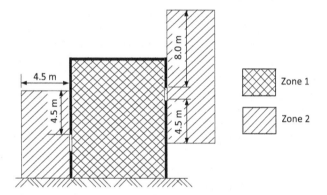

FIGURE 10.3   Gases lighter than air – source of hazard in enclosed premises.

the horizontal span of the hazardous area is reduced from 16 to 8 m in all directions (refer to Figure 10.4 as an example for the case of a distillation column along with pipe rack and associated technological structure).

If, however, there is a possibility of a large release of volatile products, the horizontal span of the hazardous area is further extended from 16 to 32 m in all directions up to a height of 0.63 m from the ground.

Zone 1: If the source of hazard falls under Zone 1, then the extent of area covered under Zone 1 shall be the same as described for Zone 2 above.

When a source of hazard gives rise to a hazardous area falling under Zone 2 and is located within an enclosed building, the whole of the inside of the building shall be classified as Zone 1 instead of Zone 2.

In case an enclosed premise does not contain a source of hazard, but is situated within either a Zone 1 or Zone 2 area, the inside of the premises shall be classified as Zone 1 unless separated by a fire wall.

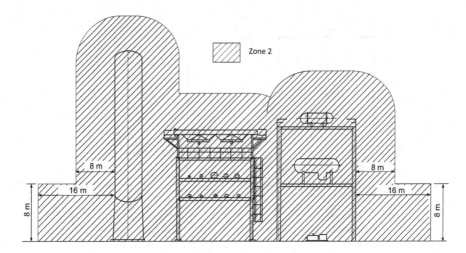

FIGURE 10.4   Extent of hazardous area – heavier than air gases.

In addition, any pit or trench below ground level and located in an area covered under Zone 2 shall be classified as Zone 1 area.

For a floating roof tank, the space above the roof and within the shell of the tank shall be classified as Zone 1. The area surrounding the tank shall be classified as Zone 2 to the extent of 3 m vertically and horizontally from the tank (Figure 10.5). If there is a dike around the tank, then Zone 2 shall extend up to the dike and the vertical extension from the ground level shall be the same as the height of the dike.

In case of fixed roof tanks which breathe to the atmosphere, the area surrounding the tank shown in Figure 10.5 as Zone 2 shall be classified as Zone 1. The extent of Zone 1 shall extend 3 m vertically and horizontally from the tank. If, however, the fixed roof tank breathes into a closed system and not to the atmosphere (inbreathing and outbreathing), the area surrounding the tank shall be classified as Zone 2.

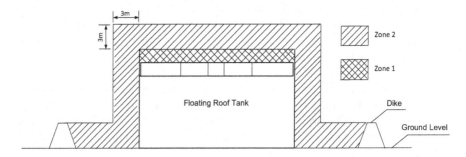

FIGURE 10.5   Extent of hazardous area for floating roof tanks.

10.7 SELECTION OF ELECTRICAL EQUIPMENT FOR HAZARDOUS ATMOSPHERES

Once an area has been classified with regard to the degree and extent of hazard, the next step is to select the type of protection required for electrical equipment located in such hazardous areas. While a process engineer is not directly involved in selecting the type of protection, it is important for him to have the basic knowledge of the same.

Flame-Proof Enclosure (Marking "Ex d"): In this type of protection, the enclosure is capable of withstanding the pressure of internal explosion of a flammable gas within it and preventing the ignition of such a gas outside the enclosure. Further, the enclosure would operate at such a temperature that surrounding explosive gases or vapors will not get ignited [7].

Pressurized Enclosure (Marking "Ex p"): In this type of protection, the enclosure is filled with a gas (air, inert gas or other suitable gas) at a pressure above that of atmospheric pressure, such that ingress of flammable gas or vapor from surrounding atmosphere is prevented [8].

Powder-Filled Enclosure (Marking "Ex q"): Powder-filled Enclosure is filled with a fine powder in such a way that if arcing occurs inside the enclosure, it will not be able to ignite a potentially explosive atmosphere outside the enclosure [9].

Oil-Filled Enclosure (Marking "Ex o"): In such an enclosure, parts that could ignite a potentially explosive mixture are immersed in oil or a non-combusting liquid such that gas or vapor present above the liquid or outside the enclosure cannot be ignited by sparks from below the surface of the liquid [10].

Increased Safety (Marking "Ex e"): In this method of protection, additional measures are taken beyond those adopted in industrial practice to achieve increased security. This ensures prevention of excessive temperatures both on the inside and outside of the equipment which in normal operation do not produce unusually high temperature or sparks [11].

Intrinsically Safe (Marking "Ex i"): Such equipment contain intrinsically safe circuits, i.e., circuits that under normal or abnormal conditions do not generate sufficient thermal or electrical energy to cause ignition of a hazardous gas or vapor [12].

REFERENCES

1. NFPA 30, *Flammable and Combustible Liquid Code*, National Fire Protection Association, 2003.
2. OISD-STD-113, *Classification of Areas for Electrical Installations at Hydrocarbon Processing and Handling Facilities*, Oil India Safety Directorate, 2001.
3. IEC 60079, Part 10: *Classification of Hazardous Areas,* International Electrotechnical Commission.

4. NFPA 70, National Electric Code – Article 500 and 505, National Fire Protection Association, 2011.
5. NFPA 497, *Recommended Practice for Classification of Flammable Liquids, Gases, or Vapors and of Hazardous (Classified) Locations for Electrical Installations in Chemical Process Areas*, 2012.
6. IEC 60079-0, Part 0: *Equipment – General Requirements*, International Electrotechnical Commission, 2017.
7. IEC 60079-1, *Explosive Atmospheres - Part 1: Equipment protection by flameproof enclosures "d"*, International Electrotechnical Commission, 2014.
8. IEC 60079-2, *Explosive Atmospheres - Part 2: Equipment protection by pressurized enclosures "p"*, International Electrotechnical Commission, 2014.
9. IEC 60079-5, *Explosive Atmospheres - Part 5: Equipment protection by powder filling "q"*, International Electrotechnical Commission, 2015.
10. IEC 60079-6, *Explosive Atmospheres - Part 6: Equipment protection by liquid immersion "o"*, International Electrotechnical Commission, 2015.
11. IEC 60079-7, *Explosive Atmospheres - Part 7: Equipment protection by increased safety "e"*, International Electrotechnical Commission, 2015.
12. IEC 60079-11, *Explosive Atmospheres - Part 11: Equipment protection by intrinsic safety "i"*, International Electrotechnical Commission, 2011.

11 Instrumentation and Controls

11.1 INTRODUCTION

The term "instrumentation" in a process plant refers to a set of devices each of which measures, indicates and records parameters such as temperature, pressure, flow and level. Analyzers also fall in this category which measure the concentrations of one or more constituents of a process stream in gaseous or liquid phase.

11.2 FLOW MEASUREMENT

11.2.1 DIFFERENTIAL PRESSURE TYPE – ORIFICE FLOWMETERS

In an orifice flowmeter, the pressure drop across an orifice installed in the flow path is used to estimate the flow (Figure 11.1). P_1 is the pressure upstream of the orifice, P_2 is the pressure at the *vena contracta* where the velocity is the maximum, and consequently the pressure P_2 is the minimum.

$$P_1 + \tfrac{1}{2}\rho v_1^2 = P_2 + \tfrac{1}{2}\rho v_2^2 \tag{11.1}$$

Assuming that the velocity profiles upstream and downstream of the orifice are uniform, the equation of continuity can be expressed as

$$Q_L = v_1 A_1 + v_2 A_2 \tag{11.2}$$

Combining Equations (11.1) and (11.2), and assuming $A_2 < A_1$ we arrive at the following equation:

$$Q_L = \frac{A_2 \left[2\left(P_1 - P_2\right)/\rho \right]^{0.5}}{\left[1 - \left(\dfrac{A_2}{A_1}\right)^2 \right]^{0.5}} \tag{11.3}$$

Equation (11.3), however, gives only a theoretical value of the flow. The area at the *vena contracta* A_2 is not known since this is not easy to measure. Therefore, the area A_2 is more conveniently taken as the area of the orifice itself. Furthermore, while P_2 is the pressure at the *vena contracta*, the measured value of P_2 may not exactly reflect the actual pressure at this point. In addition, losses may not be negligible as we have assumed.

DOI: 10.1201/9780429284656-11

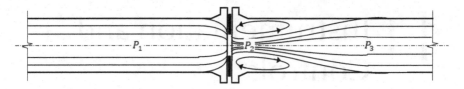

FIGURE 11.1   Orifice meter.

TABLE 11.1

Discharge Coefficients – Orifice Meters

Pipe Diameter, in	β Ratio D_2/D_1	Reynolds Number			
		10^4	10^5	10^6	10^7
2	0.3	0.6082	0.6030	0.6018	0.6015
	0.5	0.6196	0.6082	0.6051	0.6040
	0.7	0.6421	0.6164	0.6090	0.6060
6	0.3	0.6048	0.5995	0.5983	0.5980
	0.5	0.6178	0.6062	0.6031	0.6021
	0.7	–	0.6120	0.6044	0.6014
10	0.3	0.6049	0.5997	0.5985	0.5982
	0.5	–	0.6063	0.6032	0.6021
	0.7	–	0.6105	0.6029	0.5998
30	0.3	–	0.5999	0.5987	0.5984
	0.5	–	0.6065	0.6034	0.6023
	0.7	–	0.6095	0.6019	0.5988

Source: Adapted from ASME MFC-3M-2004, by permission of The American Society of Mechanical Engineers. All Rights Reserved.

To compensate for this error, Equation (11.3) is modified by adding a discharge coefficient, C_d. Table 11.1 provides values of C_d at different Reynolds numbers [1].

$$Q_L = \frac{C_d\, A_2 \left[2\left(P_1 - P_2\right)/\rho \right]^{0.5}}{\left[1 - \left(\dfrac{A_2}{A_1}\right)^2 \right]^{0.5}} \tag{11.4}$$

The ratio of the diameters D_2/D_1 is called the β ratio. Substituting this in Equation (11.4), we get

$$Q_L = \frac{C_d\, A_2 \left[2\left(P_1 - P_2\right)/\rho \right]^{0.5}}{\left(1 - \beta^4\right)^{0.5}} \tag{11.5}$$

β ratios are kept in the range of 0.3–0.7. This is the range that gives the best performance.

In such meters, velocity of the fluid is at its highest and the pressure is at the lowest at the *vena contracta*. Thereafter, the velocity will decrease to the same level as before the orifice. The pressure will also recover from P_2 to P_3 (refer to Figure 11.1). Thus, the permanent pressure loss is $P_1 - P_3$. For example, for a β ratio of 0.5, the permanent pressure loss is about 70–75% of the orifice pressure loss $P_1 - P_2$. In reality, the orifice calculation is carried out in the following way.

From the process engineer's point of view, it is the system hydraulics which is important, apart from the flow measurement. The process engineer therefore specifies the permanent pressure loss $P_1 - P_3$ which he can accept across the orifice for the given flowrate. The process data sheet, specifying a permanent pressure loss of $P_1 - P_3$ is passed on to the instrument engineer. The instrument engineer, on the other hand, needs the orifice pressure loss $P_1 - P_2$ since this is what is needed according to the formula (Equation (11.5)). The usual practice is that he takes the conservative route. He assumes $P_1 - P_3$ as the orifice pressure loss and carries out his calculations for the orifice diameter. The orifice thus gets sized for a pressure loss of $P_1 - P_3$. The actual permanent pressure loss will therefore always be lower than $P_1 - P_3$. In this way, the instrument engineer ensures that the permanent pressure loss as specified by the process engineer is never exceeded.

***

EXAMPLE 11.1

Calculate the orifice size required to measure a flowrate of 28 m³/hr. The line size is 100 mm and the permanent pressure loss is 0.3 kg/cm². The density of the fluid is 990 kg/m³.

Data:

$$D_l = 100 \text{ mm} = 0.10 \text{ m}$$

(in reality, the inner diameter of the line is not exactly 0.10 m but depends upon the pipe class. But, for the sake of example, assuming a value of 0.10 m is fine.)

C_d	= 0.6 (valid for a wide range of Reynolds numbers. Refer to Table 11.1 for details)
$P_1 - P_2$	= 0.3 kg/cm²
	= 0.3 × 10000 × 9.81 N/m²
	= 29430 N/m²
Q_L	= 28 m³/hr
	= 28 /3600
	= 0.0078 m³/s

$$0.0078 = \frac{0.6 \times 3.142 / 4 \times D_2^2 \left[2 \times (29430) / 990 \right]^{0.5}}{\left(1 - \beta^4 \right)^{0.5}}$$

or,

$$0.002146 = \frac{D_2^2}{\left[1 - \left(D_2 / 0.1 \right)^4 \right]^{0.5}}$$

Solving for D_2,

D_2	= 0.046 m
	= 46 mm

***

11.2.2 Differential Pressure Type – Venturi Flowmeters

In a venturi meter, instead of an orifice, the fluid flows through a converging cone with an angle of approximately 15–20°, followed by a throat portion of constant diameter, and a gradually diverging portion (Figure 11.2). The diverging cone has a comparatively smaller angle, typically 5–7° where most of the kinetic energy is converted back to pressure energy. The pressure recovery is significantly better compared to that for orifice meters. Contrary to the case of orifice meters, in this case, because of the gradual reduction in the area, there is no *vena contracta* [2].

The difference in pressure between the upstream side of the cone and the throat is measured and is used to estimate the flowrate based on equations similar to that used for orifice type flowmeters. In the absence of data, a discharge coefficient $C_d = 0.975$ could be taken as a standard over a wide range of Reynolds numbers. *However, at Reynolds numbers below 2×10^5 the value falls noticeably. For accurate values of C_d,* the reader is advised to refer to ASME [1]. While the venturi flowmeter is also a differential pressure type of flowmeter, the pressure drop is considerably lower compared to that in orifice type flowmeters.

The measurement principle in differential pressure type of flowmeters is well understood in the industry and is easy to calibrate at site. The system is comparatively cost effective. However, there is a pressure loss in the system, particularly in case of orifices. Also, at times, the assembly is difficult to install since it calls for a certain minimum upstream and downstream straight run pipe length [3]. Refer to the video link below for a better understanding of these two types of flowmeters:

YouTube link:

The Differential Pressure Flow Measuring Principle (Orifice-Nozzle-Venturi) - YouTube.

11.2.3 Rotameters

The rotameter is a variable area flowmeter. It consists of a tapered metering tube (smaller end at the bottom) and is mounted vertically. It is equipped with a float that is free to move up and down the length of the tube. The fluid to be measured enters from the bottom, passes in the upward direction and finally exits from the tube from the top.

In the absence of any flow, the float remains at the bottom of the tube. As the fluid enters the metering tube, the float starts to rise until the upward hydraulic forces

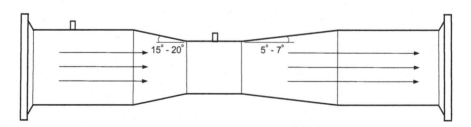

FIGURE 11.2 Venturi meter.

acting on it are equaled by the weight minus the buoyant force. The float thus moves up and down the tube in direct relation to the fluid flowrate [4].

While the cost of rotameters is low, they provide good accuracies at low and medium flowrates. However, with opaque fluids, the float may not be visible. Further, glass tube types are subject to breakage. As a result, rotameters find limited applications in large process units like refineries or petrochemicals industries.

11.2.4 ELECTROMAGNETIC FLOWMETERS

Electromagnetic flowmeters work on the basis of the Faraday's Law of Induction. The law states that when a conductor moves at right angles to a magnetic field, a voltage will be induced. The conductor in this case is the liquid. Inside the flowmeter, there are two coils that generate the magnetic field, over the cross-section of the pipe, and two electrodes that capture the induced voltage. When the liquid flows through the pipe, the induced voltage is proportional to the relative velocity of the conductor and the magnetic field [4].

Such flowmeters have the advantage of negligible pressure drop in the system and a high turndown. The instrument is also capable of handling large particles. However, the instrument cannot measure flowrates of gases and hydrocarbons. Further, a typical minimum fluid conductance of 5 microsiemens is mandatory [3]. Refer to the following video link for better understanding:

YouTube link: The Electromagnetic Flow Measuring Principle - YouTube.

11.2.5 ULTRASONIC FLOWMETERS

In ultrasonic flowmeters, the operation is based on the principle of measuring the velocity of sound as it passes through the fluid flowing in a pipe. The instrument is of two types: (1) Doppler-effect meter and (2) transit-time meter. In both these categories, the flowrate is deduced from the effect of the flowing process stream on sound waves introduced into the process stream [5].

The instrument is capable of handling both liquids and gases with a fairly high turndown. In addition, there is no pressure drop in the system. Maintenance costs are low. However, partially filled pipes may cause signal losses leading to errors in measurement. Fluctuations in operating temperatures can affect sound velocities and ultrasonic beam angles [3]. Refer to the following video link for better understanding:

YouTube link: The Ultrasonic Flow Measuring Principle - YouTube

11.2.6 CORIOLIS FLOWMETERS

A basic dual-flow Coriolis meter consists of two curved tubes through which the fluid flows. An electromagnetic drive system causes the tubes to vibrate. When there is no fluid flowing through the tubes, they vibrate towards and away from each other at their resonant frequency and the outputs of the upstream and downstream motion sensors are in phase. However, as material flows through the tubes, the Coriolis effect causes the downstream side of the loop to slightly lead the upstream side, thus creating a slight twist in the loops of the tubing. The amount of twist, and hence the phase difference between the outputs of the upstream and downstream sensors, varies linearly with the mass flowrate through the tubes. Phase is converted to time, and time delay is directly proportional to mass flowrate [6]. Refer to the following video link for better understanding:

YouTube link: The Coriolis Flow Measuring Principle - YouTube.

The principle of Coriolis flowmeter applies regardless of whether the fluid is a liquid, gas or slurry. In addition, it is also suitable for handling fluids of high viscosity. Among all the flowmeters, the Coriolis flowmeter offers the highest accuracy and also the highest turndown. However, the instrument has limited applications in large line sizes (>12"). Further, the instrument offers a certain pressure drop across it and is also susceptible to serious system noise [3].

Table 11.2 illustrates, in a nutshell, a guide to the selection of flowmeters and is self-explanatory [7, 8].

11.2.7 VORTEX FLOWMETER

Vortex flowmeters work on the principle of vortex shedding. When a non-streamlined obstruction, called the "shredder," is placed in between a flowing fluid, vortex formation occurs. As the fluid flows, it is divided into two paths by the shredder. The high-velocity fluid parcels flow past the low-velocity parcels in the vicinity of the element to form a shear layer. After a certain length of travel, the shear layer breaks down into well-defined vortices [5].

This measuring principle is based on the fact that turbulence forms downstream of obstacles in the flow. Inside each vortex flowmeter, a bluff body is therefore located in the middle of the pipe. As soon as the flow velocity reaches a certain value, vortices form behind this bluff body, are detached from the flow and transported downstream. The frequency of vortex shedding is directly proportional to mean flow velocity and thus to volume flow. The detached vortices on both sides of the bluff body generate pressure oscillations around the buff body. The devices are equipped with piezoelectric or the capacitance-type sensors which detect these oscillations and generate a primary digital, linear signal. Refer to the following video link for better understanding:

YouTube link: The Vortex Flow Measuring Principle - YouTube.

11.2.8 TURNDOWN OF FLOWMETERS

The turndown of a flowmeter is the low end of a measurement range expressed as a fraction of the high end. In other words, if a flowmeter has a full-scale rating of 10 m³/hr, and it can accurately measure down to 2.5 m³/hr, its turndown is 4:1. Such turndown ratios are typical of orifice type flowmeters. There are other flowmeters which have higher turndown ratios. Refer to Table 11.2 for turndown ratios of various flowmeters [7, 8].

TABLE 11.2
Flowmeter Selection

Service	Orifice Type	Venturi Type	Variable Area	Electromagnetic	Vortex	Ultrasonic Doppler	Ultrasonic Transit Time	Coriolis
Clean liquids, gases	yes	yes	yes	yes	yes	no	yes	yes
Dirty liquids	yes	yes	yes	yes	yes	yes	no	yes
Viscous liquids	no	yes	yes (special calibration)	yes	no	yes	yes	yes
Corrosive liquids	no		yes	no	yes	yes	yes	yes
Accuracy, ±	2–3% (of scale)	1% (of scale)	2–4% (of scale)	0.50% (of reading)	±1% (of reading)	5% (of scale)	1–5% (of scale)	0.05–0.15% (of reading)
Turndown	4:1	4:1	10:1	40:1	10:1	10:1	20:1	100:1
Pressure drop	medium	low	medium	nil	medium	nil	nil	low
Typical cost	low	medium	low	medium	medium	high	high	high

Source: Adapted from [7, 8].

11.2.9 Flow meter Selection and Accuracy

When we talk of accuracies of flowmeters, there are two types of accuracies: accuracy of reading and accuracy of scale.

Accuracy of reading: This type of accuracy is related to the true value of the reading. For example, if a flowmeter has an accuracy of reading of ±0.5%, then for a flow of 10.0 m³/hr, the error would be 0.05 m³/hr, and hence the reading could be off by ±0.05 m³/hr.

Accuracy of scale: In this type of accuracy, the error is a fixed value (in terms of m³/hr) and is related to the full scale of the flowmeter. For example, for a flowmeter with a range of 0–10 m³/hr, if the accuracy of scale is 0.5%, the error would be ±0.05 m³/hr. This would be fixed even if we move away from the full-scale capacity. Thus, for a flow of 5.0 m³/hr, the error would still be ±0.05 m³/hr (1%). Likewise, for a flow of 1 m³/h, the error would be 0.05 m³/hr (5%). In other words, as we move away from the full scale, the error becomes a larger percentage of the flow. Figure 11.3 provides a pictorial representation of these two types of accuracies.

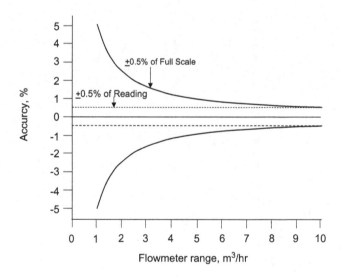

FIGURE 11.3   Flowmeter accuracy.

11.3   LEVEL MEASUREMENT

Level is one of the widely used parameters measured in the chemical process industry. There is enough literature available in the public domain. Hence, this section will focus on what in the opinion of the author is more important for a process engineer to know. Further details can be accessed from literature.

Fundamentally, there are two types of level measurement: direct and indirect. In the direct method, the varying level of the liquid is directly used as a means of obtaining the measurement. In the indirect method, the liquid level is not directly measured. Rather, the level is indirectly estimated from measurement of some other parameter, viz., hydrostatic pressure. Let us discuss some of the important ones.

11.3.1   DIRECT MEASUREMENT

Sight Glass: The sight glass consists of a graduated tube of toughened glass which is connected to the bottom of the process vessel (viz., column bottom, reflux drum, vapor-liquid separator, etc.), for which the level indication is required. The instrument, although a simple measuring device, is a common sight in chemical process industries. Figure 11.4 is a typical illustration of a sight glass.

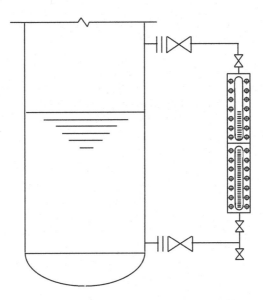

FIGURE 11.4   Level measurement – sight glass.

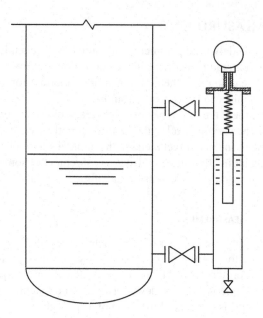

FIGURE 11.5   Level measurement – displacer type.

Displacer Type: According to the Archimedes Principle, an object, when submerged in liquid, will experience an upward thrust which is proportional to the volume of liquid displaced. The object in question here is called the "displacer." A level change in a vessel or column will cause a change in the position of the displacer, which is transferred to a torque tube assembly. As the liquid level changes, the torque tube assembly rotates. This rotary motion creates a magnetic field that is converted to an electronic signal by a sensor. The transmitter develops a 4–20 mA signal proportional to the dc amplifier voltage output. Figure 11.5 is a typical illustration of a displacer type level instrument.

Displacers work well with clean liquids and are accurate and adaptable to wide variations in fluid densities. However, once set up, the fluid measured must maintain its density. In case the density changes, then the weight of the displaced liquid changes, and this changes the calibration. Displacers are recommended only for relatively non-viscous, clean fluids. They are best used for short spans, i.e., up to 1 m. At higher spans, the instrument becomes prohibitively expensive. Cost of installation is high and many refineries are now replacing displacers on account of the inaccuracies that arise due to changes in the densities of the process liquids [9].

11.3.2   INDIRECT MEASUREMENT

Differential Pressure: A differential pressure (DP) type of level measurement is based on the determination of pressure difference generated by the height of the column of fluid. This pressure difference is a direct function of the level in the process vessel. The pressure transmitter sends a 4–20 mA signal.

***

EXAMPLE 11.2

We illustrate here how a DP type of level measurement is carried out. Consider a vessel which is equipped with a DP type of level measurement. There are two legs: HP leg and LP leg (Figure 11.6). Depending upon the location of the DP sensor, a correction has to be carried out. For this, a reference liquid is filled in the low-pressure tubing. The dimensions based on the location of the DP sensor X and Y are self-explanatory.

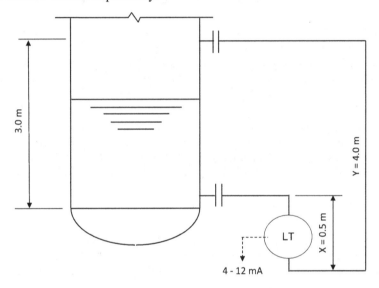

FIGURE 11.6   DP type level measurement.

Level at vessel full of liquid: 3 m
Density of process liquid: 800 kg/cm³
Density of reference liquid: 800 kg/cm³

At vessel full condition:
DP = (3 × 800) + (0.5 × 800) − (4 × 800)
     = − 0.04 kg/cm²

At vessel empty condition:
DP = (0 × 800) + (0.5 × 800) − (4 × 800)
     = − 0.28 kg/cm²

***

DPs are generally primarily useful for clean liquids. They are not to be used for liquids that solidify at higher concentrations. The main benefit of a DP type of level instrument is that it can be easily installed on a vessel. It can be easily removed after isolating by closing the block valves. However, such instruments give erroneous results if there is density variation of the liquid caused by temperature changes or change of product. These variations must always be compensated in order to get accurate measurements [9].

Radar: In this method, high-frequency radar pulses are emitted from an antenna and reflected by the surface of the medium. The time difference between transmission of the emitted pulse and reception of the reflected pulse is measured and analyzed by the instrument. The level is inferred from this time of flight.

This is a non-contact type measurement and is free of wear and tear. Therefore, it can be used in extreme process conditions. It provides highly accurate measurements in storage tanks and some process vessels. The added advantage is that vapors or dusty media have no effect on the measurement. The primary drawback is the cost of the instrument. Further, these devices cannot measure interface levels [9, 10].

Ultrasonic: Such devices work on the principle of sending a sound wave from a piezo-electric transducer to the contents of the vessel. These waves are reflected from the medium surface by the density change between air and the medium. The device measures the time between transmission and reception of the pulse, analyses it and provides a direct value for the distance between the sensor membrane and the medium surface.

The advantage of such a technique is that accuracy is unaffected by the properties of the fluid, e.g., dielectric constant, density or moisture. The transducer does not contain any moving parts, nor does it come in contact with the process fluid. In view of the single topside tank penetration, leaks are not likely to occur. The drawback is that dust, heavy vapors, surface turbulence, foam and ambient noise can affect the returning signal [9, 10].

YouTube link: Time-of-Flight measuring principle animation - YouTube.

Capacitance: The principle of capacitance level measurement is based on capacity change. An insulated probe and the tank form a capacitor whose capacitance depends on the product level. As the level rises in the tank, the capacitance increases. The increase in capacitance is directly proportional to the level increase.

The capacitance method is by far the most versatile technology for continuous level measurement. The method can handle a wide range of process conditions (from cryogenic temperatures to 540°C, and from vacuum to 680 kg/cm^2 pressure). There are no moving parts and there is only a single tank penetration point [5, 9, 10].

11.4   CONTROL VALVES

A control valve is a valve used to control fluid flow to varying degrees between minimum and maximum flow, by varying the size of the flow passage as directed by a signal from a controller. This enables the direct control of flowrate and as a consequence, the control of process parameters such as pressure, temperature, flow and level. The control valve is often referred to as the "final control element." Refer to Figure 11.7.

The capacity of a control valve is normally measured by the flow delivered by the valve against a given pressure drop. The capacity of a valve is defined by the term C_v. For liquids, the C_v is the number of gallons per minute of water at 60°F that a valve would deliver against a pressure drop of 1 psi. Thus, for example, a valve with a C_v of 15 would deliver 15 gpm against a pressure drop of 1 psi.

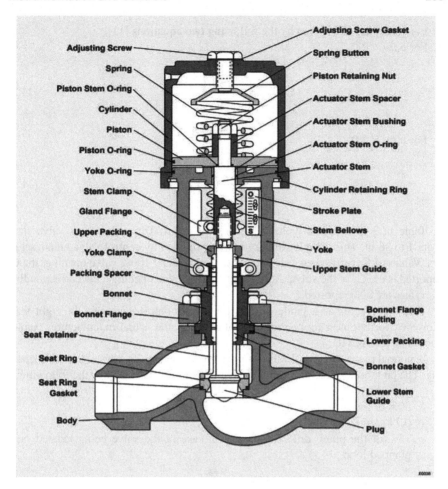

FIGURE 11.7   Control valve and its accessories.

Source: Used with permission from Flowserve Corporation.

The C_v for liquid flow is given by the following equation:

$$C_v = Q \left(\frac{G_L}{\Delta p} \right)^{0.5}$$

(11.6)

Converting the above Equation (11.6) with units consistent with this book, the equation for C_v is as follows:

$$C_v = 1.1674 \, Q_L \left(\frac{G_L}{p_1 - p_2} \right)^{0.5}$$

(11.7)

For gases, the C_v is given by the following two equations [11].
For $p_1/p_2 < 2.0$,

$$C_v = \frac{Q_G}{274} \left(\frac{G_G T}{p_1^2 - p_2^2} \right)^{0.5}$$
(11.8)

For $p_1/p_2 > 2.0$,

$$C_v = \frac{Q_G}{232\, p_1} \left(G_G T \right)^{0.5}$$
(11.9)

Table 11.3 illustrates a typical control valve chart. The table covers valve sizes from 1 to 36 in. This table is similar to those provided by control valve manufacturers. Values of C_v for various valve sizes are provided. At 100% valve opening, the C_v reported is the C_v of the valve. At reduced control valve openings, the corresponding C_v values are also reported.

The table serves as a guide on how to select control valves of the right size. However, for engineering or professional services, the actual manufactures' tables should be referred to.

In normal practice, control valves are operated between 20–80% of the valve opening. The pressure drop across control valves should be the greater of the following:

- 50% of the total frictional loss excluding the control valve
- 0.7 kg/cm²
- 15% of the pump differential head in case of the valve being located on a pumped loop.

***

EXAMPLE 11.3

A process liquid needs to be transferred on flow control from a vessel at 8.0 kg/cm²g to a vessel at 6.0 kg/cm²g. The density of the liquid is 700 kg/m³. The size of the line is 6". Prepare the specification of the control valve to meet this duty and also estimate the C_v of the valve. The normal flowrate is 30 m³/hr with a 20% turnup and 50% turndown (i.e., 36 and 15 m³/hr, respectively).

Refer to Table 11.4. There are two tables, the one on the left (pipe section upstream of the valve) and the one on the right (pipe section downstream of the valve).

Consider the pipe section upstream of the valve (the table to the left). At 30 m³/hr we calculate a pressure drop of 0.35 kg/cm² (based on the flowrate, line size, line length and number of fittings). Similarly, at flowrates of 36 and 15 m³/hr, we arrive at pressure losses of 0.50 and 0.09 kg/cm², respectively. The upstream source pressure being 8.0 kg/cm²g and the line losses provided in column 2, we arrive at the pressures at the inlet to the valve (column 3).

TABLE 11.3

Valve Characteristics

Valve Size, in	Cv of Valve at Percentage Opening									
	10	20	30	40	50	60	70	80	90	100
1	0.35	0.5	0.7	1.1	1.9	3.6	5.6	7.5	9.1	10
1 1/2	0.74	1.1	1.5	2.3	4	7.6	12	16	19	22
2	2.9	4.2	6.5	10	12	18	28	34	42	48
3	3.3	4.7	6.6	11	18	35	54	72	84	95
4	5.9	8.5	12	18	32	61	96	128	153	175
6	13	18	25	40	68	130	203	271	324	360
8	36	59	78	115	194	231	340	534	603	648
10	48	66	105	198	258	338	444	768	848	903
12	56	80	123	221	298	410	537	982	1012	1244
14	69	98	145	280	326	532	680	1167	1312	1592
16	78	110	167	290	386	580	746	1242	1545	2251
18	92	129	191	320	417	610	899	1332	1970	2890
20	120	176	245	351	512	752	1102	1627	2384	3526
24	165	243	360	531	786	1162	1718	2540	3756	5550
30	245	362	534	789	1168	1727	2554	3776	5584	8257
36	326	483	714	1056	1562	2309	3416	5050	7469	11044

Similarly, consider the pipe section downstream of the valve. Here we work backwards. The destination pressure being 6.0 kg/cm²g, and the line losses furnished in column 5, we arrive at the pressures downstream of the valve (column 6). In the third table below, column 8 presents the pressure drop across the valve against the three flowrates provided in column 7. The figures once again confirm the famous slogan for control valves "high flow --> low pressure drop, low flow --> high pressure drop".

Continuing with the third table, in column 9 we present the C_v of valves calculated using the Equation (11.7) given above for C_v of liquids. The C_v values are in the range 11.6–37.6. We now refer to the control valve chart illustrated in Table 11.3. It appears that a 2" valve with a C_v of 48 would be suitable. However, after a closer look, it appears that at the maximum flow of 36 m³/hr, for the calculated C_v of 37.6, the control valve would be 84.5% open. As explained earlier, control valves are operated between 20–80% of valve opening. Thus, we need to choose the next higher size valve. We select a 3" valve with a C_v of 95. With this valve, the range of operation lies between 40.9–61.4% of valve opening and is thus acceptable (column 13).

***

Table 11.5 illustrates the process data sheet of the control valve illustrated in Example 11.3.

TABLE 11.4
Control Valve Calculations

1 Flow rate, m³/hr	2 Line loss, kg/cm²	3 Press at valve inlet, kg/cm²g		4 Flow rate, m³/hr	5 Line loss, kg/cm²	6 Press at valve outlet, kg/cm²g
36	0.50	7.50		36	0.50	6.50
30	0.35	7.65		30	0.35	6.35
15	0.09	7.91		15	0.09	6.09

8 kg/cm²g   6 kg/cm²g

7 Flow rate, m³/hr	8 Valve ΔP kg/cm²	9 Calc Cv 1st trial	10 Cv Valve of 2"	11 % opening Valve of 2"	12 Cv Valve of 3"	13 % opening Valve of 3"
36	1.00	37.6	46	84.5	95	61.4
30	1.30	27.5	46	75.3	95	51.3
15	1.82	11.6	46	48.0	95	40.9

TABLE 11.5
Control Valve Process Data Sheet

Tag Number			FCV - 101	
Service			Flow Control	
Number required			1	
P&ID No.				
Fluid			Liquid Hydrocarbon	
		Minimum	Normal	Maximum
Flowrate	m³/hr	15	30	36
Pressure at inlet	kg/cm²g	7.91	7.65	7.50
Differential pressure	kg/cm²	1.82	1.30	1.00
Design temperature	°C		65.0	
Design pressure	kg/cm²g		10.0	
Density at T_{op}	kg/m³		700	
Viscosity at T_{op}	mPa-s		0.51	
Vapor pressure at T_{op}	kg/cm²a		0.015	
Pour point			-	
Critical pressure			25.35	
Compressibility factor			-	
Isentropic exponent			-	
Molecular weight			-	
At instrument air failure			Closed	
Leakage class			IV	

***

EXAMPLE 11.4

Determine the size and percentage opening for a control valve delivering a liquid hydrocarbon stream with the following data:

Q_L = 35 m³/hr
p_1 = 10 kg/cm²g = 11.033 kg/cm²a
p_2 = 9 kg/cm²g = 10.033 kg/cm²a
G_L = 0.8
T = 50°C (323°K)

From Equation (11.7),
we have

$$C_v = 1.1674 \times 35 \left(\frac{0.8}{11.033 - 10.033} \right)^{0.5} = 36.5$$

From Table 11.3, the C_v for a 2' valve is 48, while that of a 1 ½" valve is 22. Hence, we select a 2" valve. The percentage opening at the calculated C_v of 36.5 works out to 83.1% (after interpolation). However, an opening of 83.1% is greater than the recommended value of 80%.

Now we select a 3" valve (C_v of 95). At a calculated C_v of 36.5, the percentage opening of the valve from Table 11.3 is 60.8%. This is an acceptable value.

***

***

EXAMPLE 11.5

Determine the size and percentage opening for a control valve delivering medium pressure steam with the following data:

Mass Flowrate: 14000 kg/hr

p_1 = 13 kg/cm²g = 14.033 kg/cm²a
p_2 = 10 kg/cm²g = 11.033 kg/cm²a
G_G = 0.62
T = 478°K

Q_G works out to 17422 Nm³/hr
$p_1/p_2 = 1.27$ (< 2.0), hence Equation (11.8) will be applicable in this case.

$$C_v = \frac{17422}{274}\left(\frac{0.62 \times 478}{14.033^2 - 11.033^2}\right)^{0.5} = 126$$

From Table 11.3, the C_v for a 4" valve is 175. The percentage opening at a C_v of 126 works out to 79.4% (after interpolation).

***

EXAMPLE 11.6

Determine the size and percentage opening for a control valve delivering a methane gas stream with the following data:

Q_G = 124000 Nm³/hr
p_1 = 134 kg/cm²g = 15.033 kg/cm²a
p_2 = 3.5 kg/cm²g = 4.533 kg/cm²a
G_G = 0.555
T = 303°K
$p_1/p_2 = 3.3$ (>2.0), hence Equation (11.9) will be applicable in this case.

$$C_v = \frac{124000}{232 \times 15.033}\left(0.555 \times 303\right)^{0.5} = 461$$

From Table 11.3, the C_v for a 6" valve is 360. Since the calculated C_v in this case is 461, we have to select an 8" valve. From Table 11.3, the C_v for this 8" valve is 648. The percentage opening at a C_v of 461 works out to 76.2% (after interpolation).

***

11.5   CONTROLS IN A PROCESS PLANT

A process variable is controlled by means of a control loop. A typical control loop consists of a sensor, transmitter, controller and control valve. The sensor measures the process variable and the transmitter transmits the information to the controller located in the control room. The job of the controller is to check that the process value sent by the transmitter is in line with the desired value (called the "set point").

If not, then it will send a signal to the control valve based on which it will either close or open until the measured process value matches with the set point. This continues in a loop and hence the name.

The various controls like temperature, pressure, level and flow would best be understood through two examples, one for a reactor and the other for a distillation column. All these controls are covered in these two examples.

11.5.1 Controls in a Distillation Column

A distillation column needs to be operated with care to ensure that the desired product qualities are met, while achieving the specified production quantities.

Pressure Control: Pressure fluctuations in a distillation column change vapor loads and temperature profiles and, thus, make control more difficult. Pressure fluctuations change relative volatilities and affect performance. Effective pressure control enables more reliable operation close to the unit's design capacity. Sloley [12] has provided an in-depth discussion on 19 different column pressure control schemes for different cases.

Pressure Control (Net overhead vapor rate ≥ 0, Case 1): In this configuration, apart from an overhead liquid product, there is an uncondensed vapor product as well. In such a case, the control valve is located in the vapor outlet line. The controller controls the column top pressure by regulating this valve (refer to Figure 11.8).

Pressure Control (Net overhead vapor rate ≥ 0, Case 2): This is similar to Case 1 above; however, the mechanism here is through a split range pressure control. When column overhead pressure drops, initially at 0–50% controller output, inert gas is introduced through a control valve. Thereafter, any remaining excess pressure is controlled by releasing the gases to flare (refer to Figure 11.9).

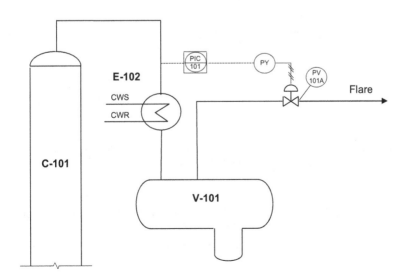

FIGURE 11.8 Pressure control (net overhead vapor rate ≥ 0, Case 1).

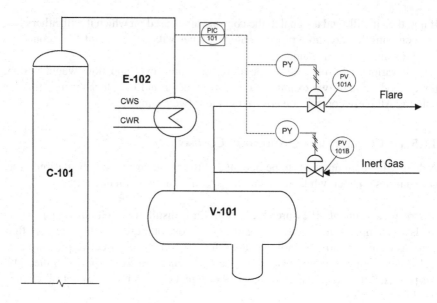

FIGURE 11.9   Pressure control (net overhead vapor rate ≥ 0, Case 2).

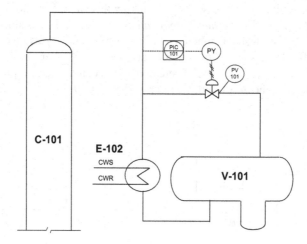

FIGURE 11.10   Pressure control (net overhead vapor rate = 0, Case 1).

Pressure Control (Net overhead vapor rate = 0, Case 1): In this case, there is only a liquid overhead product. This method of pressure control is called the "hot-vapor bypass method." Refer to Figure 11.10. Suppose the pressure of the column over-heads is higher than the set pressure. The control valve on the bypass line opens, and hot vapors flow directly into the reflux drum bypassing the condenser. The hot vapors

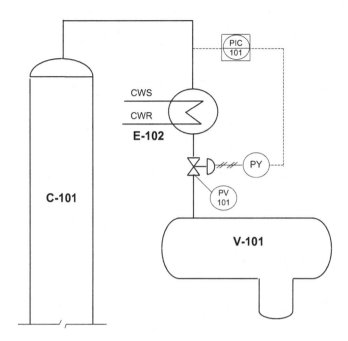

FIGURE 11.11 Pressure control (net overhead vapor rate = 0, Case 2).

must condense; however, the liquid in the reflux drum is in equilibrium and is not cold enough to absorb the latent heat of condensation. The vapors will start accumulating in the reflux drum and increase the pressure in the drum. This higher pressure in the drum will cause the liquid to back up slightly in the condenser reducing the surface area for condensation. The liquid on the submerged tubes will start to subcool. This sub-cooled liquid will absorb the latent heat of condensation and condense the accumulated vapors in the reflux drum. A new equilibrium will thus be established. Refer to Lieberman [13] for further details.

Pressure Control (Net overhead vapor rate = 0, Case 2): In the method, a part of the condenser is flooded directly by a control valve located downstream of the condenser. Refer to Figure 11.11. When pressure in the column overhead decreases, the control valve partially closes the line downstream of the condenser. As a result, the liquid backs up slightly in the condenser reducing the surface area for condensation. Therefore, less vapors will now condense, restoring the column pressure.

Temperature Control: Let us study this by means of temperature control in a reboiler in a typical distillation column. Temperature control of the column bottoms, or for that matter any critical tray is carried out by either controlling the flow of steam into the reboiler or controlling the flow of condensate outlet from the reboiler (Figure 11.12).

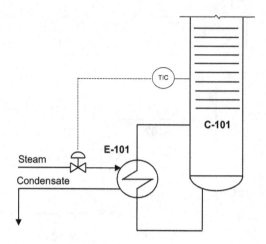

FIGURE 11.12 Temperature control by regulating flow of steam.

***

EXAMPLE 11.7   CONTROLLING THE FLOW OF STEAM

Steam header pressure:	5.5 kg/cm²g
ΔP across the control valve:	0.80 kg/cm²
Steam side pressure:	4.7 kg/cm²g
Steam condensing temperature:	156.2°C
Steam flow:	m kg/hr (m depends upon the heat duty. The higher the heat duty, the higher the value of m)
Process side temperature:	125°C (i.e., tube side temperature)
Δt:	31.2°C

Suppose the temperature in the column bottoms is such that the control valve has to increase the steam flow by 11%. The Δt should also increase by 11%, i.e., to 34.6°C (see explanation below). The condensing steam temperature should therefore be 159.6°C. This corresponds to a steam pressure of 5.2 kg/cm²g. The new process conditions would be as follows:

Steam header pressure:	5.5 kg/cm²g
ΔP across the control valve:	0.30 kg/cm²
Steam side pressure:	5.2 kg/cm²g
Steam side temperature:	159.6°C
Steam flow:	1.11 m kg/hr
Process side temperature:	125°C (i.e., tube side temperature)
Δt:	34.6°C

In the above example, the steam control valve would open to such an extent that the pressure drop across the valve comes down to 0.3 kg/cm². The equations that describe the two cases are as follows:

$$\text{At steam flow of m}: \quad m\lambda = UA\Delta t \tag{11.10}$$

$$\text{At steam flow of } 1.1\text{m}: (1.11\text{m})\lambda = UA(1.11\Delta t) \tag{11.11}$$

It is assumed that the latent heat of condensation and the overall heat transfer coefficient remains constant in both cases. While it may be argued that these properties may be different under different pressures, the difference would be marginal. The above example conveys the dynamics of how the temperature control in a column bottom functions.

It may be summarized that the rate of steam flow to the reboiler is not controlled directly. The control valve increases or decreases the steam condensing pressure and hence temperature in the reboiler, and this is how the heat transfer rate is regulated. Refer to Lieberman [13] for further details.

***

EXAMPLE 11.8   CONTROLLING THE FLOW OF CONDENSATE

When using low pressure steam, a control valve on the steam line would further reduce the steam pressure inside the reboiler and hence the temperature. In such cases, it is advisable to place the control valve on the condensate line. The flow of steam can then be regulated by raising or lowering the condensate level in the shell as shown in Figure 11.13. In this system, the level in the reboiler shell is determined by the level in the drum. As the level in the drum falls, the number of tubes exposed to the steam increases. When the level in the shell drops down to the bottom tubesheet, the level controller over-rides the temperature controller and prevents loss of condensate seal.

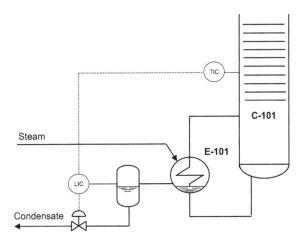

FIGURE 11.13   Temperature control by regulating flow of condensate.

***

11.5.2 Controls in a Reactor System

For the sake of illustration, a hydroprocessing unit is selected. Hydrotreating and hydrocracking are two hydroprocessing units employed in oil refineries. Both use high-pressure hydrogen to catalytically remove contaminants from petroleum fractions. In hydrotreating units, reactions that convert organic sulfur and nitrogen into H_2S and NH_3 also produce light hydrocarbons. The main chemical reactions are hydrodesulphurization (HDS), hydrodenitrification (HDN), aromatic and olefin saturation.

"Hydrocracking" is a process to convert larger hydrocarbon molecules into smaller molecules under high hydrogen pressure and elevated temperature. It is commonly applied to upgrade the heavier fractions of the crude oils to produce higher value transportation fuels.

Process flow schemes for hydrotreating and hydrocracking are similar. Figure 11.14 illustrates the reactor system. Most hydroprocessing reactions are exothermic. The heat released in naphtha and kerosene hydrotreaters is relatively low, so units

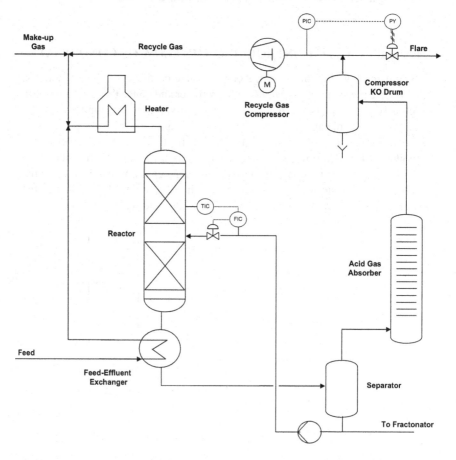

FIGURE 11.14 Reactor temperature control.

designed for these feeds may use just one reactor that contains a single catalyst bed. However, for heavier feeds and/or feeds that contain large amounts of sulfur, aromatics or olefins, the total increase in temperature can exceed 100°C. It is unsafe to allow that much temperature rise in a single bed of catalyst. To divide the heat release into smaller, safer portions, commercial units use multiple catalyst beds with cooling in between. A unit can have one bed per reactor or multiple beds in each reactor with quench zones in between. In Figure 11.14, a portion of the reactor effluent is separated and cooled via heat exchange in order to be introduced between beds as quench streams. Generally, the recycle stream used as a quench comes from the bottom of the high- or low-pressure separators. In this way, the portion of the heaviest treated fraction also has a second-pass opportunity through the reaction system.

11.5.3 Dynamics in a Distillation Column

The primary objective of distillation columns is to maintain the specified composition of the top and bottom products (or any side streams, if any). In addition, certain parameters like pressure and levels in the system also need to be controlled. The control scheme in a distillation column thus controls five variables, namely:

- Distillate composition
- Column bottoms composition
- Column top pressure
- Reflux drum level
- Column sump level.

These variables are called "controlled variables."

The disturbances that could jeopardize the product compositions in the column could be variations in:

- Feed flowrate, composition and temperature
- Steam and cooling water conditions
- Weather conditions.

In a simple column configuration having top and bottom products, there are five variables which are manipulated to achieve the desired specifications of the controlled variables:

- Reboiler duty
- Condenser duty
- Reflux flowrate
- Distillate flowrate
- Bottom product flowrate.

These five variables are called "manipulated variables."

Figures 11.15 through 11.18 show various control configurations in distillation columns. In each set of configurations, two types of schemes are shown: one in which

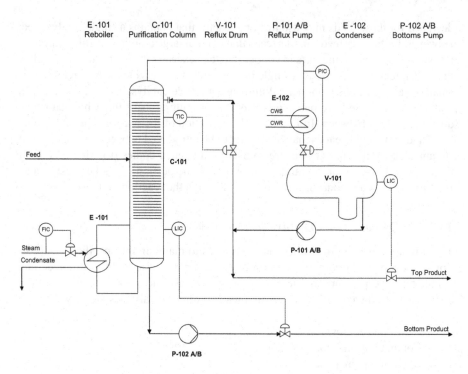

FIGURE 11.15 Control of top product purity by varying the reflux ratio (net vapor rate – 0).

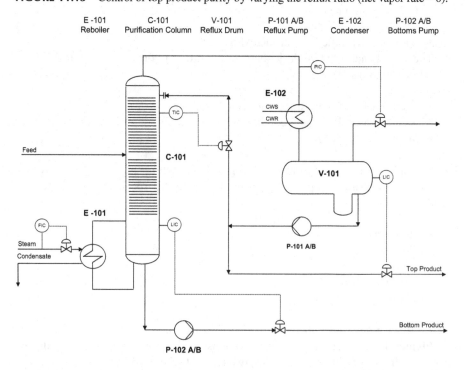

FIGURE 11.16 Control of top product purity by varying the reflux ratio (net vapor rate > 0).

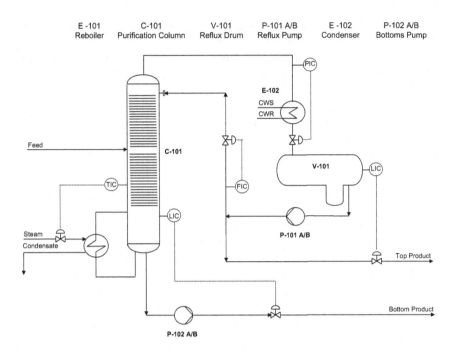

FIGURE 11.17 Control of bottoms product purity by varying the boil-up rate (net vapor rate = 0).

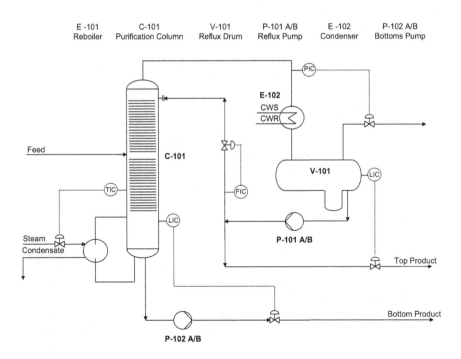

FIGURE 11.18 Control of bottoms product purity by varying the boil-up rate (net vapor rate > 0).

the top product is only liquid, and the other in which the column top generates both a liquid product and a non-condensable vapor product.

If the overhead product purity is the target, the usual practice is to control the top temperature by varying the reflux flowrate (Figures 11.15 and 11.16). Likewise, if the purity of the column bottoms is the target, the bottom temperature is controlled by varying the boil-up rate (Figures 11.17 and 11.18).

Consider Figure 11.15. The desired parameter is the overhead product purity. There is no vapor product. Suppose the concentration of condensable lights increases in the feed to the column. The following will be the sequence of events:

- Light ends will increase at the column top
- Temperature transmitter will sense a drop in tray temperature and will reduce the reflux flow
- Column pressure will rise due to accumulation of non-condensables
- The column pressure control will increase condensation in the condenser. The non-condensables will be removed from the top of the reflux drum
- Level in the reflux drum will increase. The level controller will increase the flow of distillate
- Reduced reflux will reduce the level in the column sump
- Bottoms level controller will reduce the column bottoms flowrate.

A new balance is established whereby the increased quantity of light ends in the feed, increases the top product and reduces the bottoms product. Kister [14] provides detailed discussions on such mass balance control for distillation columns.

Consider Figure 11.16. The desired parameter is the overhead product purity. There is a vapor portion in the overhead product. Suppose the concentration of lights increases in the feed to the column. A part of these lights could be non-condensables. The following will be the sequence of events:

- Light ends will increase at the column top
- Temperature transmitter will sense a drop in tray temperature and will reduce the reflux flow
- Column pressure will rise due to accumulation of non-condensables
- The column pressure control will increase condensation in the condenser. The non-condensables will be removed from the top of the reflux drum
- Level in the reflux drum will increase on account of condensation of the additional condensables. The level controller will increase the flow of distillate
- Reduced reflux will reduce the level in the column sump
- Bottoms level controller will reduce the column bottoms flowrate.

A new balance is established whereby the increased quantity of light ends in the feed increases the top product and reduces the bottoms product. This was however a subjective explanation. The exact quantity of the distillate would depend upon the quantity of the non-condensables.

Consider Figure 11.17. The desired parameter is the column bottoms product purity. There is no vapor component in the product. Suppose the concentration of heavy ends increases in the feed to the column. The following will be the sequence of events:

- Temperature transmitter will sense a rise in tray temperature and will reduce the steam flow to the reboiler
- Column pressure will decrease
- Column pressure control will decrease condensation in the condenser
- Level in the reflux drum will decrease. The level controller will decrease the flow of distillate
- Reduced boiling in the reboiler will increase the level in the column sump
- Bottoms level controller will increase the column bottoms flowrate.

A new balance is established whereby the increased quantity of heavy ends in the feed decreases the top product and increases the bottoms product.

Finally, consider Figure 11.18. The desired parameter is the column bottoms product purity. There is a vapor component in addition to the overhead liquid product. Suppose the concentration of heavy ends increases in the feed to the column. The following will be the sequence of events:

- Temperature transmitter will sense a rise in tray temperature and will reduce the steam flow to the reboiler
- Column pressure will decrease
- Column pressure control will decrease condensation in the condenser. The non-condensables will be removed from the top of the reflux drum
- Level in the reflux drum will decrease. The level controller will decrease the flow of distillate
- Reduced boiling will increase the level in the column sump
- Bottoms level controller will increase the column bottoms flowrate.

A new balance is established whereby the decreased quantity of light ends in the feed decreases the top product and increases the bottoms product. This was again a subjective explanation. The exact quantity of distillate would depend upon the quantity of the non-condensables.

11.6 CONTROL CONFIGURATION AND HARDWARE

11.6.1 DISTRIBUTED CONTROL SYSTEM

The distributed control system (DCS) is a specially designed automated control system that consists of geographically distributed control elements over the plant or control area, but there is a central operator supervisory control. Each process element, machine or group of machines is controlled by a dedicated controller. The DCS consists of a large number of local controllers in various sections of plant control area and are connected via a high-speed communication network.

The process data is transmitted from the process units and displayed on computer screens. The plant operators monitor the process data and adjust the set points of the process parameters using the keyboard. The control system has, in principle, three functional levels:

1. Level 0 consists of field devices such as transmitters for temperature, pressure, flow, level, etc. It also consists of final control elements such as the control valves and ON/OFF valves.
2. Level 1 is the controlling unit of the DCS and consists of input/output modules and the associated processors. The processors receive information from input modules in the form of 4–20 mA variable input or a discreet input. The processed information is signaled to the output modules also as 4–20 mA output or a discreet output. Such control units are often located in the so-called "Instrument Cabinet Room" or "Rack Room" close to the field devices.
3. Level 2 is the production control level. Monitoring of production and targets is done here. Records of inventory are maintained, as well as energy consumptions of various process units.

Figure 11.19 is a typical schematic diagram of a DCS system. Figure 11.20 illustrates a typical diagram of a control loop.

DCS is most suited for large-scale processing or manufacturing plants wherein a large number of continuous control loops are to be monitored and controlled. The main advantage of dividing control tasks for distributed controllers is that if any part of DCS fails, the plant can continue to operate irrespective of failed section.

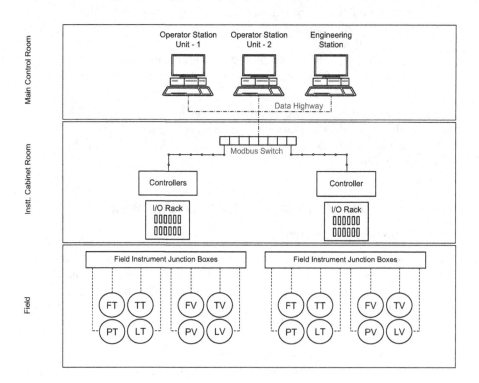

FIGURE 11.19   Schematic diagram of a DCS system.

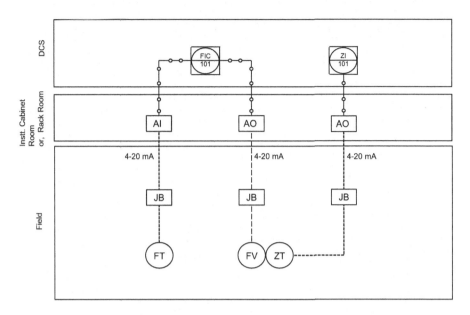

FIGURE 11.20 Schematic diagram of a control loop.

Information Displays: Suppliers of DCS systems use various types of displays.

Group Displays: In group displays, the operating parameters of a group of control loops such as four, eight or twelve loops are shown arranged in rows. Each control loop is represented by a rectangle with bars showing the process variable, the set point and the output signal [15]. Figure 11.21 illustrates a typical four-loop group display.

Graphic Displays: Graphic display allows a flowsheet to be shown on the screen so that the operator can have a more detailed look at a portion of the process rather than watching a row of bar graphs [14]. Refer to Figure 2.4 in Chapter 2 which is a

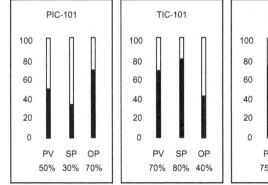
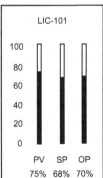
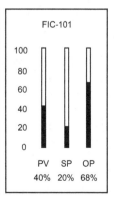

FIGURE 11.21 Group display.

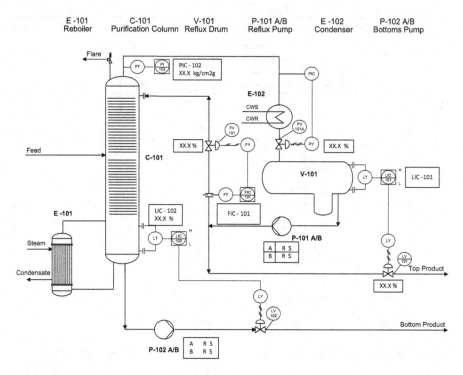

E-101   C-101   V-101   P-101 A/B   E-102   P-102 A/B
Reboiler   Purification Column Reflux Drum   Reflux Pump   Condenser   Bottoms Pump

FIGURE 11.22   DCS graphic display.

typical PFD. The corresponding DCS graphics are conceptualized by the process engineer based on this PFD. Figure 11.22 is the DCS graphics developed using the PFD. It needs to be emphasized that unlike P&ID, a PFD does not show open loops such as a pressure indicator or a level indicator. However, all transmitters are supposed to be a part of the DCS system. Therefore, the PFD used for developing the graphics needs to be suitably elaborated in the sense that it should show all instruments similar to a P&ID, while not covering all the piping components.

Overview Displays: The overview display shows the essentials of a number of groups, each group covered in a separate rectangle. The operator seeing the overview display can observe at a glance the condition of all loops in an operating area, and can easily identify any loop that is out of control. For further details, the reader is referred to Singh [15].

Trend Displays: In trend displays, the profiles of various process variables are shown over a period of time. The time duration could be 1 min, 1 hour or 24 hours. In some displays, several trend graphs are displayed together, allowing a comparison of the history of performance [15].

11.6.2 Emergency Shutdown System

The Emergency Shutdown System is a facility that is equipped to handle certain operations in the process that could lead to unwanted events such as harm to personnel, facilities or environment. The system is designed to rapidly slow down or cease certain operations such that the likelihood of such unwanted events is prevented from continuing or escalating.

Take Figure 11.23, for example. It is a P&ID equipped with ESD features. The top of the column is equipped with a PT-103. Upon a high pressure alarm, the Interlock I-101 shuts the ON/OFF valve on the steam line to the reboiler E-101, thereby preventing the unwanted event to escalate. Similarly, the column C-101 bottom is equipped with an LT-104. Upon a very low level in the column bottoms, the interlock I-102 gets triggered and stops pump P-102 A/B. This prevents a potential damage to the pump by cavitation. Likewise, LT-103 carries out a similar function with respect to the reflux drum V-101 and the pumps P-101 A/B.

It needs to be mentioned that while the level control in the reflux drum V-101 is controlled by a control loop consisting of the level transmitted LT-101, the level controller LIC-101 and the level control valve LV-101, it is important to note that the ESD system is equipped with instrumentation that is dedicated for initiating a shutdown process. Thus, for the ESD system, there is a dedicated level transmitter LT-103 which triggers the pump trip. Similarly, there is a dedicated ESD instrumentation for tripping of column bottoms pump P-102 A/B. Likewise, there is a dedicated ESD system in case of an unusual pressure build-up in the column C-101.

11.6.3 Cause and Effect Diagram

The cause and effect diagram is the preliminary document used for the developing process interlocks applicable for initiating emergency shutdowns or process shutdowns. Literally, "cause" means something that makes something else happen, and "effect" is what happens as a result of the cause. For example, in process control, cause could be a tank high liquid level alarm and the effect could be to start the tank outlet pump.

The cause and effect diagram is presented in the form of a matrix. All causes and effects associated with a specific unit of a process is shown on the cause and effect (C&E) diagram. Several process units may also be grouped on the same C&E diagram.

The causes are defined by the "tag number" and by the service description and are listed in the left section. The effects are defined by the "tag number" and by the type of action and are listed in right section of the diagram. Interlock numbers are indicated in the diagram as per P&ID as a base for logic identification. Table 11.6 illustrates the cause and effect diagram for the P&ID in Figure 11.23.

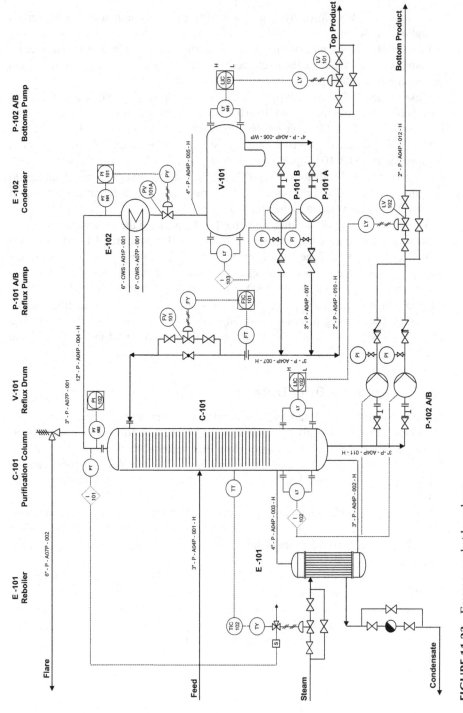

FIGURE 11.23　Emergency shutdown loops.

TABLE 11.6

Cause and Effect Diagram

Cause		Interlock No.	Effect	
Description	Tag No.		Description	Tag No.
High Pressure in C-101	PAHH-102	I-101	Close Steam Valve to E-101	TV-102
Low Level in V-101	LALL-103	I-103	Stop Pump P-101 A/B	MI-101 A/B
Low Level in C-101	LALL-103	I-102	Stop Pump P-102 A/B	MI-102 A/B

Symbols

β	Beta ratio
C_v	Capacity of the control valve
D_1	Pipe diameter (line size), m
D_2	Orifice diameter, m
G_L	Specific gravity of liquid
G_G	Specific gravity of gas (where air at 20°C and 1.033 kg/cm²a equals 1.0)
P_1	Pressure upstream of the orifice, N/m²a
P_2	Pressure at *vena contracta*, N/m²a
P_3	Pressure after recovery, N/m²a
p_1	Pressure upstream of control valve, kg/cm²a
p_2	Pressure downstream of control valve, kg/cm²a
Δp	Pressure drop across the control valve, psi
T_{op}	Operating temperature, °C
ρ	Density of the fluid, kg/m³
v_1	Velocity upstream of the orifice, m/s
v_2	Velocity at *vena contracta*, m/s
Q	Volumetric flowrate of the liquid, gpm
Q_L	Volumetric flowrate of the liquid (for flow orifices), m³/s
Q_L	Volumetric flowrate of the liquid (for control valves), m³/hr
Q_G	Volumetric flowrate of the gas (for control valves), Nm³/hr
A_1	Area of upstream pipe, m²
A_2	Area at *vena contracta*, m²
C_d	Discharge coefficient
T	Gas temperature, K
Δt	Difference in temperature between steam side and process side, °C.

REFERENCES

1. American Society of Mechanical Engineers, Measurement of Fluid Flow in Pipes Using Orifice, Nozzle and Venturi, ASME MFC-3M-2004.
2. https://www.engineeringtoolbox.com/orifice-nozzle-venturi-d_590.html

3. Emerson Flow Expert, Comparing Flow Meter Technology: Choosing the Meter, https://flowsolutionsblog.com/blog/comparing-flow-meter-technology-choosing-the-meter/

4. Hughes, T. A., *Measurement and Control Basics*, Instrument Society of America, 1988.

5. Considine, D. M., *Process Industrial Instruments & Controls Handbook*, 4th edition, McGraw Hill Inc., 1993.

6. O'Banion, T., "Coriolis: The Direct Approach to Mass Flow Measurement", *Chemical Engineering Progress*, March 2013.

7. Flow meter Selection, *Chemical Engineering - Facts at your Fingertips*, July 2009.

8. Forbes Marshall, *Engineering Data Book*, Pune, India, 2014.

9. Bahner, M., A Practical Overview of Level Measurement Technologies, Drexelbrook Engineering Company, date unknown. http://gilsonengineering.net/reference/Levelpap.pdf

10. Endress+Hauser, Level Measurement – Product Overview for Applications in Liquids and Bulk Solids, date unknown. https://portal.endress.com/wa001/dla/5001075/3449/000/12/FA00001F00EN2319.pdf

11. Ferguson Enterprises Inc., Virginia, USA, www.fnwvalve.comtech_AboutCv.pdf, 2012.

12. Sloley, A. W., Effectively Control Column Pressure, *Chemical Engineering Progress*, January 2001.

13. Lieberman, N. and Lieberman, E.T., *"A Working Guide to Process Equipment"*, McGraw Hill Inc., 3rd edition, 2008.

14. Kister, H. Z., *"Distillation Operation"*, McGraw Hill Inc., 1990.

15. Singh, S. K., *"Industrial Instrumentation and Control"* 3rd Edition, Tata McGraw Hill Education Private Limited, New Delhi, 2007.

12 Safety and Relief System Design in Process Plants

12.1 NEED FOR PRESSURE RELIEF SYSTEMS

The relieving of pressure from a process system arises from a number of reasons. It may be needed so that a system is not allowed to pressurize beyond its maximum allowable working pressure. It may be needed for emergency depressurization of high pressure or high inventory systems. In case of an external fire, the maximum allowable yield stress of the metal reduces significantly due to increased temperature. Relieving the pressure under these situations allows the actual stresses to be reduced below the lowered maximum allowable stresses, thereby preventing failure. It is also needed to take care of thermal expansions when a pipeline or equipment containing a liquid is blocked in and subsequently heated.

The chapter begins with a description of the important terminology in pressure relief systems. The causes of over-pressure are discussed. The flare system in a process plant is described. Methods are described to estimate the relieving rates for various scenarios. Inlet and outlet pipe sizing is discussed in addition to the procedure for sizing of pressure relief valves. Finally, blowdown systems are also briefly discussed. Let us first begin with the definitions of some of the important terms.

12.2 DEFINITION OF TERMS

A relief system in a process plant is a fairly complicated system. Before we proceed with the intricacies of such a system, there are number of terms that the reader should be familiar with. These are discussed below:

Pressure Relief Valve: A device actuated by the inlet static pressure and designed to open during an emergency or abnormal condition to prevent a rise in internal fluid pressure in excess of a specified valve [1].

Set Pressure: The inlet gauge pressure at which the pressure relief valve is set to open under service condition.

Over-pressure: The pressure increase over the set pressure of the relieving device expressed as a percentage.

Relieving Pressure: The sum of the valve set pressure and the over-pressure.

Superimposed Back Pressure: The static pressure that exists at the outlet of a pressure relief device at the time the device is required to operate. It is the result of the pressure in the discharge system coming from other sources and may be constant or variable.

DOI: 10.1201/9780429284656-12

Built-up Back Pressure: The increase in pressure in the discharge header that develops as a result of flow after the pressure relief device opens.

Back Pressure: The pressure that develops at the outlet of a pressure relief device after the pressure relief device opens. It is the sum of the superimposed and the built-up back pressures and may be constant or variable.

For further details refer to [1].

12.3 CAUSES OF OVER-PRESSURE

Over-pressure occurs due to a deviation in normal flow of liquid or vapor, a disturbance in the balance of energy input or both to build up in parts of the system. Causes of over-pressure may range from a single contingency to a combination of events. Not all causes will occur simultaneously. Therefore, the pressure relief devices should be sized for the condition that requires the largest relief area. Such devices are not necessarily required on every vessel but on a vessel or a group of vessels or a sub-system that could get isolated by closure of control valves. The following section deals with a brief description of the various causes of over-pressure [1].

Blocked Outlet: This is a case that arises due to the closure of a block valve or a control valve at the outlet of a process vessel while the plant is in operation, resulting in built-up pressure in excess of the maximum allowable working pressure. For control valves, this could also happen in the case of failure of instrument air, if the failsafe position of the control valve is the "closed position." However, such an event would lead to the opening and closing of a more than one control valve and is discussed later in this section.

Failure of Control Valve: Such a situation arises when a control valve which has a high pressure on the upstream side inadvertently opens resulting in the build-up of pressures higher than the maximum allowable working pressure in the downstream side. This could typically happen in the case of failure of instrument air (for those cases where the failsafe position of the control valve is the "open position").

Cooling Water Failure: Failure of cooling water in heat exchangers such as column overhead condensers, etc. will result in vapors leaving the condenser in vapor form, whereas the plant section downstream of the condenser is designed to handle only the liquid phase. This would result in a gradual build-up of pressure due to accumulation of uncondensed vapors in the system.

Power Failure: Failure of electric power will result in stoppage of column overhead air fin coolers. This will result in build-up of pressure due to accumulation of uncondensed vapors upstream. It also leads to stoppages of pumps supplying cooling water or process fluids for cooling/quench leading to pressure build-up.

Instrument Air Failure: Failure of instrument air in the plant would normally lead to the closing or opening of control valves depending on their failsafe positions. In a system or a vessel, it could so happen that some of the control valves on the outlet lines get fully closed, while those on the inlet lines get fully open upon failure of instrument air, based on their failsafe positions. This represents a case similar to that blocked outlet and would lead to pressure build-up.

Heat Exchanger Tube Failure: In shell-and-tube heat exchangers, the tubes are subject to failure from a number of cases, including thermal shock, vibration and corrosion. The result is the possibility that the high pressure stream will over-pressure the system on the low pressure side.

External Fire: Any pressure vessel in an operating plant that handles or processes inflammable liquids or gases may be exposed to fire at some point in time. This hazard may exist even though the contents of the vessel themselves are not flammable.

In case of fire on the wetted surface of a vessel, radiation to the wetted surface will be absorbed as sensible heat and the liquid temperature will rise. At the boiling point, the radiation will be absorbed almost totally as latent heat. The subsequent vapor generation will cause the pressure in the vessel to rise.

When the vessel is filled only with vapor, there is no latent heat to mitigate the rise in metal temperature. As a result, the wall temperature may rise rapidly accompanied by vapor expansion.

12.4   PRESSURE RELIEF SYSTEMS

A pressure relief system in a typical process plant consists of many pressure relief valves (PRVs) which discharge through the outlet pipes to the atmosphere (for cases where the relieving fluid is non-hazardous, viz. steam, water, air, etc.). If, however, the relieving fluid is hazardous, viz. hydrocarbons, the tail pipes from the PRVs are routed to a common flare header, which is connected to a flare stack where ultimately the hydrocarbons are burnt.

Open Disposal System: In cases where the discharge gases or vapors consist of steam, air, nitrogen or similar substances which do not pose a potential hazard, the discharge can be released to the atmosphere.

Closed Disposal System: If the fluid relieved is toxic or flammable, or could result in the formation of a flammable mixture upon discharge, it is mandatory to discharge the gases through a closed disposal system such as the flare. The flare system converts the flammable vapors to less objectionable compounds by combustion.

12.5   COMPONENTS OF A CLOSED DISPOSAL (FLARE) SYSTEM

Pressure Relief Valve Outlet Piping: The flare system consists of outlet pipes from the various PRVs of a unit. These pipes are routed to a common unit flare header. The unit flare header finally routes the flare gases via the unit flare knock-out (KO) drum and the main flare header to the flare stack (Figure 12.1).

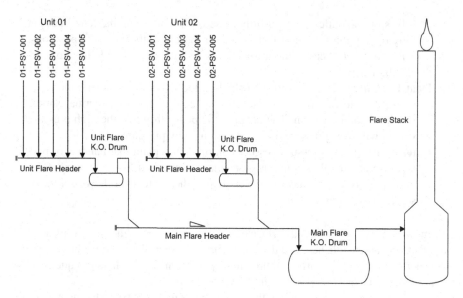

FIGURE 12.1   Flare network in a process complex.

Source: Adapted from *CHEMICAL ENGINEERING*, Copyright© September 2008, by Access Intelligence, Rockville, MD 20850.

Unit Flare Header: Discharge pipes from the pressure relief valves in individual units are connected to the respective unit flare headers. The unit flare headers are either connected directly to the main flare header or are routed to the respective unit flare KO drums, which are in turn connected to the main flare header (Figure 12.1). A minimum slope of 1:500 is recommended. All unit flare headers are continuously purged from the upstream end towards the respective KO drums to maintain a positive pressure and thus avoid ingress of air into the system. Fuel gas, or nitrogen, is typically used as purge gas.

Unit Flare Knock-Out Drum: A unit flare knock-out (KO) drum is mandatory in cases where the discharge from a particular unit is expected to contain appreciable quantities of liquids, especially liquids which are corrosive, fouling or congealing. Another reason why such a drum may be required is when the main flare KO drum is located far away (as in the case of large refineries and petrochemical units), and therefore it is not feasible to maintain a continuous slope of all headers towards the main flare KO drum. In such a case, the unit flare headers are sloped towards the unit flare KO drums. The vapors from these drums are taken from the top of the unit flare KO drums as a result of which the header gains some elevation. Thereafter, the gases are routed to the main flare header [2].

Typically, unit flare KO drums are sized to separate droplets falling in the range of 300–600 microns. In addition, a hold-up time of liquid discharge of 5–10 minutes from a single contingency is also considered. The liquids collected in such drums are drained by gravity to the unit closed blowdown drum. In case the discharged liquid is the congealing type, heat tracing is normally provided in these drums, or the drums are provided with steam coils [2].

Mail Flare Header: The main flare header receives discharge either directly through individual unit flare headers or through unit flare KO drums if such drums are provided. If discharge is from the unit flare KO drums, then these drums collect all the liquid discharged from the individual units. In such a case, the main flare header needs to be sized only for vapor flow. The flare header should be devoid of any pockets and should typically have slope of 1:500 towards the main flare KO drum [2].

Although flare headers are normally sized based on pressure drop criteria, the velocity criteria cannot be ignored. A Mach number in the range of 0.2–0.5 is recommended. Another criterion that should be checked is the change in density of flare gas as it passes through the length of the flare header. In many cases, where flare discharges are at high temperatures, on account of the length of the flare header, the flare gases cool down leading to an increase in the density and, correspondingly, a decrease in the flowrates. Thus, while estimating pressure drops in such flare lines, it is better to divide the header into sections and estimate the pressure drops for each section separately.

Main Flare Knock-Out (KO) Drum: Irrespective of the provision of unit flare KO drums, a main flare KO drum is mandatory and should be installed close to the flare stack. It takes care of condensation in the header on account of atmospheric cooling. Similar to the unit flare KO drums, these drums are also sized to separate out liquid droplets of 300–600 microns in size, and a hold-up of 20–30 minutes [1] of liquid release. The KO drum is provided with pumps to remove the liquid collected to a safe location. The pumps should be selected such that the liquid hold-up could be emptied out in 15–20 minutes. KO drums handling congealing liquids should be provided with steam coils [1].

Seal Drum: Seal drums are provided close to the fare stack or are even sometimes integral with the flare stack. These drums protect against flame flashback from the flare tip. The seal drum should have a diameter of at least two times that of the flare pipe diameter [1].

Flare Stack: Flare stacks are usually elevated structures designed to burn out flammable vapors.

12.6 OVER-PRESSURE AND RELIEVING RATES

12.6.1 BLOCKED OUTLET

This phenomenon occurs when all outlets of a vessel or system are closed. Sources of over-pressure include pumps, compressors, high pressure supply lines connected to the system, etc. For liquids, the maximum liquid pumping rate is considered. In case of centrifugal pumps, pressure relief protection is not required if the connected equipment and piping are designed for the pump shut-off pressure. However, in case of reciprocating pumps, such a protection is normally required. For steam and vapors, it shall be the maximum incoming rate.

The relieving rate should be determined at the relieving pressure and temperature. In addition, the effect of frictional pressure drop between the source of over-pressure and the system being protected should also be considered in determining the relieving rate.

12.6.2 Failure of Control Valve

One of the most well-known emergency situations is that of a steam control valve installed on the steam inlet line to a column reboiler (Figure 12.2). If the steam control valve fails to the open position, there will be a substantially higher steam flow to the reboiler. As a result, the reboiler operation will shift towards its design duty. This will result in higher vaporization from the reboiler. The resistance to flow in the downstream equipment and piping remaining unchanged, this will result in the column pressure approaching the design pressure of the column. At this point, the pressure relief valve will open, releasing the pressure. To calculate the relieving rate of the pressure relief valve, here is an equation which can be used as a simple guideline [3]:

$$\text{Relieving rate} = \left[\frac{Q_R - Q_C}{\lambda_R} \right] \tag{12.1}$$

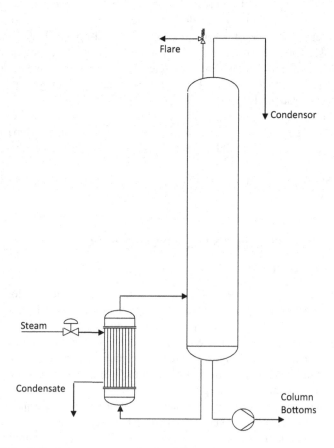

FIGURE 12.2 Relieving rate on control valve failure.

***

EXAMPLE 12.1

Consider a distillation column separating a hydrocarbon from water from a hydrocarbon-water mixture. The column reboiler is fed with medium pressure steam. The control valve on the steam line has a C_v of 606. The column operates at a pressure of 3.5 kg/cm²g. The safety valve of the column is set at a pressure of 5.5 kg/cm²g.

At normal operating condition:

Steam flowrate to the reboiler is 14000 kg/hr (at a control valve pressure drop of 4.5 kg/cm²). The reboiler duty at this condition is 6880900 Kcal/hr.

At control valve full open condition:

Steam flow is estimated as 23500 kg/hr (at a control valve pressure drop of 1.0 kg/cm²). The reboiler duty at this condition is 11363000 Kcal/hr.

At the relieving pressure of 5.5 kg/cm²g, the latent heat of the column bottom liquid is 495 Kcal/kg.

Pressure relief valve relieving rate = (11363000 − 6880900)/495 = 9055 kg/hr.

***

12.6.3 EXTERNAL FIRE

The following formulae are used to estimate the relief load for a vessel on external fire subject to the condition that there are prompt firefighting efforts and drainage of flammable materials away from the vessel [1]:

$$Q = 43,000\,FA^{0.82} \tag{12.2}$$

Where adequate drainage and firefighting equipment do not exist, the following formula is valid:

$$Q = 70,900\,FA^{0.82} \tag{12.3}$$

F has a value of unity for uninsulated vessels. For insulated vessels, F has a value of 0.3 for an insulation conductance of 22.71 W/m².K, and 0.026 for an insulation conductance of 1.87 W/m².K. For further details of F for various values of insulation conductance, refer to [1].

In such calculations, hydrocarbon fires are considered only up to a height of 7.62 m (25 ft). According to experience, pool fires are not likely to impinge for long durations above this height [1].

***

EXAMPLE 12.2

Consider the same distillation column as in Example 12.1. Consider a pool fire in the vicinity of the column. The column has the following particulars:

Vessel bottom T/L w.r.t grade, h_1	m	: 3.00
Liquid level in vessel, h_2	m	: 5.00
Fire source elevation, h_3	m	: 0.00
Diameter of column, D_C	m	: 4.00
Heat of vaporization at relieving condition, λ	Cal/kg : 495	

The wetted surface can be given as:

Wetted cylindrical surface area, A_1 $= \pi \, L \, D_C$

Wetted head area, A_2 $= 1.375 \times \pi \, (D_C)^2/2$

Total wetted area, A $= A_1 + A_2$

Effective liquid level, L $= 7.62 - h_1$

  $= 4.62$ m

Cylindrical surface area, A_1 $= \pi \, L \, D_C$

  $= 3.142 \times 4.62 \times 4$

  $= 58.06$ m^2

Wetted head area, A_2 $= 1.375 \times \pi \, D^2/4$

  $= 1.375 \times 3.142 \times (4.0)^2/4$

  $= 17.28$ m^2

Total wetted area, A $= A_1 + A_2$

  $= 58.06 + 17.28$

  $= 75.34$ m^2

Heat input to the vessel, Q $= 70{,}900 \, F \, A^{0.82}$ (F = 1, uninsulated vessel)

  $= 70{,}900 \times 1 \times (75.34)^{0.82}$

  $= 2.454 \times 10^6$ W

  $= 2.454 \times 10^6 \times 3600/1000$ kJ/hr

  $= 8834400$ kJ/hr

  $= 12113493$ kCal/hr

Relieving rate, F $= Q / \lambda$

  $= 2113493 / 495$

  $= 4270$ kg/hr

Table 12.1 shows the process data sheet of a pressure relief valve that caters to the control valve failure and external fire cases as illustrated in Examples 12.1 and 12.2, respectively.

TABLE 12.1
Pressure Relief Valve Data Sheet

Tag No.				PSV - 101	
Quantity				2 (1 Opn + 1 Std.by)	
Service				Hydrocarbon-Water Mixture	
P&ID No.				20122-PR-002	
Inlet Line No.				3"-P-A07P-003	
Protected Vessel Design Pressure			kg/cm^{2}g	5.5	
Protected Vessel Design Temperature			°C	177	
Relieving Rate (Control Valve Failure)			kg/hr	9055	
Relieving Rate (External Fire)			kg/hr	4270	
Operating Pressure	kg/cm^{2}g	Set pressure	kg/cm^{2}g	3.5	5.5
Over-pressure			%	21	
Back Pressure		Back Pressure			
(Constant)	kg/cm^{2}g	(Variable)	kg/cm^{2}g	0.5	0.4
Total Back Pressure			kg/cm^{2}g	0.9	
Molecular Weight			kg/kmol	19.2	
Viscosity at relieving temperature			cP	0.015	
Cp/Cv				1.28	
Compressibility Factor					

***

12.6.4 HEAT EXCHANGER TUBE FAILURE

10/13 Rule: In a shell-and-tube heat exchanger, there are fluids flowing through the shellside and tubeside. Depending upon the operating pressures of these fluids, each of these sides has a specified design pressure. In case of a tube rupture, the fluid from the high pressure side flows into the low pressure side. It might so happen that the low pressure side gets pressurized to an unacceptable level. The 10/13 rule helps us to decide whether or not a pressure relief valve is required in the low pressure side of the heat exchanger.

Refer to Table 12.2. The high pressure side on a shell-and-tube heat exchanger is the tubeside with a design pressure (P_{des}) of 20 kg/cm²g. On the shellside, there are two cases. In Case 1, the P_{des} is 15.4 kg/cm²g. According to ASME Section VIII, Division 1, vessels designed for internal pressure shall be subjected to a hydrostatic test pressure (P_{test}) at minimum 1.3 times the maximum allowable working pressure MAWP (this further comes with a rider that the MAWP may be assumed to be the same as P_{des} when calculations are not made to determine the MAWP). With this rider, let us consider that the P_{test} of the shellside has been carried out at 1.3 times the P_{des}, i.e., at 20 kg/cm²g.

Let us assume that there is a tube rupture and that the tubeside is at its worst operating condition, i.e., at its design pressure of 20 kg/cm²g. There will be flow of fluid through the rupture portion and the pressure in the shellside will build-up. The maximum pressure the shellside could reach is 20 kg/cm²g. According to Table 12.2, the shellside has already been tested at a pressure of 20 kg/cm²g. Thus it can be safe to assume that the shellside would be able to withstand the pressure build-up due to the tube rupture and, thus, no PRV is required. The ratio of P_{des} low pressure side to P_{des} high pressure side is 15.4/20 = 0.77 = 10/13.

Consider now Case 2, where the P_{des} on the shellside is 12.0 kg/cm²g. The test pressure is 1.3 times the, P_{des}, i.e., 15.6 kg/cm²g. In case of a tube rupture, the maximum pressure the shellside could reach is 20.0 kg/cm²g, which is higher than the P_{test}. Therefore, shellside would not be able to withstand the pressure build-up due to the tube rupture and, thus, a PRV would be required.

Summary: This is the essence of the 10/13 rule. If P_{des} of the low pressure side is $\geq$10/13 of the P_{des} of the high pressure side, the low pressure side would be able to withstand the pressure build-up upon tube rupture, and therefore no PRV would be required.

Equations and Guidelines to Be Used: Critical flow pressure for expansion of a compressible gas or vapor across a nozzle, or an orifice is given by [4]:

TABLE 12.2
Tube Rupture Scenarios

Parameter	Unit	Tubeside	Shellside Case 1	Shellside Case 2
Design Pressure (P_{des})	kg/cm²g	20.0	15.4	12.0
Test Pressure (P_{test})	kg/cm²g	26.0	20.0	15.6

$$P_{cf} = P_1 \left[\frac{2}{k+1} \right]^{\frac{k}{(k-1)}}$$ (12.4)

Crane Co. [5] provided equations for flow of compressible fluids through a sharp-edged orifice. Accordingly, the relieving flowrate for vapor, through an orifice on account of a tube rupture is given as:

$$W = 2407.7\, C\, A_t\, Y \sqrt{\Delta P . \rho_v}$$ (12.5)

If $P_{CF} > P_2$, $\Delta P = P_1 - P_{CF}$.
If $P_2 > P_{CF}$, $\Delta P = P_1 - P_2$.

For flow from tubeside to shellside, the orifice coefficient is typically 0.74. The net expansion factor Y is expressed as:

$$Y = 1 - 0.4 \frac{\Delta P}{P_1}$$ (12.6)

For flow from shellside to tubeside, the orifice coefficient is typically 0.60. The net expansion factor Y is expressed as:

$$Y = 1 - 0.317 \frac{\Delta P}{P_1}$$ (12.7)

For tube rupture cases, only one cross-sectional area is considered.

For further details, viz., liquid flow and mixed flow, refer to [6, 7].

***

EXAMPLE 12.3

Consider a heat exchanger with conditions as in Table 12.2. For shellside, we consider Case 2. The P_{des} of the shellside, i.e., 12.0 kg/cm²g is lower than 10/13 the P_{des} of the tubeside, i.e., 20.0 kg/cm²g. Therefore, a PRV is required on the shellside. The process data is as follows (refer to Figure 12.3):

Pressure at high pressure side P_1: 20 kg/cm²g (299.2 psia)
Shellside design pressure: 12 kg/cm²g
Pressure at low pressure side P_2: 12 + 10% over-pressure = 13.2 kg/cm²g
                                      (202.4 psia)

(The shellside is connected to the pressure relief valve. Hence, at relieving condition of the PRV, the pressure at low pressure side P_2 would be the shellside design pressure + the over-pressure.)

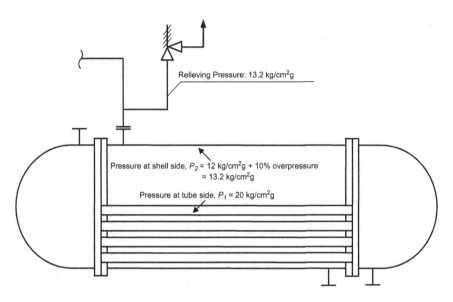

FIGURE 12.3 Pressure conditions at tube rupture.

k :	1.13
ρ_v:	31.92 kg/m^3 = 1.99 lb/ft^3
Tube diameter:	19.05 mm = 0.75 in

Total tube rupture area A_t: = 2 x 3.142/4 x (0.75)^2 = 0.883 in^2

Critical pressure P_{cf}:

$$= P_1 \left[\frac{2}{1.13+1} \right]^{\frac{1.13}{(1.13-1)}} = 173.1 \text{ psia}$$

Here $P_2 > P_{CF}$; therefore,

$\Delta P = 299.2 - 202.4 = 96.8$ psia

Flow is from tubeside to shellside, hence from Equation (12.5),

Y	$= 1 - 0.4\dfrac{96.8}{299.2} = 0.87$
C	$= 0.74$
W	$= 2407.7 \times 0.74 \times 0.883 \times 0.87 \sqrt{96.8. \times 1.99}$
	$= 18997$ lb/hr $= 8617$ kg/hr

***

12.6.5 COOLING WATER FAILURE

Cooling water failure affects mainly column overhead condensers. In cases of cooling water failure, the difference between the total vapor entering and leaving the condenser at the relieving conditions should be considered as the relieving load. However, one should bear in mind that at relieving conditions, more vapor might condense because of a higher pressure compared to the normal operating condition. Hence, a rigorous simulation should preferably be carried out to avoid oversizing the pressure relief valve [8].

12.6.6 POWER FAILURE

In case of a total power failure in a plant, the consequences are as follows:

- Motor driven equipment will stop, thereby leading to stoppage of feed input, product output and reflux input
- Air coolers, combustion air, etc. will stop
- Cooling towers will stop (refer to relevant section below)
- Motor-operated valves will fail
- Instrument air will be interrupted (refer to relevant section below)

In oil and gas industries, stand-by pumps are installed. Critical stand-by pumps are equipped with automatic start-up facility through an emergency power supply. Normally, credit is not given for such a facility. However, if a High Integrity Protective System (HIPS) with triple redundant signals [8] is installed for such an automatic start-up, and time difference between power failure and the stand-by drive coming into operation is not significant for process conditions approaching an over-pressure condition, it might be possible to take credit for such an installation.

12.6.7 INSTRUMENT AIR FAILURE

This event could lead to the opening and closing of several control valves in a system, both on the inlet and outlet lines. Over-pressure could occur, if there is a net flow into the system. The required relieving capacity in such a case is the difference between the maximum expected inlet flow from the valves that are fully open and the normal expected outlet flow from the valves that are fully closed. The flow should be calculated at the relieving condition. In oil and gas industries, appropriate backup facilities are provided such as stand-by air compressors, automatic cut-in of spare machine, buffer vessels that hold a backup of 10–15 minutes of instrument air supply after the air compressor has shut down, etc.

In general, failure of instrument air will cause all control valves to take their fail-safe positions (air-fail-open or air-fail-close). Therefore, if a plant is well designed, an instrument air failure holds limited risk of over-pressure [8].

12.7 GROUPING OF RELIEVING LOADS

The individual relieving loads from various sources estimated above should be grouped together for various contingencies in order to design the relieving system components downstream of these relief valves. A table listing such loads should be prepared for each of the units in the complex. From this table, the total governing load for the largest single contingency for each plant should be estimated [2]. Table 12.3 shows such a tabulation.

TABLE 12.3
Grouping of Relieving Loads

Tag No.	Power Failure			Cooling Water Failure			External Fire		
	Relief Load, tonnes/hr	Temperature, °C	Mol. Weight, kgmol/kg	Relief Load, tonnes/hr	Temperature, °C	Mol. Weight, kgmol/kg	Relief Load, tonnes/hr	Temperature, °C	Mol. Weight, kgmol/kg
PSV - 001							12.6 (Note 1)	79	53.4
PSV - 002	3.8	82	55.4	7.3	83	54.3			
PSV - 003							3.2	84	56.2
PSV - 004							5.8 (Note 1)	87	57.3
PSV - 005	7.3	91	59.2	3.5	92	60.1			
PSV - 006							6.7 (Note 1)	92	61.2
PSV - 007							9.1	89	59
PSV - 008	3.6	98	65.2						
PSV - 009							8.2	102	67.1
PSV - 010				4.8	103	68.9	15.7	104	67.9
PSV - 011	7.4	105	68.7	7.6	104	69.0			
PSV - 012							12.5	106	69.00
Total	**22.1**	**95.3**	**62.7**	**23.2**	**95.4**	**63.0**	**25.1**	**84.3**	**56.4**

Note: Note 1 is the fire zone which produces the governing load.

12.8   PIPING CONSIDERATIONS

In case of an over-pressure in a vessel, the relevant pressure relief valve will start to open at the set pressure. At this moment, the downstream pressure at the valve is the superimposed back pressure of the system. The valve keeps opening as the over-pressure builds up. The resultant flow creates a built-up back pressure on the discharge pipe. As long as the built-up back pressure is less than the over-pressure after the valve opens, the valve will remain open and perform satisfactorily. If, however, the built-up back pressure is at a rate greater than the over-pressure, the balance of forces will tend to close the valve. Therefore, the sizing of discharge pipes is very important in such systems.

Relief Device Inlet Piping: Excessive pressure loss at the inlet of a pressure relief valve can cause rapid opening and closing of the valve, or chattering. Chattering will result in lower capacity and damage to the seating surfaces.

When a pressure relief valve is installed on a line directly connected to a vessel, the total permanent pressure loss between the protected equipment and the pressure relief valve should not exceed 3% of the set pressure of the valve except as permitted for pilot-operated pressure relief valves. This pressure should be calculated using the rated capacity of the pressure relief valve. The nominal size of the inlet piping must be the same as or larger than the nominal size of the pressure relief valve inlet flange connection. When a rupture disk device is used in combination with a pressure relief valve, the pressure-drop calculation must include the additional pressure drop developed by the disk [1].

Relief Device Outlet Piping: The basic criterion for sizing the discharge piping and the relief manifold is that the back pressure which develops as a result of the flow does not reduce the relieving capacity below the amount required to protect the vessel from over-pressure. Where conventional safety relief valves are used, the relief manifold system should be sized to limit the built-up back pressure to approximately 10% of the set pressure of each pressure relief valve. In addition, the effect of superimposed back pressure from other valves that may be releasing simultaneously also should be considered. Thus, while tailpipes from individual relief devices are sized based on the rated capacity of the device, common headers and manifolds in multiple device installations are sized based on the worst case cumulative required capacities of all the devices that may be expected to relieve simultaneously through that header in a single over-pressure event. Typical contingencies include cooling water failure, power failure and instrument air failure.

With balanced bellows type pressure relief valves, higher back pressures may be used. The functioning of such valves is not dependent upon the back pressure. However, their capacities gradually decrease as the back pressure increases from 30 to 50% of the set pressure.

In summary, selecting the flare header size is a tradeoff between the size of the header and the type of pressure relief valve. A high outlet Mach number results in a smaller header size, but with a higher back pressure, and may even result in the requirement of balanced bellows type of relief valve which is more expensive. It is common practice to limit the outlet Mach number to 0.5 corresponding to the peak flow [1].

The isothermal flow equation based on outlet Mach number is given by API 521 [1]. This method calculates pressure build-up backwards up to the outlet of relief valves, and thus avoids the trial and error method of calculation:

$$\frac{fl}{D} = \frac{1}{M_2^2}\left[\left(\frac{P_1}{P_2}\right)^2 - 1\right] - \ln\left(\frac{P_1}{P_2}\right)^2 \qquad (12.8)$$

The Mach number at the outlet of each pipe section is given by [1]:

$$M_2 = 3.293 \times 10^{-7} \left(\frac{W}{P_2 D^2}\right)\left(\frac{Z.T}{kM_w}\right)^{0.5} \qquad (12.9)$$

***

EXAMPLE 12.4

Consider a process unit having nine safety valves relieving to the flare system. There are two relief scenarios: cooling water failure and external fire. Table 12.4 summarizes the relief loads of all the pressure relief valves along with the relieving case. It can be seen that the total relieving load for external fire case is much higher than that for cooling water failure case. This gives the impression that the main flare header should be sized based on external fire case.

However, the external fire case occurs only at localized areas where the fire occurs. The extent of this fire depends on the size of the catch basin for oil spillage. Figure 12.4 illustrates this in detail. It can be seen that oil spillage in a particular area of a plant runs into the nearest catch basin. The paved areas are sloped towards the respective catch basins. It is therefore clear that any hydrocarbon spillage is restricted to a particular zone and is therefore localized. This is why in Table 12.4, there are sub-totals for external fire case, i.e., each sub-total stands for fire in a localized area.

Failure of cooling water on the other hand affects the whole plant. The plant wide relief load for cooling water failure is 67220 kg/hr. In contrast, the relief loads from the various plant sections are significantly lower (Table 12.4). Hence, here cooling water failure case has been considered as the governing case.

The flare network is divided into segments as shown in Figure 12.5. Segment 1 is the section of the flare header between battery limit A and point B. Similarly, segment 2 is the section of the flare header between points B and C. The following calculations estimate the main flare line size and the branch header sizes.

Calculations: (**between point A and point B**)

Flowrate:	67220 kg/hr
Superimposed back pressure at battery limit (point A):	1.4 kg/cm²a (i.e., P_2)

TABLE 12.4
Summary of Relieving Loads

| Plant Section | PSV Tag No. | Relieving Rates | | Temperature, °C | Mol Wt., kg/kg-mol | Viscosity, cP | Comp. Factor, Z |
		External Fire, kg/hr	CW Failure, kg/hr				
1	PSV-001	9500	10120	90	80	0.02	0.99
	PSV-002	9800	8100	88	78	0.02	0.99
	PSV-003	6000	9900	95	84		
	Sub-total	**25300**					
2	PSV-004	10500	0	86	74	0.02	0.99
	PSV-005	12100	9800	78	68	0.02	0.99
	Sub-total	**22600**					
3	PSV-006	12500	11000	78	68	0.02	0.99
	PSV-007	15000	0	84	72	0.02	0.99
	Sub-total	**27500**					
4	PSV-008	8900	9800	76	64	0.02	0.99
	PSV-009	8200	8500	72	58	0.02	0.99
	Sub-total	**17100**					
	Total	**92500**	**67220**				

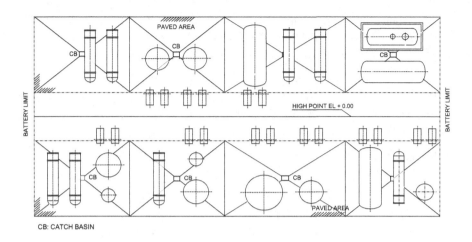

FIGURE 12.4 Drainage areas in the plot plan.

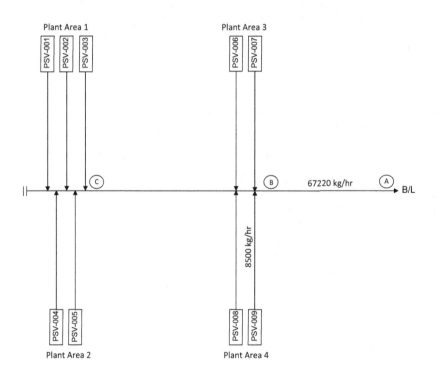

FIGURE 12.5 Flare network divided into segments.

Temperature of relieving vapors:	90°C
Average molecular weight of relieving gases:	80
Compressibility factor:	0.99
Ratio of specific heats, C_p/C_v:	1.4
Density of vapors (calculated):	2.986 kg/m³
Reynolds number:	560148
Equivalent length of pipe (assumed for calculations):	40 m
Viscosity:	0.00001 kg/m.s (0.01 cP)
Roughness factor:	0.2 mm
Diameter of the pipe (assumed):12" (300 mm)	
N_{Re}	7923351
f	0.018
Mach number (from Equation (12.9))	0.36
P_2, i.e., pressure at point A:	1.40 kg/cm²a
Hence P_1, i.e., pressure at point B (from Equation (12.8))	1.63 kg/cm²a

The Mach number is rather low, i.e., we can go with a higher Mach number. So, let us try with a line size of 10" (i.e., 250 mm)

N_{Re}	9510410
f	0.019
Mach number based on a 10" line size Equation (12.9))	0.51

This appears reasonable.

Therefore, we finalize a line size of 10"

P_1, i.e., pressure at point B:	1.97 kg/cm²a

Next we need to size and measure the pressure drops across the branch lines. Let us calculate the branch line size between point B and PSV-009.

Calculations: (between points B and PSV-009) 1st Trial

Set pressure of PSV-009:	5.0 kg/cm²g
Flowrate (refer to Table 12.4)	8500 kg/hr
Pressure at point B:	1.97 kg/cm²a (i.e., P_2)
Temperature of relieving vapors:	90°C
Average molecular weight of relieving gases:	80
Compressibility factor:	0.99
Ratio of specific heats, C_p/C_v	1.4
Equivalent length of pipe (assumed for calculations):	30 m
Viscosity:	0.00001 kg/m.s (0.01 cP)
Roughness factor:	0.2 mm
Diameter of the pipe (assumed):	3" (80 mm)
N_{Re}	3757412
f	0.025
Mach number calculated (from Equation (12.9)):	0.45
P_2, pressure at point B:	1.97 kg/cm²a
P_1, pressure at PSV-009 (from Equation (12.8)):	3.55 kg/cm²a

The back pressure at PSV-009 is 3.55 kg/cm²a or 2.52 kg/cm²g. The set pressure of PSV-009 is 5.00 kg/cm²g. The ratio of back pressure to set pressure is 50.4%, which exceeds the upper limit recommended for balanced bellows type pressure relief valves.
So, let us go for a line size of 4" (100 mm).
Calculations: (between points B and PSV-009) 2nd Trial

Set pressure of PSV-009:	5.0 kg/cm²g
Flowrate:	8500 kg/hr
Pressure at point B:	1.97 kg/cm²a (i.e., P_2)
Temperature of relieving vapors:	90°C
Average molecular weight of relieving gases:	80
Compressibility factor:	0.99
Equivalent length of pipe (assumed for calculations):	30 m
Viscosity:	0.00001 kg/m.s (0.01 cP)
Roughness factor:	0.2 mm
Diameter of the pipe (assumed):	4" (100 mm)
N_{Re}	3006165
f	0.025
Mach number calculated (Equation (12.9)):	0.29
P_2, pressure at point B:	1.97 kg/cm²a
P_1, pressure at PSV-009 – by iteration:	2.50 kg/cm²a

Hence, the back pressure at PSV-009 is 2.50 kg/cm²a or 1.47 kg/cm²g. The set pressure of PSV-009 is 5.00 kg/cm²g. The ratio of back pressure to set pressure is 29.4% which is acceptable for balanced bellows type pressure relief valves. The Mach number of 0.29 is also acceptable.

***

Refer to [9] for a similar example.

12.9 SIZING OF PRESSURE RELIEF VALVES

Depending upon the flow being critical or sub-critical, equations for sizing of pressure relief valves in vapor or gas service fall under two different categories. For this, let us first understand the definitions of critical and sub-critical flow.

Critical Flow: "Critical flow" is also referred to as "choked flow" or "sonic flow." It can occur at a restriction in a line such as a relief valve orifice, where piping goes from a small branch into a larger header, where pipe size increases.

Critical flow pressure P_{cf} is the absolute pressure at the nozzle exit at sonic velocity. Further, suppose the upstream relieving pressure is P_1. The ratio P_{cf}/P_1 is called the "critical pressure ratio." As long as the pressure downstream of the relief valve is less than or equal to the critical flow pressure P_{cf}, the throughput will vary with the inlet pressure and be independent of outlet pressure. This is called "critical flow." If, on the other hand, the downstream pressure exceeds the critical flow pressure P_{cf}, the flow would be sub-critical [4].

The critical pressure ratio is given by:

$$P_{cf} = P_1 \left[\frac{2}{k+1} \right]^{\frac{k}{(k-1)}} \tag{12.10}$$

Pressure relief devices in gas or vapor service that operate at critical flow conditions may be sized using the following equation:

$$A_0 = \frac{W}{C K_d P_1 K_b K_c} \sqrt{\frac{tZ}{M_w}} \tag{12.11}$$

where the coefficient C is given by:

$$C = 0.03948 \sqrt{k \left[\frac{2}{k+1} \right]^{\frac{(k+1)}{(k-1)}}} \tag{12.12}$$

***

EXAMPLE 12.5 CALCULATE THE ORIFICE FOR A PRESSURE RELIEF VALVE WITH THE FOLLOWING PROCESS DATA

Gas	Propane
Relief rate, W	30000 kg/hr
Relieving temperature, t	80°C (= 353°K)
Set pressure	10 kg/cm²g
Over-pressure	10%
Back pressure, P_2	1.3 kg/cm²g [127.5 kPa(a)]

We calculate as follows:

Relieving pressure, P_1	= (10 x 1.1) + 1.033 = 12.033 kg/cm²a = 1180.3 kPa(a)
For propane, k	= 1.13

The critical flow pressure works out to

P_{cf}

$$= 1180.3 \left[\frac{2}{1.13+1} \right]^{\frac{1.13}{(1.13-1)}}$$

$$= 682.7 \text{ kPa(a)} = 581.4 \text{ kPa(g)}$$

Back pressure   = 1.3 kg/cm²g = 127.5 kPa(g)

Since the back pressure is less than the critical flow pressure P_{cf}, the relief valve sizing would be based on critical flow.

For calculating the orifice area, the parameters are as follows:

W   = 30,000 kg/hr

C

$$= 0.03948 \sqrt{1.13 \left[\frac{2}{1.13+1} \right]^{\frac{(1.13+1)}{(1.13-1)}}}$$

	= 0.0251
K_d	= 0.975 (for the case of PRV)
K_b	= 1.0 since the ratio of back pressure to set pressure is <30%
K_c	= 1.0 since no upstream rupture disk is installed
t	= 353°K
Z	= 0.9
M	= 44.09 (for propane)

The orifice area is calculated as:

A_0

$$= \frac{30,000}{0.0251 \times 0.975 \times 1180.3 \times 1 \times 1} \sqrt{\frac{353 \times 0.9}{44.09}}$$

$$= 2788 \text{ mm}^2 = 4.32 \text{ in}^2$$

***

Sub-critical Flow: When the back pressure of the safety valve exceeds the critical flow pressure P_{cf}, the flow through the pressure relief device would be sub-critical. Equation (12.13) can be used to calculate the required discharge area for the orifice [4].

$$A_0 = \frac{17.9W}{F_2 K_d K_c} \sqrt{\frac{tZ}{M_w P_1 (P_1 - P_2)}} \tag{12.13}$$

where,

$$F_2 = \sqrt{\frac{k}{k-1} r^{\left(\frac{2}{k}\right)} \left(\frac{1-(r)^{\left(\frac{k-1}{k}\right)}}{1-r}\right)}$$

r = ratio of back pressure to the upstream relieving pressure

***

EXAMPLE 12.6   CALCULATE THE ORIFICE FOR A PRESSURE RELIEF VALVE WITH THE FOLLOWING PROCESS DATA

Gas	Propane
Relief rate, W	25000 kg/hr
Relieving temperature, t	80°C (= 353°K)
Set pressure	4.0 kg/cm²g
Over-pressure	10%
Back pressure, P_2	2.8 kg/cm²g [376.0 kPa(a)]

We calculate as follows:

Relieving pressure, P_1 $= (4.0 \times 1.1) + 1.033 = 5.433$ kg/cm²a
$= 532.9$ kPa(a)

For propane, k $= 1.13$

The critical flow pressure
works out to

$$P_{cf} = 532.9 \left[\frac{2}{1.13+1} \right]^{\frac{1.13}{(1.13-1)}}$$

$= 308.3$ kPa(a) $= 207.0$ kPa(g)

Back pressure $= 2.8$ kg/cm2g $= 274.6$ kPa(g)

Since the back pressure is greater than the critical flow pressure P_{cf}, the relief valve sizing would be based on sub-critical flow.

For the orifice area, the parameters are as follows:

W $= 25,000$ kg/hr

r $= 376.0/532.9$
$= 0.706$

F_2

$$= \sqrt{ \frac{1.13}{1.13-1} (0.706)^{\left(\frac{2}{1.13}\right)} \left(\frac{1-(0.706)^{\left(\frac{k-1}{k}\right)}}{1-0.706} \right) }$$

$= 0.791$

K_d $= 0.975$ (for the case of PRV)

K_c $= 1.0$ since no upstream rupture disk is installed

t $= 353°$K

Z $= 0.9$

M $= 44.09$ (for propane)

The orifice area is calculated as:

A_0

$$= \frac{17.9 \times 25000}{0.791 \times 0.975 \times 1} \sqrt{ \frac{353 \times 0.9}{44.09 \times 532.9 \times (532.9 - 376.0)} }$$

$= 5387$ mm² $= 8.35$ in²

***

Standard Orifice Areas: API 526 [10] lists standard effective orifice areas which relief valves need to meet. Therefore, once the orifice areas have been calculated using Equations (12.11) or (12.13), they need to be corrected to the next standard orifice areas. Pressure relief valves are manufactured with these standard orifice

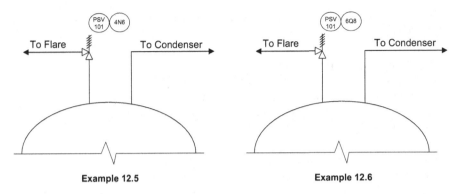

FIGURE 12.6 Relief valve orifice sizes.

areas. Each orifice area has a designation. For example, a PRV designated with an orifice "D" implies an orifice size of 0.110 in. An orifice "E" implies an orifice size of 0.196 in, and so on. For details, the reader may refer to API 526 [10].

In addition, once an orifice area has been specified, and material of construction and the inlet and outlet pressure ratings have been finalized, standard tables are available in API 526 [9] which provides further details. This information is useful for process engineers in adding expanders and reducers at the inlet and outlet of pressure relief valves in the P&IDs.

Example 12.5 shows the required orifice area of 4.32 in². This corresponds to an "N" type orifice with an area of 4.34 in². The pressure and temperature ranges in Example 12.5 correspond to an inlet flange rating of 150#. Hence, we go for the combination 4N6. The corresponding pressure relief valve with the "N" type orifice as typically is illustrated in a P&ID is shown in Figure 12.6.

Example 12.6 shows the required orifice area of 8.35 in². This corresponds to a "Q" type orifice with an area of 11.05 in². The pressure and temperature ranges in Example 12.6 correspond to an inlet flange rating of 150#. Hence, we go for the combination 6Q8. The corresponding pressure relief valve with the "Q" type orifice as typically illustrated in a P&ID is shown in Figure 12.6.

12.10 BLOWDOWN SYSTEMS

Blowdown [11] is a liquid stream containing water, oil or other chemicals, or a combination of these which are required to be drained from various process equipment in a process plant under various operating conditions such as normal operation, start-up, shutdown or emergencies. Since these streams are generally waste products, the handling and disposal of these often do not receive enough importance in plant design and operation. However, handling these blowdown streams in enhancing the safety of a process plant is equally important. The following section recognizes the various blowdown streams commonly encountered in the process industry and provides guidelines for their safe handling and disposal.

Aqueous Blowdown: These streams contain mainly water, with small amounts of hydrocarbons. Such streams are generally encountered during the normal operation of the units. These streams may be continuous or intermittent in nature. Some typical examples are:

- Water from the boots of reflux drums
- Water draw-off from oil separators
- Stripped water withdrawal from sour water strippers
- Water draw-off from crude oil tanks and other product tanks
- Boiler blowdown

Hydrocarbon Blowdown: These streams mainly contain hydrocarbons and may be encountered during normal operation, start-up, normal shutdown and emergency conditions. The streams are generally intermittent. Some typical examples are:

- Draining of process equipment during normal operation
- Drains from pumps, pipelines and manifolds
- Drains from sampling of products from equipment and piping
- Draining of equipment during planned shutdown
- Emergency draining of a process equipment like furnace, tower, vessel, etc.

Chemical Blowdown: These streams contain aqueous solutions of chemicals and may also contain small quantities of hydrocarbons. Such streams may be continuous or intermittent in nature. Some typical examples are:

- Caustic drains from treating plants
- Drains from chemical cleaning operations
- Effluents from water treatment plants
- Water draw-off from sour water strippers which may contain sulfides, H_2S, phenols, etc.
- Cooling tower blowdowns

12.11   HANDLING BLOWDOWN STREAMS

Aqueous Blowdown: All aqueous blowdowns which are continuous in nature and originate from pressure vessels are disposed to the oily water sewer through a level control valve. Figure 12.7 illustrates a typical scheme. Such a disposal method is valid for blowdowns which contain non-volatile hydrocarbons.

However, if the continuous aqueous blowdown is form a vessel operating at atmospheric pressure, automatic level indicator-cum-controller may not be necessary. Instead, a U-seal with a siphon breaker can be provided. Water draining from barometric condenser drums and water seal vessels in flare headers are some typical examples of this arrangement. Figure 12.7 also depicts a typical arrangement of such a scheme.

In cases where the aqueous blowdown is drawn from light ends reflux drums, there is a possibility of dissolved hydrocarbons going to the sewer and creating hazardous conditions. Under such conditions, the water draw-off should first be separated from the vapors in a vapor disengaging drum. The vapors released from the top of this drum are connected to the flare header and the water from the bottom of the drum, free from hydrocarbons, is disposed to the oily water sewer. Figure 12.8 illustrates a typical scheme.

Boiler Blowdown: Continuous blowdown from a boiler or a steam drum is usually at high temperatures and pressure. Therefore, such streams should be handled carefully to avoid hazards to operating personnel. If the boiler blowdown stream is at

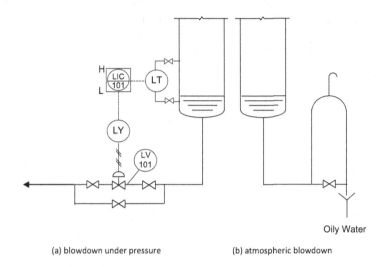

(a) blowdown under pressure   (b) atmospheric blowdown

FIGURE 12.7   Aqueous blowdown.

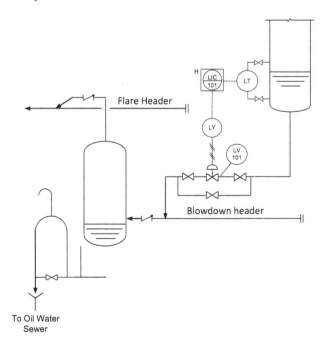

FIGURE 12.8   Aqueous blowdown containing volatile hydrocarbons.

high pressure, it is normally flashed into a lower pressure steam system before being disposed. Figure 12.9 illustrates a typical scheme.

Closed Hydrocarbon Blowdown: These blowdown streams contain predominantly hydrocarbons. They are intermittent in nature and are required to be handled during draining operations while the plant is under operation and also during planned or emergency shutdowns or start-up activities. The blowdown drum is located underground and vented to the flare. Different vessels/equipment are drained to this drum under gravity through a closed piping system. A submersible pump routes the contents of the drum to feed tanks, or to slops or even back to the process system from where they were drained. Figure 12.10 illustrates a typical scheme.

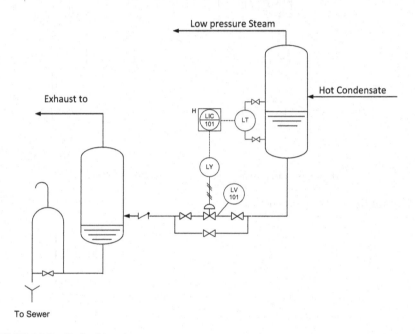

FIGURE 12.9   Boiler blowdown.

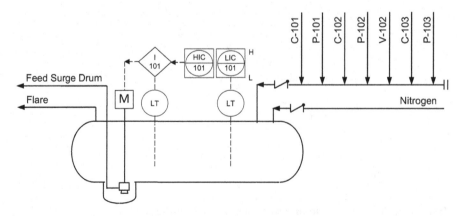

FIGURE 12.10   Closed hydrocarbon blowdown.

Symbols

A	Total wetted surface, m^2
A_1	Cylindrical wetted surface area, m^2
A_2	Head wetted surface area, m^2
A_o	Orifice area of pressure relief valve, mm^2
A_t	Total tube rupture area, in
C	Coefficient determined from an expression of the ratio of specific heats
D_c	Column diameter, m
D	Pipe inside diameter, m
f	Moody's friction factor
F	Environment factor, dimensionless
F_2	Coefficient for sub-critical flow
h	Wetted height, m
K_b	Capacity correction factor due to back pressure. For conventional and pilot-operated valves, $K_b = 1$, for balanced bellows valves, $K_b = 1$ as long as ratio of back pressure/set pressure < 30%
K_d	Effective coefficient of discharge = (0.975 for pressure relief valves)
K_c	Combination of correction factor for installations in combination with an upstream rupture disk (= 1.0 when no rupture disk is installed, = 0.9 when an upstream rupture disk is installed and the combination does not have a certified value)
k	ratio of C_p/C_v at relieving condition
l	Equivalent length of pipe, m
L	Wetted length of vessel, m
M_1	Mach number at inlet of pipe
M_2	Mach number at outlet of pipe
M_w	Molecular weight
N_{Re}	Reynolds number
P_{cf}	Critical flow pressure, psia (for heat exchanger tube rupture calculations)
P_{cf}	Critical flow pressure, kPa(a) (for pressure relief valve sizing calculations)
P_1	Pressure at high pressure side, psia (for heat exchanger tube rupture calculations)
P_1	Upstream relieving pressure, kPa(a) (for pressure relief valve sizing calculations)
P_1	Pressure at pipe inlet, $kg/cm^2(a)$ (for relief valve outlet piping calculations)
P_2	Pressure at low pressure side, psia (for heat exchanger tube rupture calculations)
P_2	Back pressure, kPa(a) (for pressure relief valve sizing calculations)
P_2	Pressure at pipe outlet, $kg/cm^2(a)$ (for relief valve outlet piping calculations)
P_{des}	Design pressure
P_{test}	Test pressure
Q	Total heat absorbed by the wetted surface of the vessel, W
Q_R	Reboiler duty at relieving condition, Kcal/hr
Q_O	Reboiler duty at operating condition, Kcal/hr
r	Ratio of backpressure to the upstream relieving pressure
T	Absolute temperature, °K

t	Relieving temperature of the gas or vapor, $^{\circ}K$
Z	Compressibility factor at relieving conditions, dimensionless
W	Relief rate of the pressure relief valve, kg/hr
α	Angle, radians
λ	Enthalpy of vaporization, Kcal/kg
λ_R	Latent heat of vaporization at relieving condition, Kcal/kg
π	Constant (= 3.14159)

REFERENCES

1. American Petroleum Institute, "Guide for Pressure-Relieving and Depressuring Systems," API 521, 6th edition, January 2014.
2. Oil Industry Safety Directorate, Government of India, Process Design and Operating Philosophies on Pressure Relief and Disposal System, OISD-Standard-106, August1999.
3. Mofrad, S. R., "Relief Rate Calculation for Control Valve Failure", *Hydrocarbon Processing*, January 2008.
4. American Petroleum Institute: "Sizing, Selection and Installation of Pressure-Relieving Devices, Part-1 Sizing and Selection," API 520, 9th edition., July 2014.
5. Crane Co., "Flow of Fluid", Technical Paper No. 410, 2018.
6. Wong, W. Y., "PRV Sizing for Exchanger Rupture", *Hydrocarbon Processing*, February 1992.
7. Aspen Technology Inc., "Equations and Example Benchmark Calculations for Emergency Scenario Required Relief Loads", *While Paper V8.8: Control Valve Failure, Heat Exchanger Tube Rupture, Hydraulic Expansion and Fire*, 2015.
8. Wong, W. Y., "Protect Plants against Overpressure", *Chemical Engineering*, June 2001.
9. Mukherjee, S., "Pressure-Relief System Design", *Chemical Engineering*, November 2008.
10. American Petroleum Institute: "Flanged Steel Pressure Relief Valves," API 526, 5th ed., 2002.
11. Oil Industry Safety Directorate, Government of India, Process Design and Operating Philosophies on Blowdown and Sewer System, OISD-Standard-109, November 1988.

13 Hazard and Operability Study

13.1 INTRODUCTION

The Hazard and Operability Study, also known as the HAZOP Study, is a detailed examination of a plant or a system to determine what would happen if the plant or the system were to operate in a way that is different from its normal design. In this step-by-step examination process, a systematic questioning is carried out for every part of the system to establish how deviations from design intent can arise, what could be the likely negative consequences (if any) in terms of hazard or operability of the system, what safeguards are provided in the system to take care of such consequences, and finally what actions need to be taken to remedy the situation. It is a technique in which a systematic investigation conducted by the HAZOP team leads to a brainstorming process. A team of experts from various engineering disciplines participates in the study [1, 2].

The concept of HAZOP was invented by Imperial Chemical Industries (ICI) in the United Kingdom. It gained widespread acceptance in the chemical process industry after the Flixborough disaster in 1974, in which 28 people were killed and hundreds of others injured.

13.2 DEFINITIONS AND METHODOLOGY

The following are some of the definitions of various terms used in the text:

- **Hazards:** Any operation or incident that could possibly cause a release of toxic, flammable or explosive chemicals or any action that could result in injury to personnel or loss of life and/or property.
- **Operability:** Any operation inside the design envelope that may cause an operational disruption or shutdown, possibly leading to a violation of environmental, health or safety regulations or negatively impacting profitability.
- **Intention:** How a process is expected to operate within the node.
- **Guidewords:** In the course of the questioning process and analysis, certain "guidewords" are used to identify deviations from the design intents of key process parameters. Examples of guidewords that are typically used in a HAZOP study include: no, less, more, reverse, other than, higher, lower, etc.

DOI: 10.1201/9780429284656-13

- **Deviations:** These are departures from design intentions. The above guide-words are used in conjunction with certain process parameters, viz. flow, pressure temperature, level, phase, etc. This leads to the generation of potential deviations. Typical examples of deviations include the following:
  - No Flow
  - Less Flow
  - More Flow
  - Reverse Flow
  - As Well as Flow
  - Other than Phase
  - High Pressure
  - Low Pressure
  - High Temperature
  - Low Temperature
  - High Level
  - Low Level.

Causes: Possible reasons why the deviation may occur.

Consequences: Results of the deviations which will affect either the operability or safety of the plant.

Safeguards: Features that could either prevent the cause or mitigate the consequences arising out of a particular deviation.

Methodology: The HAZOP team applies all relevant guidewords to the system in question and identifies the possible deviations from the design intent. The possible causes, effects and corrective/preventive measures are listed in the HAZOP report.

13.3 PREPARATORY WORK

Forming the HAZOP Team: The HAZOP team should be headed by a chairman, or study leader, who will be the person in charge of the study. The chairman should be fully conversant with HAZOP methodology and should have experience of HAZOP studies for similar plants. Ideally, he should not be directly associated with the project under study but should be capable of providing a wide-angle view. There should be a HAZOP secretary, whose main job is to record the HAZOP notes; therefore, he too needs to be a technical person.

The rest of the HAZOP team should ideally consist of engineers from various disciplines. In addition, it invariably should also consist of members from the client's organization. One representative should be present from plant operations having first-hand experience of day-to-day operations of a similar plant. In case the plant under review is in the design and engineering phase, the presence of the process lead engineer is necessary. To summarize, the team should be so selected that all questions raised during the meeting can be answered immediately rather than resorting to outside expertise, which is a time-consuming process.

The HAZOP team should consist of the following personnel:

- Chairman
- Secretary
- Project Manager (as required)
- Representative from the client's organization
- Process Lead Engineer
- Instrument Lead Engineer
- Process Operations Engineer
- Instrumentation Operations Engineer
- Piping Engineer (as required)
- Mechanical Engineer (as required).

Assembling Documents: Before beginning the study, all relevant documentation should be collected and made available. These should typically consist of:

- Process Description
- Process Flow Diagrams
- Piping and Instrumentation Diagrams (P&IDs)
- Equipment Layout Drawings
- Process Data Sheets of Equipment
- Process Data Sheets of Control Valves and Pressure Relief Valves
- Cause and Effect Diagrams
- Vendor Documentation of Package Units, viz., compressors, refrigeration units, etc.

Sequence of Study: It is not practically possible to carry out a HAZOP study of a plant as a whole. Therefore, the plant or system is split into sections referred to as "nodes." This is further discussed in Section 13.4.

The sequence of action for the HAZOP can be represented as below:

- Start with a particular node
- Select a process parameter (viz., flow)
- Apply a guideword describing deviation (viz., high flow)
- Discuss the possible causes
- Discuss the possible consequences
- Recommend remedial measures
- Decide upon actions to be taken
- Repeat the procedure with another guideword (viz., low flow).

Once all the guidewords have been applied, repeat the procedure with other parameters (viz., temperature, pressure, level, etc.). Also, once all the parameters have been covered, repeat the procedure with the other nodes.

Sub-dividing the Plant and Marking the P&IDs: As mentioned above, it is not practically possible to carry out a HAZOP Study of any unit as a whole. The P&IDs are therefore divided into sections referred to as "nodes." Nodes are numbered and marked in the order in which the process takes place, i.e., from upstream to downstream. Nodes also represent sub-systems within a P&ID. For example, a pump with its suction, discharge and spill-back lines may be grouped as a node. In a vessel, the inlet/outlet lines, including control valves, level instruments and safety valves may be grouped as a node.

Making the List of Relevant Guidewords: Having divided the plant into nodes and having marked the P&IDs, it should now be relatively easy to select the relevant set of guidewords. The process lead engineer or the operations engineer should check that the guidewords selected cover all aspects of the system.

Carrying out the Study: The stage is now set to start the study. Copies of the marked up P&IDs are distributed to the team members. It is important to mention certain guidelines at this stage as follows:

- Discourage random marking on the master P&IDs by team members.
- Discourage cross-talk and private discussions between team members.
- Try to involve team members who are keeping silent by prompting them with questions.
- If the discussion loses focus or direction, try to bring back the team to the heart of the problem.
- Sometimes, a specific topic leads to a series of discussions and results in a substantial amount of time devoted to it. In such a case, to avoid delay in the overall schedule, the overall responsibility of collecting additional data/information is best placed on a particular individual to be carried out outside the HAZOP meeting, allowing the meeting to proceed smoothly.
- The chairman should be an independent member and should not be perceived as favoring a particular section of the team. This is of particular importance when members from both the client and engineering teams are present.
- If the schedule is slipping, in order to catch up on time, certain team members tend to create shortcuts and dictate to the HAZOP secretary the deviations, causes, consequences, etc. instead of following the prescribed HAZOP procedure. This practice must be discouraged, as it is better to add a few hours to complete the study than compromise the quality of the final output.

13.4 SAMPLE HAZOP

Consider the P&ID of the Purification Column (Figure 2.9) described in Chapter 2. At the top, it consists of a condenser, reflux drum, reflux line back to the column and an overhead product line. At the bottom, it consists of a reboiler which sends vapors back to the column, as well as a bottom product line.

The Reflux Drum and reflux from Reflux Pump P-101 to Column C-101 is designated as Node 1.

We begin with the first deviation, i.e., "No Flow." When could there be a no flow situation? The tripping of the pump P-101 A/B could be one such cause. So, what could be the consequence? The first consequence could be an upset in the column in the form of the trays running dry. The second consequence could be a level rise in the Reflux Drum V-101. Are there any safeguards? Yes, a high-level alarm is provided in V-101, so the panel operator sees this alarm and gives suitable instructions to the field operator in case of a pump trip. What recommendation could be given? The first recommendation could be to provide a low flow alarm in the reflux line to the top of the column C-101. The other one could be to check that there is at least 5 minutes hold-up time between NLL and HLL in V-101 so that the operator can start the stand by pump before the vessel gets flooded.

TABLE 13.1
HAZOP Worksheet

Deviations	Causes	Consequences	Safeguards	Recommendations
No Flow	Pump P-101 trips	Upset in column C-101. Level rises in reflux drum V-101.	High level alarm provided in V-101.	Provide low flow alarm in reflux line.
Less Flow	Pump P-101 strainer chokes	Upset in column C-101. Level rises in reflux drum V-101.	High level alarm provided in V-101.	Include procedures for periodic cleaning of pump strainers in operating manual.
More Flow	Reflux line control valve stuck open	Possible flooding in column C-101. Level falls in reflux drum V-101.	Low level alarm provided in V-101.	Consider high differential pressure drop indicator across column.
As well as flow	Leakage in tubes of the condenser.	Not likely. Shellside is at higher pressure.		
Reverse Flow	Unlikely			
Low Level	Refer to More Flow.			
High Level	Pump P-101 trips.	Flooding in reflux drum and liquid backup into the condenser. Also refer to No Fow.	High level alarm provided in V-101.	Check that there is at least 5 minutes hold-up time between NLL and HLL in V-101 so operator can start the stand by pump before the vessel floods.
High Pressure	Flow control valve in reflux line stuck closed.	Pump P-101 goes to shut-off. Pump discharge pressure becomes 22 kg/cm²g. Pump discharge line is only 150# rating, i.e., A04P.		Change pump discharge line from 150# to 300# rating (i.e., from A04P. to B04P).
Low Pressure	Pump P-101 trips.	Refer to High Level and No Flow.		

Note: Node is Reflux Drum and reflux from Reflux Pump P-101 to Column C-101.

This completes the first deviation. Similarly, all other possible deviations would have to be examined along with the causes, consequences, safeguards and recommendations. This needs to be carried out for all nodes. Table 13.1 illustrates a sample HAZOP worksheet.

13.5   HAZOP STUDY CLOSEOUT

After the HAZOP is completed, time is given to the engineering team to work on the recommendations. These could be carrying out changes in the P&IDs and the process data sheets of equipment or instruments. If the technology is a matured one, chances are that the P&IDs have been subjected to HAZOP studies earlier and the changes are generally to ensure conformity with specific requirements and safe operating procedures within the company. However, if the technology is relatively new, there are changes that could be related to hardware and its design. In either case, after the completion of the time allotted to the engineering team, a HAZOP closeout is conducted where a review is carried out to ensure that all recommendations are incorporated in the engineering documents.

This marks the closeout of the HAZOP Study.

REFERENCES

1. Lihou, M., Hazard & Operability Studies, Lihou Technical & Software Services, 2017. www.lihoutech/com/hazop1.htm
2. Rausand, M., *System Reliability Theory*, 2nd Edition, Wiley, 2004.

14 Revamp Engineering and Capacity Augmentation

14.1 INTRODUCTION

In Chapter 1 we described the various aspects under basic engineering and detail engineering which are carried out for Greenfield projects. However, there is another category of engineering that relates to existing chemical process plants, i.e., revamp engineering. A chemical process plant is set up after due consideration of many factors such as market demand for the product and hence the plant capacity, plant cost, operating cost, space availability, space for future expansion, etc. However, once a plant has been built and has been in operation for a few years, if the demand for the product is rising, the owner starts pondering whether the capacity of the plant could be increased to meet the growing demand. In other words, he is looking at revamping his plant to augment capacity. In this chapter, we will talk about revamp engineering.

A plant could be revamped to achieve some of the following objectives [1]:

- Change feedstock
- Apply more energy-efficient technology
- Extend the life of the plant
- Increase the capacity of the plant.

In this chapter we will focus on the last list item: increase of plant capacity. The first step in capacity augmentation of a plant involves identifying and removing bottlenecks. In this connection, it is worth mentioning that several equipment and bulk components installed in the plant have design margins. Distillation trays (and hence column diameters) are normally specified with a 10% margin. Heat exchangers are also designed with margins of 5–10% on the surface area. Similarly, pumps are specified with a 10% margin. Likewise, control valves also have additional margins [2].

Going one step further, there are sometimes differences between what is specified and what is finally procured in the course of the project execution. While a 10% design margin on area is a standard practice for heat exchangers, during project execution, depending on standard tube lengths and diameters available, standard shell diameters and based on project and client standards, we might end up procuring exchangers with even higher margins.

DOI: 10.1201/9780429284656-14

The point to emphasize here is that there are inherent margins available in various equipment beyond what is required to run the plant at the design load. These are the margins that are utilized while increasing the plant throughput.

During a revamp activity, the process plant in question is studied in its entirety and the roadblocks that come in the way of increase in capacity are identified. The bottleneck could be a pipe that has a diameter smaller than recommended for the increased flow. It could be also a column where the revamp vapor liquid-traffic gives rise to an unacceptable pressure drop, or it may be a heat exchanger which has a surface area that falls short of the required future heat load prediction. It could even be a control valve which allows only a portion of the intended flow even at the full open condition. In most cases, it is a combination of some of these factors. In the worst case, most of the above bottlenecks could be present.

14.2   LINE HYDRAULICS AND PUMPS

In any capacity augmentation exercise, flows will increase and so will frictional losses. Losses can occur across lines, control valves, flow orifices and heat exchangers. Nobody would replace a heat exchanger just to minimize frictional losses. It is now a trade-off between the line, control valve and flow orifice. Experience shows that within a process unit, frictional losses across lines normally fall in the range of 0.1–0.2 kg/cm^2. For an increase in flow by 120%, the additional losses on account of increased flows work out to approximately 0.04–0.09 kg/cm^2. Such increase in frictional losses can easily be controlled. Once the flow increases, the opening of the control valve in the line also increases. So, the first step in arriving at a solution would be to see that the percentage opening of the control valve does not increase beyond the recommended 80%. If this happens, then the next step would be to replace the flow orifice with that of a higher diameter. The check here would be that the β ratio does not exceed the recommended 0.70. If this also fails, then the last resort would be to focus on the pump. Normally, process lines in the unit could have lengths of 40–50 m and replacing lines, apart from the cost of the hardware, would also incur costs for installation. Hence, for small revamps up to the order of 120–130%, line sizes are normally kept unchanged.

Debottlenecking a pump comprises of two types of changes: the flow and the head. Let us take a case study.

***

EXAMPLE 14.1

Consider the Feed Pump P-301 sending feed to a Pre-Fractionator from a Feed Surge Drum V-301 (Figure 14.1). The existing pump has the following specifications:

Rated flow   : 40 m^3/hr
Head   : 43 m (3.57 kg/cm^2)
Liquid density   : 830 kg/m^3
Liquid viscosity : 0.35 cP

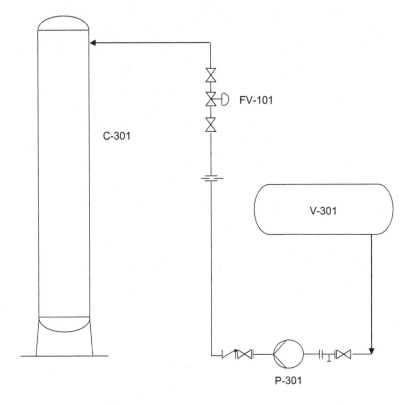

FIGURE 14.1   Debottlenecking a pump.

Refer to the pump curve illustrated in Figure 14.2. This is the same pump model illustrated in Figure 5.6, in Chapter 5, and this model fits with the requirements. The operating point is marked in the figure. In the existing pump, the above specifications are met with an impeller diameter of 192 mm, at a pump efficiency of 64%. The power consumption for the pump at shaft is given by (refer to Equation (5.11)):

$$kW_{shaft} = \frac{QH}{36.7\eta}$$

where,

kW_{shaft} = Shaft Power (kW)
Q = Pump Flowrate (m³/hr)
H = Pump Differential Head (kg/cm²)
η = Pump Efficiency

The shaft power of the pump works out to

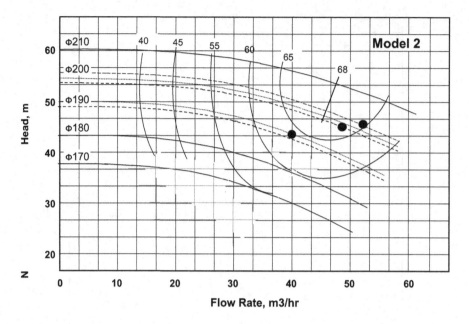

FIGURE 14.2 Pump characteristic curve (Model 2).

$$kW_{shaft} = (40 \times 3.57)/(36.7 \times 0.64) = 6.08\,kW$$

The shaft power is less than 22 kW; hence, the motor power should be 125% of the shaft power, i.e., 7.6 kW. The nearest standard motor rating is 9.3 kW, and the same was selected.

***

While executing a capacity augmentation, the easiest way to proceed is to go for a higher diameter impeller. This takes care of both the increase in flow as well as the head. The repercussions of this change could lead to the following:

- Change in only the impeller diameter
- Increase in motor rating, in view of a higher power consumption (pump model remaining unchanged since the same model accommodates a higher impeller diameter)
- Change in the pump model as well (in case the existing pump model is not able to accommodate the higher impeller diameter).

In order to keep the costs under control, attempt should be made to replace only the impeller while keeping the motor and pump unchanged. This could possibly be achieved in the following way.

In revamp activities, increase in flow requirements is inevitable. However, efforts should be made to minimize the increase in the head requirements in order to avoid unnecessary increases in power consumptions. One way to achieve this is to decrease the frictional losses. Frictional losses in pipes in typical process plants are of the order of 0.10–0.15 kg/cm². Existing pipes in the plant can normally take higher flows with acceptable increase in frictional losses (unless the existing pipes fall on the upper end of allowable velocities). The other losses are those across control valves. The allowable pressure drop across the control valve FV-101 (Figure 14.1) was checked. It was found that in the existing unit, an allowable pressure drop of 1.0 kg/cm² was specified. This could still be brought down to 0.7 kg/cm² in the revamp case (with a higher valve opening), and this is quite acceptable according to industry practice.

***

EXAMPLE 14.2

In line with the above approach, we have managed to balance the increase in the line losses in the pipe by reducing the allowable pressure drop across the control valve. The revamp pump will now have the following specifications under the revamp flow of 48 m³/hr:

Rated flow   : 48 m³/hr
Head   : 45 m (3.74 kg/cm²)
Liquid density   : 830 kg/m³
Liquid viscosity : 0.35 cP

It can be seen that the flow has increased by 20%, while the head has increased marginally. The other parameters have remained unchanged.

We find that under the revamp conditions, the flow and head can be met with an impeller of diameter of 198 mm at a pump efficiency of 66.5%. Hence, in order to satisfy the revised specifications, we would need to procure a new impeller of diameter 200 mm and get it trimmed to 198 mm (refer to Figure 14.2).

The shaft power of the pump under the revamp conditions works out to (refer to Equation (5.10))

$$kW_{shaft} = \frac{48 \times 3.74}{36.7 \times 0.665}$$
$$= 7.36\,kW$$

The shaft power is less than 22 kW, hence the motor power should be 125% of the shaft power, i.e., 9.2 kW. The existing motor is 9.3 kW. Hence, we need to change only the impeller, and the motor remains unchanged for this revamp case.

***

EXAMPLE 14.3

Now consider a revamp case with 30% additional flow. Under such revamp conditions, the revamp specifications of the pump have been worked out as:

Rated flow   : 52 m³/hr
Head   : 48 m (3.98 kg/cm²)
Liquid density   : 830 kg/m³
Liquid viscosity : 0.35 cP

We find that under the revamp conditions, the flow and head can be met with an impeller of diameter of 202 mm at a pump efficiency of 65.5%. Hence, we would need to procure a new impeller of diameter 210 mm and get it trimmed to 202 mm (refer to Figure 14.2).

The shaft power of the pump under the revamp conditions works out to (Equation (5.10)):

$$kW_{shaft} = \frac{52 \times 3.98}{36.7 \times 0.655}$$

$$= 8.61\,kW$$

The shaft power is less than 22 kW, hence the motor power should be 125% of the shaft power i.e., 10.8 kW. The existing motor is 9.3 kW. Hence in this case, we need to change not only the impeller but the motor as well.

The reader may note that we are now at the upper end of the pump curves, and further increase in capacity might call for a change of pump model.

***

14.3 COLUMNS

With the examples of revamp cases covered above, the reader would by now be in a position to understand the ways in which a column could be debottlenecked. First, let us understand the parameters where the hydrodynamics of a column is evaluated.

The process performance in terms of the separation and product quality is guaranteed by the process designer or process licensor (in case of a licensed process). However, the hydrodynamic performance of the column is guaranteed by the tray manufacturer or tray vendor (or the packing vendor, as the case may be). There are a number of parameters which contribute towards tray hydrodynamics. However, there are three groups of parameters which need to be satisfied to guarantee proper tray performance, namely:

- The total pressure drop across the column should not be more than the pressure drop specified by the process licensor.
- A proper choice of the tray active area, number of valves, downcomer area, downcomer clearance and weir height should ensure the desired pressure drop.
- The jet flood and the downcomer flood should also be within certain limits.

We will now discuss three specific examples below.

***

EXAMPLE 14.4

Let us take the case of a final distillation column in a petrochemical unit, C-205 (refer to Table 14.1). We will consider here the tray section 30–60 for our study. Refer to Case 1(a) for the existing vapor liquid traffic and the existing tray geometry. The column diameter is 2250 mm with a tray spacing of 450 mm. The existing trays in the column are equipped with 315 floating valves. The allowable pressure drop in this tray section is 0.20 kg/cm².

TABLE 14.1
Tray Geometry and Hydraulics (Example 14.4)

Item		C-205 Final Distillation Column								
Designation		1(a)			1(b)			1(c)		
Case No.										
Case		Original Traffic			Revamp Traffic			Revamp Traffic		
Tray Geometry		Original Geometry			Original Geometry			Higher Valve Count		
Column Section No.		1			1			1		
Tray Section		30–60			30–60			30–60		
Type of Tray		Floating Valves			Floating Valves			Floating Valves		
Percentage of Tray	%	50	100	110	50	100	110	50	100	110
Load										
Vapor to Tray	kg/hr	23296	46590	51250	27955	55908	61500	27955	55908	61500
Density	kg/m³	10.64	10.64	10.64	10.64	10.64	10.64	10.64	10.64	10.64
Vapor Viscosity	mPa-s	0.02	0.02	0.02	0.02	0.02	0.02	0.02	0.02	0.02
Liquid from Tray	kg/hr	22784	45568	50125	27341	54682	60150	27341	54682	60150
Density	kg/m³	586.3	586.3	586.3	586.3	586.3	586.3	586.3	586.3	586.3
Surface Tension	dyne/cm	10.44	10.44	10.44	10.44	10.44	10.44	10.44	10.44	10.44
Liquid Viscosity	mPa-s	0.113	0.113	0.113	0.113	0.113	0.113	0.113	0.113	0.113
Column										
Diameter	mm	2250			2250			2250		
Tray Spacing	mm	450			450			450		
Tray passes		1			1			1		
Column Area	m²	3.977			3.977			3.977		
Active Area	m²	2.915			2.915			2.915		
Flow Path Length	mm	1390			1390			1390		
Downcomer Chord Height top/bottom	mm	430/430			430/430			430/430		

(Continued)

TABLE 14.1 (Continued)
Tray Geometry and Hydraulics (Example 14.4)

Parameter	Units									
Weir Length	mm	1769.29			1769.29			1769.29		
No. of Valves		315			315			39.5		
Open Area	%	8.18			8.18			10.26		
Downcomer Clearance	mm	30			30			30		
Exit Weir Height	mm	45			45			45		
Weir Load	m³/hr.m	21.96	43.93	48.32	26.36	52.71	57.98	26.36	52.71	57.98
Downcomer Backup Liquid	mm	124.61	176.02	193.36	132.78	212.25	237.36	133.85	191.46	211.03
Downcomer Flood	%	17.60	35.19	38.71	21.11	42.23	46.45	21.11	42.23	46.45
Jet Flood Percent	%	32.41	61.79	67.97	38.37	74.15	81.57	38.37	74.15	81.57
Dry Tray Pressure Drop (per tray)	mm liq.	34.40	52.96	63.82	35.01	75.66	91.18	34.28	48.59	58.55
Total Tray Pressure Drop (per tray)	mm liq.	75.83	100.57	111.75	78.74	123.92	140.03	78.92	100.24	110.86
Total Tray Pressure Drop (all trays)	kg/cm²	0.133	0.177	0.197	0.138	0.218	0.246	0.139	0.176	0.195
Allowable Tray Pressure Drop (all trays)	kg/cm²	0.200			0.200			0.200		

It is now proposed to increase the capacity of this column by 20%. In other words, the vapor-liquid load would also increase in a similar proportion. The column diameter cannot be increased, since substantial costs are involved in procuring a new column. Let us think of solving this in another way. We try to investigate whether the same column diameter with identical tray geometry can handle the increased vapor-liquid traffic (refer to Case 1(b) of Table 14.1). It can be noticed that jet flood and downcomer flood are within reasonable limits. However, the pressure drop at 120% load exceeds the limit of 0.20 kg/cm².

This means that we are left with no other option but to try with a new tray geometry. We first try with an increased number of valves in the tray, i.e., with 395 valves instead of the existing 315 (refer to Case 1(c) of Table 14.1). We find that all parameters are now within the limit. The total pressure drop is now 0.195 kg/cm². This was, however, a simple case.

***

EXAMPLE 14.5

Let us take a case of a cyclopentane column in a refinery, C-105 (refer to Table 14.2). We will study the tray section 1–20 for our revamp case. Refer to Case 2(a) for the existing vapor-liquid traffic and the existing tray geometry. The diameter of the column is 2100 mm with a tray spacing of 600 mm. There are 430 floating valves in this section. The allowable pressure drop in this tray section is 0.15 kg/cm².

The revamp scenario calls for an increase in the capacity of this column by 20%. In principle, the vapor-liquid load would also increase by a similar proportion. As in Example 14.4, we first try to investigate whether the same column diameter with identical tray geometry can handle the increased vapor-liquid traffic (refer to Case 2(b) of Table 14.2). It can be seen that the downcomer flood, jet flood and tray pressure drop (at 120% load) are in excess of the specified/recommended limits.

Next, we try to address the problem with an increased number of valves per tray. Let us try with 835 fixed valves; fixed valves give a comparatively lower pressure drop compared to conventional floating valves (refer to Case 2(c) of Table 14.2). We now find that the pressure drop and the jet flood are within limits. However, the downcomer flood at 120% load is still 99.1%. This means the downcomer is not adequate to handle the increased liquid load and is therefore choking.

We now try by increasing the downcomer area. We execute this by increasing the downcomer chord height from 240 to 310 mm (refer to Case 2(d) of Table 14.2). We now find that all parameters are within the limit.

The hardware part needs to be elaborated here. In cases where only the number of valves needs to be changed in a tray, only the tray active panels need to be replaced (refer to Figure 14.3). No change is needed in the tray inlet panels since there are no valves in these panels below the downcomers. The downcomers also do not need any replacement since their geometry remains unchanged. However, when chord heights of the downcomers are to be changed, the weir lengths also get changed. As a result, in addition to the tray panels, the downcomers also need

TABLE 14.2
Tray Geometry and Hydraulics (Example 14.5)

Item	Unit	2(a) Original Case, Original Geometry			2(b) Revamp Case, Original Geometry			2(c) Revamp Case, Valve type changed			2(d) Revamp Case, Valve type changed, Downcomer area increased		
Column Section Number		1			1			1			1		
Tray Section		1–20			1–20			1–20			1–20		
Type of Tray		Floating Valves			Floating Valves			Fixed Valves			Fixed Valves		
Percentage Load	%	50	100	120	50	100	120	50	100	120	50	100	120
Vapor to Tray	kg/hr	26815	53630	64356	32178	64356	77227	32178	64356	77227	32178	64356	77226.9
Density	kg/m³	8.71	8.71	8.71	8.71	8.71	8.71	8.71	8.71	8.71	8.71	8.71	8.71
Vapor Viscosity	mPa-s	0.02	0.02	0.02	0.02	0.02	0.02	0.02	0.02	0.02	0.02	0.02	0.02
Liquid from Tray	kg/hr	22743	45486	54582	27292	54582	65498	27292	54582	65498	27292	54582	65498
Density	kg/m³	571	571	571	571	571	571	571	571	571	571	571	571
Surface Tension	dyne/cm	10.5	10.5	10.5	10.5	10.5	10.5	10.5	10.5	10.5	10.5	10.5	10.5
Liquid Viscosity	mPa-s	0.148	0.148	0.148	0.148	0.148	0.148	0.148	0.148	0.148	0.148	0.148	0.148
Column Diameter	mm	2100			2100			2100			2100		
Tray Spacing	mm	600			600			600			600		
Tray Passes		1			1			1			1		
Column Area	m²	3.464			3.46			3.46			3.46		
Active Area	m²	3.025			3.03			3.03			2.83		
Flow Path Length	mm	1620			1620			1620			1480		
Downcomer Chord Height top/bottom	mm	240/240			240/240			240/240			310/310		
Weir Length	mm	1336.26			1336.26			1336			149.93		
Number of Valves		430			430			835			835		

Parameter	Units	10.76			10.76			11.9			12.73		
Open Area	%	10.76			10.76			11.9			12.73		
Downcomer Clearance	mm	40			45			40			40		
Exit Weir Height	mm	50			50			50			50		
Weir Load	m3/hr.m	29.81	59.61	71.54	35.77	71.54	85.84	35.77	71.54	85.84	32.08	64.16	76.99
Downcomer Backup Liquid	mm	140.66	188.22	226.54	149.34	226.54	286.02	114.20	195.55	253.41	105.16	181.15	231.98
Downcomer Flood	%	34.41	68.82	82.59	41.29	82.59	99.10	41.29	82.59	99.10	28.44	56.88	68.25
Jet Flood Percent	%	30.35	59.30	71.17	35.75	71.17	85.40	31.31	62.61	75.13	32.86	65.72	78.86
Dry Tray Pressure Drop (per tray)	mm liq.	35.13	47.38	67.69	35.68	67.69	96.71	16.95	67.82	97.66	16.95	67.82	97.66
Total Tray Pressure Drop (per tray)	mm liq.	84.16	103.86	126.62	87.24	126.62	161.86	54.30	98.93	128.02	49.23	95.84	125.06
Total Tray Pressure Drop (all trays)	kg/cm²	0.096	0.119	0.145	0.10	0.14	0.18	0.062	0.113	0.146	0.056	0.109	0.143
Allowable Tray Pressure Drop (all trays)	kg/cm²			0.15			0.15			0.15			0.15

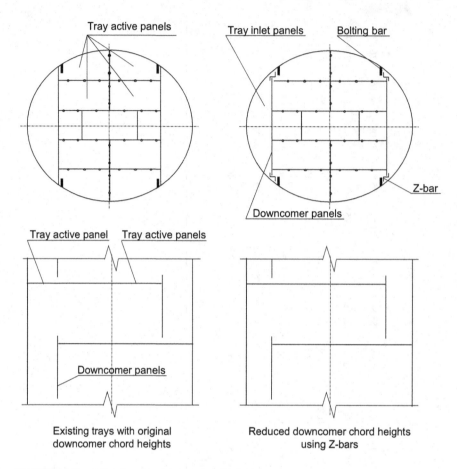

FIGURE 14.3 Adjusting downcomer chord heights.

Source: Adapted from *Indian Chemical Engineer*, October-December 2007.

to be replaced. We need to mention that the downcomers are bolted to strips of metal called "bolting bars." These bolting bars are welded to the column shell. It involves a lot of work to cut the existing bolting bars and weld new ones to become compatible with the new downcomer panels.

To overcome this problem, adaptors (also sometimes called "Z-bars") are fitted to bridge the gap between the existing bolting bars and the new down-comers. Refer to Figure 14.3 which is self-explanatory.

***

EXAMPLE 14.6

Let us take the case of a rectification column in a process plant, C-301 (refer to Table 14.3). Here, we will consider the tray section 1–30 for our study. Refer to Case 3(a) which describes the existing vapor liquid flows and the existing tray geometry. The column has a diameter of 2100 mm and a tray spacing of 450 mm. The trays are equipped with 368 floating valves. The allowable pressure drop for this tray section is 0.20 kg/cm².

TABLE 14.3

Tray Geometry and Hydraulics (Example 14.6)

Item		3(a)			3(b)			3(c)			3(d)		
Designation		C-301 Rectification Column											
Case No.		3(a)			3(b)			3(c)			3(d)		
Case		Original Case			Revamp Case			Revamp Case			Revamp Case		
Tray Geometry		Original Geometry			Original Geometry			Valve type changed			Valve type changed Downcomer area decreased		
Column Section Number		1			1			1			1		
Tray section		1–30			1–30			1–30			1–30		
Type of Tray		Floating Valves			Floating Valves			Fixed Valves			Fixed Valves		
Percentage Load	%	50	100	110	50	100	110	50	100	110	50	100	110
Vapor to Tray	kg/hr	27954	55909	61500	33545	67091	73800	33544.8	67090.8	73800	33544.8	67090.8	73800
Density	kg/m3	11.30	11.30	11.30	11.30	11.30	11.30	11.30	11.30	11.30	11.30	11.30	11.30
Vapor Viscosity	mPa-s	0.02	0.02	0.02	0.02	0.02	0.02	0.02	0.02	0.02	0.02	0.02	0.02
Liquid from Tray	kg/hr	19528	39055	42961	23434	37493	51553	23433.6	37493	51553.2	23433.6	37493	51553.2
Density	kg/m3	610	610	610	610	610	610	610	610	610	610	610	610
Surface Tension	dyne/cm	10.6	10.6	10.6	10.6	10.6	10.6	10.6	10.6	10.6	10.6	10.6	10.6
Liquid Viscosity	mPa-s	0.113	0.113	0.113	0.113	0.113	0.113	0.113	0.113	0.113	0.113	0.113	0.113
Column Diameter	mm	2100			2100			2100			2100		
Tray Spacing	mm	450			450			450			450		
Tray Passes		1			1			1			1		
Column Area	m2	3.464			3.464			3.464			3.464		
Active Area	m2	2.736			2.736			2.736			2.915		

(Continued)

TABLE 14.3 (Continued)
Tray Geometry and Hydraulics (Example 14.6)

Parameter		Unit												
Flow Path Length		mm	1420			1420			1420			1540		
Downcomer Chord Height	top/bottom	mm	340/340			340/340			340/340			280/280		
Weir Length		mm	1547.13			1547.13			1547.13			1427.73		
Number of Valves			368			368			650			650		
Open Area		%	10.18			10.18			10.24			9.61		
Downcomer Clearance		mm	30			30			30			30		
Exit Weir Height		mm	45			45			45			45		
Weir Load		m3/hr.m	20.69	41.38	45.52	24.83	39.73	54.63	24.83	39.73	54.63	26.91	43.05	59.19
Downcomer Backup Liquid		mm	120.59	168.48	184.58	128.32	180.68	225.15	92.40	150.11	193.13	98.81	157.53	204.93
Downcomer Flood		%	20.72	41.44	45.58	24.86	39.78	54.69	24.86	39.78	54.69	32.95	52.72	72.49
Jet Flood Percent		%	37.53	71.06	78.17	44.44	81.89	93.80	39.11	76.40	86.04	37.36	72.78	82.19
Dry Tray Pressure Drop (per tray)		mm liq.	33.05	50.58	60.95	33.64	72.26	87.08	17.23	68.92	83.40	17.23	68.92	83.40
Total Tray Pressure Drop (per tray)		mm liq.	73.53	97.13	107.76	76.37	114.80	134.52	42.56	89.93	107.10	43.26	91.73	109.14
Total Tray Pressure Drop (all trays)		kg/cm2	0.135	0.178	0.197	0.140	0.210	0.246	0.078	0.165	0.196	0.079	0.168	0.200
Allowable Tray Pressure Drop (all trays)		kg/cm2	0.2			0.2			0.2			0.20		

The column is undergoing a capacity augmentation by 20%. This implies that the vapor-liquid load would also increase accordingly. Refer to Table 14.3, Case 3(b). Here, a rating check has been carried out for the augmented vapor-liquid traffic with the existing column diameter and tray geometry. It can be noticed that the jet flood and the tray pressure drop (at 110% load) both exceed specified/recommended limits.

To tackle the problem, let us first try with increased number of valves per tray and 650 fixed valves (refer to Case 3(c)). After increasing the number of valves, the pressure drop across the tray section is within limits. However, the jet flood at 110% load is still a matter of concern. This means the active area is inadequate to handle the increased vapor load and is thus resulting in a high jet flood.

Our final try is therefore with an increase in the active area. We do this by decreasing the downcomer chord height from 340 to 280 mm (refer to Case 3(d)). This is achieved by the use of "adaptors" (refer to Figure 14.3). We now notice that all parameters are within limits.

***

Summary: Example 14.4 was a simple one. Only a change in the number of valves took care of the problem. In Examples 14.5 and 14.6, the open areas were increased to solve the problem of high pressure drop.

However, this was only one part of the problem. In Example 14.5, there was the additional problem of downcomer flood. In other words, the liquid load was too high for the downcomer to absorb. Therefore, the downcomer area had to be increased by increasing the downcomer chord height. Similarly, in Example 14.6, the added problem was that of jet flood. The vapor load was too high for the existing active area. The active area, therefore, had to be increased by decreasing the downcomer chord height. For further reading, the reader is referred to [3].

14.4   CONTROL VALVES AND INSTRUMENTATION

There are many interactions between a process engineer and his instrumentation counterpart during detailed engineering.

***

EXAMPLE 14.7

A plant is undergoing revamp with a capacity augmentation of 25%. The flow-rates in various sections of the plant therefore have also increased by approximately a similar proportion. We take the example of one such line feeding reflux to a column (Figure 14.4):

Present scenario
Flowrate: 40 m^3/hr
Line size: 4" (100 mm)
C_v of control valve: 95 (opening of 66.2%)
Density of liquid: 800 kg/m^3
Viscosity of liquid: 0.70 cP

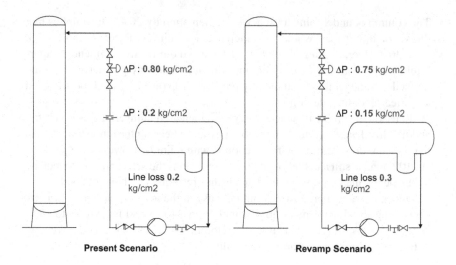

FIGURE 14.4   Revamp scenario of a reflux loop.

ΔP across control valve: 0.8 kg/cm²
ΔP across flow orifice: 0.2 kg/cm² (19620 N/m²)
Orifice diameter: 56 mm
Line loss: 0.2 kg/cm²

Revamp scenario
Flowrate: 50 m³/hr (0.01389 m³/s)
Line size: 4" (100 mm)
Density of liquid: 800 kg/m³
Proposed ΔP across control valve: 0.8 kg/cm²
Proposed ΔP across flow orifice: 0.1 kg/cm² (9810 N/m²)
Line loss: 0.3 kg/cm²

It may be noticed that the process engineer has not changed the line size since the existing 4" line is enough to convey the revamp flowrate. The intention is to make minimum changes. The pump is therefore kept unchanged. The control valve was operating at 66.2% earlier and it is expected that it could still convey the revamp flowrate with a slightly higher opening. Hence, the ΔP across the control valve has not been changed either. The line losses have gone up by 0.1 kg/cm². Hence, the process engineer has reduced the drop across the orifice to 0.1 kg/cm², in order to keep the total hydraulics unchanged.

The instrument engineer calculated the orifice area required to convey a flow of 50 m³/hr through a 4" line, at a ΔP of 0.1 kg/cm².

This is how it was calculated. From Equation (11.5), in Chapter 11, we proceeded as follows:

C_d was calculated from Table 11.1 as 0.606

$$0.01389 = \frac{0.606 \times \pi / 4 \times D_2^2 \left[2 \times (9810 / 800)\right]^{0.5}}{\left(1 - \beta^4\right)^{0.5}}$$

$$0.00589 = \frac{D_2^2}{\left[1 - (D_2 / 0.1)^4\right]^{0.5}}$$

Solving, $D_2 = 71.3$ mm

This means, the β ratio is greater than 70%, and hence is not recommended. Therefore, the process engineer has to go for a higher ΔP across the orifice in order to get a β ratio <70%. He now selects a higher ΔP of 0.15 kg/cm² (14715 N/m²). The calculations are as follows:

Repeating the same calculations, $D_2 = 65.8$ mm. The β ratio works out to 65.8% and hence is acceptable.

The line loss has gone up by 0.1 kg/cm² (0.2 to 0.3 kg/cm²) and the ΔP across flow orifice has gone down by 0.05 (0.2 to 0.15 kg/cm²).

Therefore, the deficit of 0.05 kg/cm² has to be adjusted in the ΔP across the control valve. Consequently, the ΔP across the control valve needs to be brought down from 0.80 to 0.75 kg/cm². But still, we need to check the percentage opening.

From Equation (11.6), we have the following equation:

$$C_v = 1.1674 \, Q_l \left(\frac{G_L}{p_1 - p_2}\right)^{0.5}$$

Substituting the values, we have,

$$C_v = 1.1674 \times 50 \left(\frac{0.8}{0.75}\right)^{0.5} = 60.28$$

From Table 11.3, for the calculated C_v of 60.28, the percentage opening for the existing 3" valve with a C_v of 95 works out to 73.5%, which is acceptable.

***

EXAMPLE 14.8

Another frequently encountered problem that instrument engineers face is that of β ratio in flow orifices. Consider the following example.

A flow orifice is specified to deliver a flow of 38 m³/hr. The line size is 100 mm. Due to some hydraulic constraints, the pressure drop is limited to 0.05 kg/cm². The density of the fluid is 990 kg/m³ and the viscosity 0.8 cP.

Data:

D_1 $= 100\ mm = 0.10\ m$

ΔP $= 0.05\ kg/cm^2$

         $= 0.05 \times 10000 \times 9.81\ N/m^2$

         $= 4905\ N/m^2$

Q_L $= 38\ m^3/hr$

         $= 38\ /\ 3600$

         $= 0.0106\ m^3/s$

From Equation (11.5), we proceeded as follows:

C_d was calculated from Table 11.1 as 0.606.

$$0.0106 \quad = \frac{0.606 \times \pi\ /\ 4 \times D_2^2\ [2 \times (4905\ /\ 990]^{0.5}}{\left(1 - \beta^4\right)^{0.5}}$$

$$0.00707 \quad = \frac{D_2^2}{\left[1 - (D_2\ /\ 0.1)^4\right]^{0.5}}$$

Solving for D_2 we get:

D_2 $= 0.0760\ m$

         $= 76.0\ mm$

The problem here is that the β ratio is 0.76, which is higher than the recommended value of 0.70. Since we have a limitation on the pressure drop, we cannot increase the same. The only option is to increase the line size.

We now take the next higher line size of 150 mm.

Applying the same procedure, we get:

D_2 $= 0.0822\ m$

         $= 82.2\ mm$

The β ratio is 0.548, which is acceptable.

***

This chapter provided to the readers a flavor of what goes on in the capacity augmentation of a process unit. Revamping a process plant is a different ball game and is more complicated compared to the normal design and engineering of a Greenfield unit. It calls for experienced hands to execute the changes. Having said that, this process is definitely interesting since it involves a lot of thought and brainstorming.

Symbols

β	Beta ratio
D_1	Pipe diameter (line size), m
D_2	Orifice diameter, m
G_L	Specific gravity of liquid
kW_{shaft}	Power consumption at shaft, kW
p_1	Pressure upstream of control valve, kg/cm²a
p_2	Pressure downstream of control valve, kg/cm²a
ΔP	Pressure drop across control valve, kg/cm²
ΔP	Pressure drop across orifice, N/m²
ρ	Density of the fluid, kg/m³
Q	Pump flowrate, m³/hr
Q_L	Volumetric flowrate of the liquid, m³/s
C_d	Discharge coefficient.

REFERENCES

1. Reddy, K. V., "Chemical Process Plants: Plan for Revamps", *Chemical Engineering*, December 2015.
2. Mukherjee, S., "Revamping your Process Plant," *Chemical Engineering*, April 2012.
3. Mukherjee, S., "Debottlenecking a Distillation Column for Capacity Augmentation," *Indian Chemical Engineer,* Vol. 49, No. 4, October–December 2007.

15 Interaction and Coordination with Detail Engineering Disciplines

15.1 INTRODUCTION

Engineering work cannot be executed in isolation. Basic engineering for a medium-size unit could involve 8–12 engineers at peak load. For a large process complex having multiple units, the basic engineering would involve engineers in that proportion for each of the units. Detail engineering activities for a large complex could involve several hundred engineers. Further, at the construction site, the number of workers and staff could easily be a few thousands.

Even after the process engineer has released all his deliverables, his job is far from over. There would continue to be many questions from various disciplines, particularly, piping, instrumentation and mechanical during the engineering phase. Once the taskforce has shifted for construction and commissioning activities, such queries would start coming from the construction site. This chapter gives the reader a glimpse into such queries and how resolutions are achieved.

15.2 HEAT EXCHANGERS

The heat exchanger design cycle calls for substantial interaction with the engineers of the mechanical department. The process engineer carries out the thermal calculations of the heat exchanger that satisfies the process requirements emerging from the process simulation. The output from the process engineer is a preliminary process data sheet (PDS) which also includes a preliminary tube layout drawing and a sketch of the setting plan for the heat exchanger.

The preliminary PDS issued by the process engineer is thereafter sent for a check by the mechanical engineer. During the mechanical design check, the design conditions and the geometry of the exchanger specified in the preliminary PDS are used as a basis to estimate various mechanical details such as the thicknesses of the shell, head and tubesheet. In the course of this check, the mechanical engineer also refers to various engineering standards such as TEMA, ASME, etc. The mechanical design check could also result in minor changes in the geometry, such as the number of tubes and the tube layout. This information is then fed back to the process engineer who might adjust his thermal design to suit the feedback from the mechanical design. The final PDS is then issued by the process engineer.

On the basis of the data in the final PDS, the mechanical design output is presented in the form of a mechanical data sheet (MDS). This MDS is sent to the fabricator (also sometimes called the "vendor") for detailed mechanical design followed by fabrication. The fabricator carries out the detailed mechanical calculations based on the guidelines given in the specified engineering standards. In this step, all the fabrication-related issues are taken care of. In fact, some of the details such as tube layout, inlet and outlet baffle spacing, etc. may even be adjusted by the vendor to meet the guidelines specified in the MDS and other engineering standards.

The fabricator generates the fabrication drawings (sometimes also called the "workshop drawings"). These drawings are now sent back to the mechanical engineer who now carries out a detailed check of these drawings. It is important that there is no deviation on the process specifications with regard to geometry (e.g., TEMA type, type of baffle, baffle orientation, etc.) in vendor documents. It might so happen that the vendor might have made slight adjustments in the baffle spacing (central, inlet and outlet) and the baffle cut for ease of fabrication. These need to be approved by the process engineer. If there has been a good coordination between the three parties, i.e., the process engineer, mechanical engineer and fabricator, this stage of vendor drawing review normally would go through without any concerns. If there are no further comments, then the fabricator is given a "go-ahead" for fabrication.

15.2.1 CASE STUDY

Problem Statement: Consider a hydrocarbon exchanger in an oil refinery with the following process data:

Shellside Inlet	*Shellside Outlet*
Hydrocarbon vapor-liquid mixture	Hydrocarbon vapor-liquid mixture
Mass fraction vapor: 0.075	Mass fraction vapor: 0.014
Pressure: 5.0 bar (g)	Temperature: 41.0°C
Temperature: 64.0°C	Flowrate: 119,000 kg/hr
Flowrate: 119,000 kg/hr	
Tubeside Inlet	*Tubeside Outlet*
Cooling water	Cooling water
Pressure: 5.0 bar (g)	Temperature: 41.0°C
Temperature: 31.0°C	

Allowable shellside pressure drop: 0.25 bar
Allowable tubeside pressure drop: 0.60 bar
Table 15.1 specifies the principal process parameters for this study.

Design Step 1 (preliminary thermal data sheet, TDS): Based on the process data specified in Table 15.1, thermal design was carried out by the process engineer. The output of the thermal design is illustrated in the form of a preliminary TDS, Table 15.2 which also includes an exchanger setting plan (Figure 15.1) and the tube layout (Figure 15.2).

Design Step 2 (mechanical design check): The preliminary TDS, along with the preliminary tube layout drawing generated during the thermal design, was next sent to the mechanical engineer for review and comments. Findings were as follows:

* In the preliminary TDS, the tube bundle to shell clearance was 33.4 mm (Figure 15.2). The mechanical engineer recommended 44.0 mm as a minimum based on engineering standards.

TABLE 15.1
Process Parameters

No.	Parameter	Shellside	Tubeside
1	Fluid	Hydrocarbon	Cooling Water
2	Flowrate, kg/h	1,19,000	To be calculated
3	Mass vapor fraction in/out	0.075/0.014	0.0/0.0
4	Temperature in/out, °C	64/41	31/41
5	Heat duty, MW	2.39	
6	Inlet pressure, bar (g)	5	5
7	Allowable pressure drop, bar	0.25	0.60
8	Fouling resistance, m²K/W	0.00026	0.00036
9	Material of construction	Carbon steel	Carbon steel

TABLE 15.2
Preliminary Thermal Data Sheet

1	Type of heat exchanger	AES
2	Shell inside diameter, mm	1050
3	Surface area, m²	359
4	Over-design on area, %	7.0
5	Calculated pressure drop, bar	SS:0.23/TS:0.50 (allow -0.60)
6	Tube dia. x wall thickness, mm	19.05 x 2.11
7	Tube length, m	6
8	Layout angle/pitch, deg/mm	90/25.4
9	Tube count/pass	1032/4
10	Baffle type/number	Single segmental - parallel/16
11	Baffle spacing (central/inlet/outlet), mm	316.3/468.5/595.5
12	Baffle cut, %	27

- As a result, the tube bundle diameter got reduced, and it is therefore not possible to accommodate 1032 tubes (as proposed in the preliminary TDS) within a shell of diameter 1050 mm and with the given tube layout. Only 992 tubes are possible. Figure 15.3 illustrates the 40 tubes (black shaded) that cannot be accommodated as a result of an increase in the tube bundle to shell clearance.
- In the TDS, a preliminary tubesheet thickness of 45 mm was chosen. The mechanical engineer calculated a thickness of 60 mm.

Design Step 3 (performance check based on comments from mechanical design check): Based on the recommendations from the mechanical design check, the tube count was reduced from 1032 to 992. In addition, the bundle to shell clearance was increased from 33.4 to 44.3 mm. Also, the tubesheet thickness was increased to 60.0 mm.

A final thermal design check was again carried out with the above changes. This resulted in a heat exchanger surface area of 343 m² and an over-design (based on surface area) of 2.56%. This over-design is a bit too low and hence not recommended.

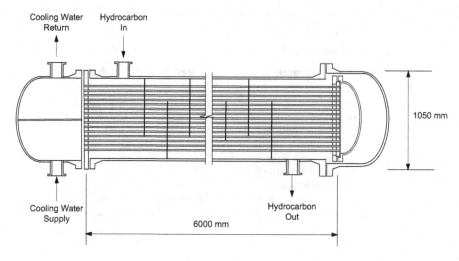

FIGURE 15.1 Exchanger setting plan.

Table 15.3 illustrates the performance of the exchanger after incorporating comments from the mechanical engineer.

Design Step 4 (fine-tuning of thermal design and release of final TDS and MDS): In order to arrive at a reasonable over-design (say, at least 5%), the shell diameter was fine-tuned from 1050 to 1075 mm, while retaining the length of the tubes as 6 m. The number of tubes as a result of the increased diameter was now 1064. After incorporating the above-mentioned changes, we arrived at an over-design of 7.8%, which is reasonable.

Table 15.4 illustrates the revised heat exchanger geometry based on the final TDS. Figure 15.4 illustrates the revised tube layout.

The final TDS, including the setting plan and the tube layout, was then released for mechanical design and generation of the MDS.

Design Step 5 (vendor design): Based upon the final MDS (which was prepared with data from the final TDS), the vendor carried out the detailed calculations. Certain dimensions, although minor, are changed by the vendor without consulting the process engineer. Table 15.5 illustrates these changes in the heat exchanger geometry. The changes are also illustrated in Figures 15.5 and 15.6, respectively, submitted in the form of drawings by the vendor.

The vendor considered the baffle type and orientation as mentioned in the MDS. However, the inlet, outlet and central baffle spacings calculated by the vendor were slightly different from those mentioned in the MDS.

This was type "AES" heat exchanger. During the thermal design calculations, a full support plate at floating head was considered (Figure 15.6). The distance between the full support plate and the floating head was assumed as 101 mm by the calculation tool. However, according to the vendor's design, the distance between the full

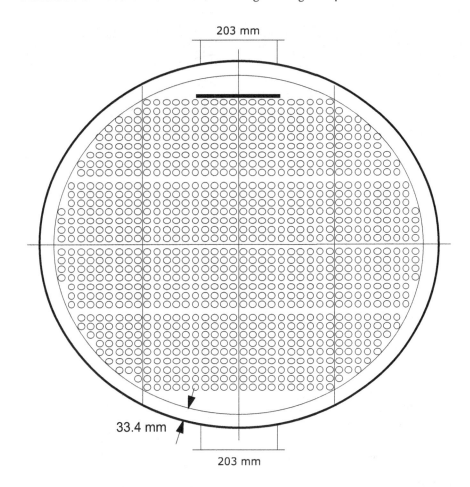

Shell inside diameter: 1050 mm   Outer tube limit: 983.2 mm
Tube type : Plain   Tube diameter : 19.05 mm
Tube pitch : 25.40 mm   Tube layout angle : 90°
Tube count : 1032   Baffle cut % : 27

FIGURE 15.2   Tube layout based on preliminary thermal design.

support plate and the floating head was 260 mm. This means that the dead zone between the floating head and the support plate was higher than that considered initially, resulting in reduced effective area for heat transfer.

Further, the baffle cut mentioned in TDS was 27%. The vendor had however considered a baffle cut of 26%. According to the vendor drawing, the baffle edge passes along the tube center line. From the fabrication point of view, this is the right way,

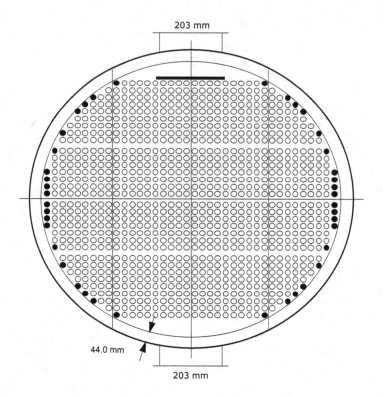

Shell inside diameter: 1050 mm   Outer tube limit: 983.2 mm
Tube type : Plain   Tube diameter : 19.05 mm
Tube pitch : 25.40 mm   Tube layout angle : 90°
Tube count : 992   Baffle cut % : 27

FIGURE 15.3   Tube layout based on comments of mechanical design check.

TABLE 15.3
Thermal Design with Comments from Mechanical Engineer

1	Type of heat exchanger	AES
2	Shell inside diameter, mm	1050
3	Surface area, m²	343 (originally 359)
4	Over-design on area, %	2.56 (originally 7.0)
5	Calculated pressure drop, bar	SS:0.23/TS:0.54 (allow -0.60)
6	Tube dia. x wall thickness, mm	19.05 x 2.11
7	Tube length, m	6
8	Layout angle/pitch, deg/mm	90/25.4
9	Tube count/pass	992/4
10	Baffle type/number	Single segmental - parallel/16
11	Baffle spacing (central/inlet/outlet), mm	314.3/468.5/595.5
12	Baffle cut, %	27

TABLE 15.4
Final Thermal Design

1	Type of heat exchanger	AES
2	Shell inside diameter, mm	1075 (originally 1050)
3	Surface area, m²	368 (originally 359)
4	Over-design on area, %	7.8 (originally 7.0)
5	Calculated pressure drop, bar	SS:0.22/TS:0.48 (allow -0.60)
6	Tube dia. x wall thickness, mm	19.05 x 2.11
7	Tube length, m	6
8	Layout angle/pitch, deg/mm	90/25.4
9	Tube count/pass	1064 (originally 1032)/4
10	Baffle type/number	Single segmental - parallel/16
11	Baffle spacing (central/inlet/outlet), mm	316.8/461/588
12	Baffle cut, %	27

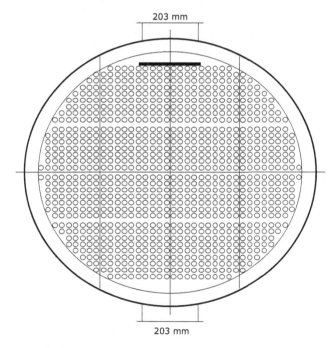

Shell inside diameter: 1075 mm   Outer tube limit: 985.0 mm
Tube type : Plain   Tube diameter : 19.05 mm
Tube pitch : 25.40 mm   Tube layout angle : 90°
Tube count : 1064   Baffle cut % : 27

FIGURE 15.4 Tube layout according to final thermal design.

TABLE 15.5

Adjusted Thermal Performance Based on Vendor Design

1	Type of heat exchanger	AES
2	Shell inside diameter, mm	1075 (originally 1050)
3	Surface area, m²	358 (originally 359)
4	Over-design on area, %	4.9 (originally 4.0)
5	Calculated pressure drop, bar	SS:0.22/TS:0.48 (allow -0.60)
6	Tube dia. x wall thickness, mm	19.05 x 2.11
7	Tube length, m	6
8	Layout angle/pitch, deg/mm	90/25.4
9	Tube count/pass	1064 (originally 1032)/4
10	Baffle type/number	Single segmental- parallel/16
11	Baffle spacing (central/inlet/outlet), mm	316.8/461/588
12	Baffle cut, %	26

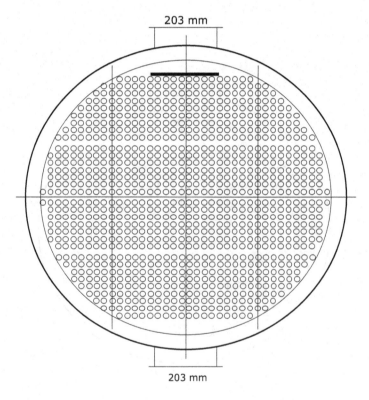

Shell inside diameter: 1075 mm   Outer tube limit: 985.0 mm
Tube type : Plain   Tube diameter : 19.05 mm
Tube pitch : 25.40 mm   Tube layout angle : 90°
Tube count : 1064   Baffle cut % : 26

FIGURE 15.5   Final tube layout by vendor.

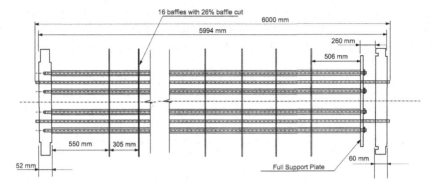

FIGURE 15.6   Tube bundle assembly drawing by vendor.

Source: Reprinted by special permission from *CHEMICAL ENGINEERING*, Copyright© March 2016, by Access Intelligence, Rockville, MD 20850.

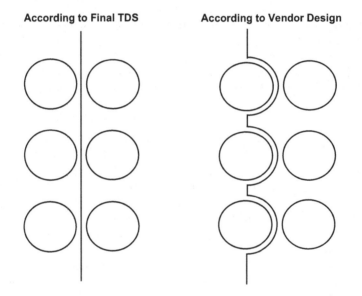

FIGURE 15.7   Baffle cut.

and the baffle edge should pass through the central line of tubes and not through the gaps between the tubes (Figure 15.7).

Design Step 6 (final process cross-check): In view of the fact that the vendor had provided a slightly different baffle cut, baffle spacings, and has changed the distance between the full support plate and the floating head, a final process cross-check was carried out to check the thermal performance. The baffle arrangement affects the shellside flow pattern and therefore the shellside heat transfer coefficient. It is therefore important that the actual baffle arrangement provided by the vendor be checked with respect to thermal performance.

A revised thermal check was carried out and the process engineer was satisfied that the changes in process performance as a result of the changes in the vendor drawings were not significant. If the design steps 1–5 are systematically carried out, this final step is normally a formality.

In summary, it is important for a process engineer to have a solid understanding of the heat exchanger fundamentals and the process requirements to ensure an effective heat exchanger selection and its thermal design. A good coordination between the process engineer, mechanical engineer and vendor is important to achieve a smooth design cycle. This avoids unnecessary last minutes changes and ultimately leads to smooth overall project execution. Refer to Mukherjee [1] for similar examples.

15.3   COLUMN TRAYS

In today's world, projects are executed on a fast-track basis. Project durations of 20–24 months from ordering to the stage of "ready for start-up" are not uncommon, even for medium-size process units. Under this scenario, it is important that activities like engineering, fabrication and delivery to site of long delivery items like columns are meticulously planned [2].

Once the client places the order for project execution on the EPC contractor, it is important that engineering and procurement activities for columns are taken up at the earliest. One of the important activities is to get a confirmation of the column diameters from prospective tray vendors. While tray vendors normally concur with the diameters specified in the basic engineering package, in a one-off case, the tray vendor may propose a diameter increase by 50–100 mm. Therefore, it is important to get this parameter confirmed at an early stage. Once this issue is settled, the activities for engineering and procurement of the columns should be taken up.

Ordering trays and internals is normally taken up after the columns are ordered. However, once these are ordered, there should be an exchange of information between the tray and the column vendor. The tray vendor designs items such as the tray support rings, seal pan, etc. These however, being welded items, form part of the column shell. Similarly, top feed/reflux arrangement may need certain welded attachments on the shell. Tray downcomers are designed and supplied by tray vendors. However, they are fastened to metallic strips called "bolting bars," which are also welded items and therefore part of the column vendor's scope. Likewise, the tray vendor designs the distributor pipes for feed and reflux; however, in order to design these pipes, he should know the orientation of the corresponding nozzles, which are fabricated along with the column shell. Figures 9.5 and 9.6 illustrate certain welded parts which are included in the column vendor's scope but are designed based on guidelines by the tray vendor. It is therefore important that the tray vendor reviews the drawings of the column vendor. Similarly, the column vendor should also review the drawings of the tray vendor in order to get an overall picture of the internals and also identify possible mistakes committed by the latter.

The EPC contractor should take the lead in coordinating this exchange of information, while also ensuring his active participation. A disciplined coordination between the EPC contractor, tray and column vendors ensures a smooth project execution, minimizing possible problems that could arise during tray erection and column commissioning.

15.3.1 CASE STUDY

Based on the process data sheets for trays and packings, the vendor submits his calculations for review/approval. Table 9.6 illustrates a typical vendor's calculation for the sieve tray column based on specifications furnished in the process data sheet in Table 9.5. The following details need to be checked by the process engineer:

- Vapor-liquid traffic as specified in the process data sheet
- Operating conditions
- Turndown/Over-design
- Column diameter
- Number of trays, tray type, tray spacing
- System factor based on foaming characteristics
- Downcomer backup
- Foam height
- Liquid height over exit weir
- Tray pressure drop

Let us begin systematically by analyzing Table 9.6 line-by-line. Tray sections 1–20 of the vendor's calculations have been chosen for analysis.

Lines 3–18: In this part of the tray calculations, the tray loading data of the process data sheet (Table 9.5) are reproduced. There is nothing specific to check here apart from possible errors committed by the vendor in entering data.

Lines 20–30: This section of tray calculations deals with tray geometry. We find that the vendor has specified a column diameter of 1300 mm against 1250 mm estimated in the process data sheet. The tray spacing and the tray passes are in line with the process data sheet.

The vendor has selected straight downcomers having a chord height of 120 mm. This fixes the weir length and flow path length (refer to Figure 15.8). From geometry, the weir length works out to 752.6 mm and the flow path length 1060 mm. The vendor's output is in line with these calculations.

The downcomer area (shaded portion in Figure 15.8) works out to 0.061 m² from geometry. The tray active area, which is the tray area minus two times the shaded portion (refer to Figure 15.8) works out to 1.20 m². The vendor's output is in line with these calculations.

At 110% load, the liquid flow rate being 13.59 m³/hr, and the downcomer area being 0.061 m², the downcomer velocity works out to 0.062 m/s. This is well below the limit of 0.12–0.21 m/s recommended for low-foaming systems and therefore is acceptable.

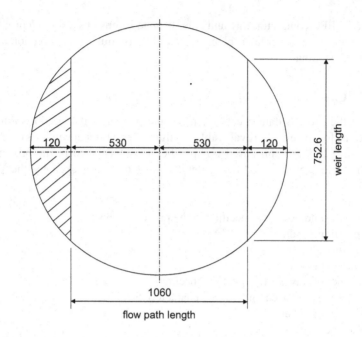

FIGURE 15.8   Tray geometry from vendor.

The downcomer area is 4.59% of the column area. The recommended range being 5–8%, there is a slight deviation here, although not significant.

Lines 32–37: This portion of the tray calculations deals with hydraulic performance data. The jet flood percent at 110% load is within the specified limit of 85%. The weir loading at 50% load is 9.03 m³/hr.m, which is just within the acceptable lower limit.

At 50% load, the total tray pressure drop is 0.0018 kg/cm². The dry pressure drop is reported at 0.0008 kg/cm². This gives the hydraulic tray pressure drop at 0.0010 kg/cm². Thus, the dry tray pressure drop and the wet tray pressure drop are fairly well matched. Therefore, even at a turndown of 50%, the column can be operated.

15.4   COLUMN BOTTOMS

The column bottoms configuration consists of the sump of the column, reboiler and associated piping. Figures 7.12 through 7.15 illustrate the various column bottoms configurations. The associated piping forms a critical part of the configuration. The distance between the column bottoms tangent line and the reboiler tubesheet (in case of a vertical thermosiphon reboiler) is specified by the process engineer in the basic engineering package. The following example illustrates the importance of the column bottoms arrangement and how the process engineer interacts with the engineering disciplines during detail engineering.

***

EXAMPLE 15.1

Refer to Figure 15.9. Figures 15.9(a) and (b) show the process requirements as specified in the basic engineering package. The distance between the column bottoms tangent line and the bottom of the reboiler tubesheet is important. Equally important is the orientation of the reboiler return nozzle to the column shell. Figure 15.9(b) shows that the reboiler return nozzle should be oriented opposite to the seal pan, i.e., at an orientation of 180° (a look at the figure will make it clear). The reason is that fluid at the outlet of the reboiler contains vapor along with the liquid. If, by mistake, the reboiler return nozzle is oriented at 0°, then while the liquid falls into the sump, the vapor gets locked between column shell and the curtain of the liquid falling down from the seal pan. This leads to upset conditions at the bottom leading to some carryover of the liquid from the liquid curtain back to the bottom tray by the vapors coming with the reboiler return fluid.

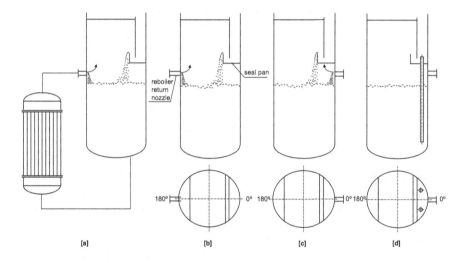

FIGURE 15.9 Column bottoms arrangement.

Normally, such mistakes should not happen with experienced engineers, but imagine a rare case that a mistake has happened and the reboiler return nozzle has been punctured in the wrong location, i.e., at an orientation of 0° (Figure 15.9(c)). The reason for this could have been that the reboiler was located in such a way that it was much easier to route the reboiler outlet line to a nozzle placed at 0° orientation. The process engineer is now consulted to suggest a solution. The solution suggested is illustrated in Figure 15.9(d). The seal pan is punctured and two dip pipes are installed running all the way down to the bottom of the column. The dip pipes should be sized to convey liquid by gravity flow down to the column sump. These pipes should also project upwards to a distance beyond the bottom of the seal pan. In this way, a liquid level is maintained above the seal plan and this acts as a seal to prevent vapors from reboiler return line from bypassing through the bottom downcomer. This may not be the preferred solution, but it is a compromise and would still work.

***

15.5 INSTRUMENTATION

A problem that is raised occasionally by the instrumentation engineer relates to the pressure drop across flow orifices. Sometimes it is seen that with the pressure drop specified, the ß ratio is difficult to meet. Consider the following example:

***

EXAMPLE 15.2

An orifice meter measures flow through a 4" line through which a liquid of density 990 kg/m³ flows. The permissible Δp across the orifice is specified as 0.1 kg/cm².

Data:

$$
\begin{aligned}
D_I &= 100 \text{ mm} = 0.10 \text{ m} \\
C_d &= 0.612 \text{ (from Table 11.1)} \\
P_1-P_2 &= 0.1 \text{ kg/cm}^2 \\
&= 0.1 \times 10000 \times 9.81 \text{ N/m}^2 \\
&= 9810 \text{ N/m}^2 \\
Q_L Q_L &= 50 \text{ m}^3/\text{hr} \\
&= 50/3600 \\
&= 0.0139 \text{ m}^3/\text{s}
\end{aligned}
$$

The problem here is that orifice diameter is calculated as 73.9 mm, which means a ß ratio of 0.739, and is thus beyond the recommended range of 0.3–0.7. A look at the data shows that in cases where a low permissible pressure drop Δp is specified, the orifice sizes are normally on the higher side, in order to meet the permissible pressure drop criteria. This is further complicated when the line is optimistically sized. A look at the flow data shows that for a flow of 0.0139 m³/s of liquid through a 4" line, the velocity works out to 1.77 m/s which is on the higher side for a 4" line. However, going for a 6" line would make the velocity 0.79 m/s which is very conservative and would increase the cost of the pipe. Hence, we stay with 4" line size. In such cases where we have an optimistic line size and a low Δp, we may encounter such problems.

The solution to this is to increase the permissible Δp to 0.25 kg/cm² (i.e., 24524 N/m²). With this increase in Δp, the orifice diameter works out to 61.7 mm which means a ß ratio of 0.617 which is within the acceptable range. This satisfies the requirements of the instrumentation engineer.

***

15.6 3D MODEL REVIEW

The 3D model of a process plant is carried out during the detail engineering phase of a project. In this exercise, the equipment layout is used to make a basic model of the plant that includes in principle all equipment, buildings, structures, ladders and staircases, platforms, pipes and piping components and instrumentation. 3D model reviews are of critical importance to the success of a project. The model review team follows the checklist available while carrying out this review so that no item is missed. The review is carried out in accordance with sound engineering practice to ensure that all the requirements with respect to the following are satisfied:

- Layout
- Construction
- Operation and maintenance
- Safety
- Process point of view

To be more elaborate, the following are modeled as a minimum:

- All equipment and machinery
- All process piping and piping components
- All underground piping
- Pipe racks and sleepers
- All structures and platforms
- Roads and access ways
- Space for construction and fabrication
- All ladders and staircases
- Electrical and instrumentation cable trays
- Battery limit interfaces

The model developed is systematically checked in three stages: 0–30%, 30–60% and 60–90%. In addition, there is a final 90–100% stage. After each stage, a check is carried out by the engineering team. They typically replace traditional squad checks that were carried out in the past. While a detailed explanation of the checks carried out at each stage is beyond the scope of this book, here is a brief description [3]:

In Stage 1 (0–30%), the review mainly focusses on the locations of major equipment and compliance with safe distances between equipment, access for construction, operation, and maintenance, roads, escape routes, structures and platforms, pipe racks, all large bore piping and those that are critical from the stress point of view, preliminary philosophy of underground piping, battery limits, main cable ducts, etc.

During Stage 2 (30–60%), modeling of all equipment, structures and pipe racks, etc. is checked. Modeling of all process and utility lines is checked. All critical instruments are covered. All small-bore pipings are covered, and the underground pipings are also checked. In addition, all firefighting equipment as well as location of safety equipment are checked.

During Stage 3 (60–90% completion), all the pending items from the previous stages are checked for closure. In addition, a check is carried out of the finalized plot plan, equipment layout, piping and supports, structures, electrical and instrumentation components. Modeling of package items is also checked. Location of eyewash stations and safety showers is finalized as well as the locations of fire and gas detectors. Location of utility headers and sub-headers is checked. Modeling of cable trays is looked into. Normally the owner also sometimes participates in this stage.

In addition to the above, the following is a typical list of checks that are carried out [4]:

- *Safety*
  - Proper escape routes have been provided
  - Safety showers and eyewash stations are located at appropriate places
  - Access has been provided for firefighting and rescue activities
  - The utility stations are provided at operating platforms
  - Location of fire and gas detectors is provided

- *Lines*
  - Piping entering and leaving the unit are grouped systematically
  - Pipe routings have been done to facilitate ease of drainage
  - Pockets in vapor lines have been avoided

- *Heat Exchangers*
  - Proper piping arrangement done for removal of shell and channel covers
  - Adequate access has been provided for pullout of tube bundles
  - Drains and vents are properly connected

- *Pumps*
  - The valves around pumps are operable
  - Care has been taken for the volume of liquid which must be drained when pump strainer is opened for cleaning
  - Provision has been kept for the pump to be safely handled during maintenance

- *Process*
  The process engineer is an important member of the 3D model review team. He carries out various checks from the process aspect, since the performance of the plant depends on how correctly all process guidelines are taken care of. In particular, the process engineer carries out the following checks:
  - Piping suction line routing (typically for minimum elbows in case of critical pumps)
  - Cooling water flow routed from bottom to top in shell-and-tube heat exchangers
  - Drain lines do not have pockets
  - There are no loops in the column outlet lines to condensers
  - Orifice meters have straight pipe lengths upstream and downstream
  - All valves are operable. If not, then suitable ace is provided (either by relocating them, or suitable step-up platforms are provided)
  - Inlet lines to pressure relief valves are short

Model review is one of the critical activities during detail engineering of a process plant. An efficient, well-coordinated review could result in few surprises later on in the project, especially once the activities shift to the project site.

15.7 MISCELLANEOUS

One of the common queries placed before a process engineer is problems related to NPSH.

Suppose the process engineer releases a pump data sheet with an $NPSH_a$ of 4.5 m of liquid. Going by the standard guideline to keep a gap of 1.0 between $NPSH_a$ and $NPSH_r$ (refer to Section 5.4.5) the mechanical engineer tries to procure a pump with an $NPSH_r$ requirement of ≤ 3.5 m liquid. In some cases, this is difficult to get. Let us say the $NPSH_r$ requirement for the nearest available pump is 3.7 m of liquid. In such a case, the matter is again referred to the process engineer. There could be two solutions to this:

- Increase the suction line size such that the suction line losses get reduced by 0.2 m of liquid
- Increase the elevation of the suction vessel by 0.2 m

In the first option, this means going for the next higher line size. It, however, also means going for higher sizes for the suction isolation valve, strainer, elbows and flanges. In addition, it also involves increasing the nozzle size of suction vessel nozzle. If procurement of the piping has already been initiated, then it needs to be seen if such a change is possible. If not, then the second option could be thought of.

In the second option, increasing the height of the suction vessel could be achieved in two ways:

- Increasing the height of the support legs or skirt (for larger vessels)
- Increasing the height of the vessel foundation

If the vessel has already been ordered, then one could think of increasing the height of the foundation.

If, however, procurement activities have not yet progressed much and multiple options are possible, then one needs to assess and seek the cheaper and most practical option.

This chapter has presented a flavor of the various interactions that take place between various disciplines in course of execution of a process plant. With proper discipline and cooperation between various engineering departments, it is possible to execute such plants with minimal hiccups.

Symbols

β Beta ratio
D_l Pipe diameter (line size), m

P_1 Pressure upstream of the orifice, $N/m^2 a$
P_2 Pressure at *vena contracta*, $N/m^2 a$
C_d Discharge coefficient

REFERENCES

1. Mukherjee, S. and Bhattacharyya, S., "Shell-and-Tube Heat Exchangers: The Design Cycle", *Chemical Engineering*, March 2016.
2. Mukherjee, S., "Tray Column Design – Keep Control of the Details", *Chemical Engineering*, September 2005.
3. http://3ddraughting.com/3d-model-review/
4. https://www.enggcyclopedia.com/2012/05/3d-model-review-process-design/

16 Pre-commissioning, Commissioning and Start-up

16.1 INTRODUCTION

While we go into the last couple of chapters of this book, let us recapitulate the various activities that take place in the execution of a process plant that we learned in Chapter 1, namely, basic engineering, detail engineering, procurement, construction, pre-commissioning, commissioning and start-up. While each of the above activities has its own importance, construction and pre-commissioning are the two crucial activities that determine a smooth start-up. Before we process further, it would be appropriate to provide definitions of certain relevant terms:

Mechanical Completion: Mechanical completion (MC) is the milestone achieved when construction activities have been completed and all equipment, instruments, piping components and electrical equipment and the related accessories are installed in accordance with the project specifications and procedures. Once this is fulfilled, the Mechanical Completion Certificate is awarded to the contractor. Refer below to the link of erection of the world's largest distillation column at Dangote Refinery, Nigeria:

YouTube link: World's largest Crude distillation column, Dangote Refinery, Nigeria - YouTube.

Pre-commissioning: Pre-commissioning activities are those that are taken up after mechanical completion but before commissioning. Typical activities include removal of rust, loading of lubricants, installation of mechanical seals, and removal of temporary bracings. Flushing and chemical/mechanical cleaning of lines are important activities under this heading. Alignment of rotating equipment is carried out. The cooling water system is charged with water and then taken into circulation. The steam lines are charged with steam. A thorough pre-commissioning is essential to ensure a smooth start-up of the plant.

Commissioning: While, in pre-commissioning, the activities are carried out for individual equipment, instruments or piping, commissioning activities involve systems. Energization of electrical equipment is carried out during this phase. This is followed by test run of rotating equipment. Testing of the control system and emergency shutdown systems and calibration and loop/functional check of instrumentation also form part of the activities in this phase. Drying out of fired heaters is carried out, if applicable. Loading of catalysts is also carried out at this stage. This is followed by the pressure/vacuum hold tests and inertization. Pressure/vacuum hold tests of the various sub-systems are carried out. Finally, the unit is purged and inertized with nitrogen.

Start-up: Start-up includes the remaining few activities before the plant goes into operation. The flare system is taken on line. The fuel gas and fuel oil lines are charged. Finally, hydrocarbon feed is taken in and circulation established. Step-by-step procedure is followed in accordance with the plant operating manual to bring the unit up to normal operation.

16.2 PRE-COMMISSIONING PLANNING AND MANPOWER REQUIREMENT

It has been established that the pre-start-up costs as a percentage of total plant investment are 5–10% for established processes, 10–15% for relatively new processes and 15–20% for novel processes [1].

In order to ensure that a plant undergoes a smooth start-up, it is important that the last phase of mechanical completion and the preparatory activities for commissioning are thoroughly performed. Proper planning is very essential in order to achieve a smooth start-up. One of the important tasks of the planning team is to determine the correct commissioning sequence of the utilities. For this, the commissioning personnel should get in touch with the construction staff at an early stage of piping erection in order to establish priorities for mechanical completion. Such an interaction could lead to important findings such as arriving at a more optimized construction schedule [2]. Therefore, when construction activities have achieved about 75% completion, the construction team should reorient its focus to meet the needs of the commissioning team [3].

The first step would be to divide the plant into various sections in the order mechanical completion is desired. For example, large compressors, if any, may have to be test run at a fairly early stage in order to identify possible problems in advance. In order to achieve this, all process lines and equipment located in the compressor loop will have to be pre-commissioned on priority. Likewise, reactors, which have to be filled with catalyst normally, have to undergo a drying process involving fired heaters in the drying loop. Therefore, equipment and piping included in the catalyst/fired heater loop also need to be pre-commissioned fairy early. Refrigeration systems, which are used to chill a process stream, also fall under this category and need to be test run on priority. It is therefore essential that a schedule is drawn well before the commencement of start-up activities taking into account such activities that fall on the priority path. Figure 16.1 illustrates a typical commissioning schedule.

The team, headed by the commissioning manager, consists of process engineers who are further supported by instrumentation, electrical, and mechanical engineering disciplines. In addition, as construction activities approach completion, certain members also shift from the construction to the commissioning team, since they are now required here. The number of engineers of each discipline depends upon the magnitude of the plant. Vendor representatives of package items, viz., refrigeration,

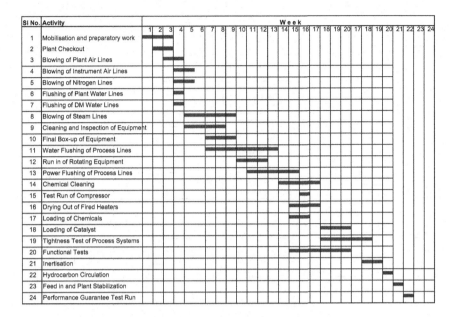

FIGURE 16.1 Commissioning schedule.

Source: Adapted from *CHEMICAL ENGINEERING*, Copyright© January 2005, by Access Intelligence, Rockville, MD 20850.

compressors, etc. are called at appropriate times during test runs of respective package units. In this way, they too form part of the commissioning team. Finally, the commission team is not complete without a dedicated skilled labor force with necessary tools and tackle. This team is particularly useful during pre-commissioning activities when a lot of temporary installations need to be installed such as erection of spool pieces, preparing and installing temporary lines, fabrication of adaptable fittings for connection of temporary hoses, removal of control valves, and similar activities. Such a labor force should typically consist of welders, gas cutters, grinders, fitters and riggers. The erection sub-contractor normally provides this labor force. Since pre-commissioning activities require such fabrication work only for short durations, the construction companies therefore hire such a team from the local market at cheaper rates. Figure 16.2 illustrates a typical commissioning organization.

16.3 PLANT CHECKOUT

Although the plant may have achieved mechanical completion, it is important that before commencing pre-commissioning and commissioning activities, all equipment and pipes are thoroughly checked to ensure that they confirm to the "approved for construction" P&IDs and the project specifications. In this manner, mistakes (if any) committed during construction could be spotted and rectified. Inspection of the plant is normally carried out for the following areas:

- Health and safety aspects
- Static equipment, viz., vessels and heat exchangers, reactors and columns

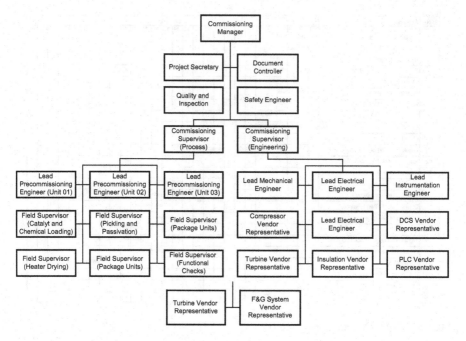

FIGURE 16.2   Commissioning organization.

- Rotating equipment, viz., pumps, compressors, agitators, etc.
- Piping systems
- Electrical and instrumentation.

Equipment on which the performance guarantees of the plant depend, such as the reactors and columns, are checked in detail by the process engineer (for open art units) or by the process licensor (for licensed processes).

Tables 16.1 and 16.2 provide indicative lists of typical checks that are essential before commencing preparations for start-up.

16.4   PUTTING UTILITIES INTO OPERATION

In order that pre-commissioning activities could commence, the utilities such as steam, cooling water, plant water, plant air, instrument air, etc. must be put in service. In addition, it is important to establish the proper sequence of commissioning of the utility systems. The systems that in turn require the minimum number of utilities should be commissioned first. Table 16.3 illustrates utility requirements for various utility systems for a typical petrochemical complex [2].

Water treatment system requires only power and raw water. Both these utilities should initially be sourced from outside, and raw water system should be commissioned. With raw water available and power sourced from outside, the filtered water system should ideally be commissioned next. With filtered water available and power

TABLE 16.1

Checklist for Plant Inspection – 1

Personnel Safety	Vessels/Heat Exchangers	Columns/Reactors
1.0 Ensure a medical check-up of all site personnel prior to commencement of work	1.0 General cleanliness	1.0 General cleanliness
2.0 Check that copies of accident prevention regulations are available	2.0 Internal baffles: type, orientation, levelness	2.0 Levelness of column trays
3.0 Check that first aid kit is available	3.0 Dip pipes correctly installed	3.0 Proper fit of all internals
4.0 Display emergency telephone numbers for accidents, fire, etc. at important places	4.0 Vortex breakers are as per specification	4.0 Liquid tightness of trays
5.0 Check availability of fire extinguishers	5.0 Demisters correctly installed	5.0 Correctness of type and height of packings
6.0 Have necessary safety clothing to be used by all personnel at site	6.0 Location and orientation of instrument nozzles	6.0 Tightness of all boltings and clips
7.0 Check availability of all safety equipment	7.0 Inside of tubes inspected to check possible fouling	7.0 Height of catalyst bed
8.0 Make available guidelines for handling of dangerous materials	8.0 Motor switches located near exchangers for air coolers	8.0 Direction/orientation of feed pipe or distributor, etc.
9.0 Ensure procedure for issue of work permits agreed with owner	9.0 Proper motor rotation	9.0 Correct type and size of vortex breakers
10.0 Ensure spark-free tools available at site	10.0 Tube fin surfaces in good condition with no construction debris	10.0 Leak test of distributor trays done
11.0 Check that gas and fire alarm systems are properly functioning	11.0 For water-cooled exchangers, check for location of thermal relief valve, vent and drain inside the outlet block valve	11.0 Support plates securely fastened
12.0 Cordon off rea demarcated for pre-commissioning and commissioning	12.0 Instrumentation easily accessible from grade	12.0 Freedom of movement of valve caps and other contact devices
13.0 Ensure safety and eye wash showers are functioning properly	13.0 Drains connected to safe location	13.0 Instrumentation easily accessible from grade
14.0 Ensure all gratings are firmly anchored	14.0 Relief valves bench tested and correctly installed	14.0 Drains connected to safe location
15.0 Ensure construction debris is removed from site	15.0 Insulation provided as specified	15.0 Relief valves bench tested and correctly installed

Source: Reprinted by special permission from *CHEMICAL ENGINEERING*, Copyright© January 2005, by Access Intelligence, Rockville, MD 20850.

TABLE 16.2
Checklist for Plant Inspection – 2

Piping

1.0	Correct gasket types installed
2.0	Insulation provided as specified
3.0	Steam/electrical tracing provided as specified
4.0	Check valves installed in correct flow direction
5.0	Slopes provided as per requirements
6.0	Requirement of no pockets adhered to as specified
7.0	Insulation provided as per specification
8.0	Field instruments visible from grade
9.0	Valves to be operated accessible from grade
10.0	Pressure relief valves bench treated and correctly mounted
11.0	Vents and drains properly located
12.0	Insulation provided wherever required
13.0	Piping requirement for special procedures correctly done
14.0	Spectacle blinds are installed in the correct position
15.0	Locked open/closed valves locked in proper position

Machinery

1.0	NPSH requirements met
2.0	Discharge pressure gauge readable from discharge block valve
3.0	Suction strainer easily removable for cleaning
4.0	Minimum flow bypass provided if required
5.0	Check valves installed in correct direction
6.0	Adequate means of venting and draining of the pump casing are available
7.0	All drains and vents routed to safe location
8.0	Pulsation dampeners provided for reciprocating pumps
9.0	Relief valves provided for reciprocating pumps
10.0	Warm-up lines provided across discharge check valve when pumping hot liquids
11.0	Suction/discharge valves and auxiliary piping and controls easily accessible and operable
12.0	Pumps and drivers correctly aligned
13.0	Cooling water to mechanical seal (centrifugal pumps) provided with sight flow indicators
14.0	Calibration of reciprocating pumps carried out
15.0	Alignments are properly done

Electrical/Instrumentation

1.0	Control system configuration correct
2.0	Correct location of instruments
3.0	Flow direction for flow instruments and control valves are correct
4.0	Measuring ranges and scales is correct
5.0	Temperature sensors are of correct length
6.0	Orifice plates are of correct bore diameter and installed in correct direction
7.0	Interlock systems are checked
8.0	Emergency shutdown systems are checked
9.0	Functional checks are carried out
10.0	Analyzers are calibrated
11.0	Emergency and redundant power supplies are checked
12.0	Insulation, screening, earthing of power lines are checked
13.0	Switching devices, motor controls, setting of protection relays are checked
14.0	Direction of rotation of all electrical drives are checked
15.0	Lightening protection and earthing systems are checked

Source: Reprinted by special permission from *CHEMICAL ENGINEERING*, Copyright© January 2005, by Access Intelligence, Rockville, MD 20850.

TABLE 16.3
Utility Requirements for Commissioning of Utility Systems

Utility System	Utilities Required for Commissioning	Remarks
Water Treatment	Power	Note 1
	Raw Water	Note 1
Filtered Water	Power	Note 1
	Raw Water	
Cooling Water	Power	Note 1
	Filtered Water	
Plant Air	Power	Note 1
	Cooling Water	
	Filtered Water	
Instrument Air	Power	Note 1
	Plant Air	
	Cooling Water	
	Filtered Water	
	Instrument Air	Note 2
Nitrogen	Power	
	Plant Air	
	Cooling Water	
	Filtered Water	
	Instrument Air	
Effluent Treatment	Power	
	Plant Air	
	Cooling Water	
	Filtered Water	
	Instrument Air	
Demineralized Water	Power	
	Plant Air	
	Cooling Water	
	Filtered Water	
	Instrument Air	
	Effluent Treatment Facility	
Captive Power	Power	
	Plant Air	
	Cooling Water	
	Filtered Water	
	Instrument Air	
	Effluent Treatment Facility	
	Nitrogen	

Source: Adapted from *CHEMICAL ENGINEERING WORLD*, August 2004.
Note: Note 1: From outside.
Note 2: One dryer must be initially commissioned manually to produce sufficient instrument air to commission the rest of the instrument air unit.

sourced from outside, the cooling water system should then be taken in line. Subsequently, plant air should be commissioned followed by instrument air, nitrogen, effluent treatment and, finally, captive power plant.

16.5   CLEANING EQUIPMENT AND PROCESS LINES

Construction activities contribute to a lot of leftovers in the form of dirt, debris, insulation material, paper, cloth pieces, etc. in equipment. According to the author's experience, even articles like grinding wheels, welding rods, lunch boxes, shoes, etc., have been found inside vessels and columns. Many of these materials, especially dirt, debris, mineral wool, weld splatter, etc. find their way into the pipes as well. The basic purpose of cleaning is to remove these foreign materials left during the construction period or formed due to piping work and/or naturally generated rust. Otherwise, if left untouched, they could cause contamination, clogging of pipes or control valves, malfunctioning of equipment, and damage to rotating equipment.

Amongst the various types of cleaning methods, the common ones include hand (mechanical) cleaning, hydraulic pressure flushing, air blowing/blasting, steam blowing, oil flushing, pigging, alkaline boiling and acid cleaning. Depending upon the criticality of the process and the requirements at site, one or a combination of these methods is applied.

Manual Cleaning: This is one of the first jobs in a pre-commissioning activity. From the construction activities, there is a carryover of a lot of foreign matter, viz., weld spatters, construction debris, mineral wool, welding rods, etc. All equipment need a round of mechanical cleaning by wire brush or cloth to ensure that they are free from dirt and filth.

Flushing/Blowing: After the equipment have been cleaned by mechanical means, a round of flushing and blowing operations is carried out to complete the cleaning cycle. In this method, the kinetic energy of fluids is used to remove finer particles like sand, rust weld slag, etc. which move along with the flowing fluid and are removed. As a general practice, water flushing is applied in the suction of pumps (usually columns and vessels). Discharge piping could also be flushed with the pump in operation. It can also be deployed to piping/equipment that could be conveniently cleaned by water flowing under gravity using a water or a fire hose. In systems where entry of water will generally not be preferred, air blowing may be used. In addition, reactors are also equipment where air blowing should take preference over water flushing.

Flushing with water: For flushing with water, one of the following methods may be adopted:

- By filling the suction vessels with water and flushing the pipes connected downstream of the pumps by running the pump with water. The pump used for such a power flushing method is a permanently installed piece of equipment.
- Pipes connected downstream of vessels can be flushed by discharging water filled in these columns/drums/vessels.
- By injecting large volumes of water with a temporary device such as a hose and a pump.

Cooling water lines should be flushed with cooling water. All equipment in the cooling water loop, viz., heat exchangers, condensers, sample coolers, lube oil coolers and cooling water to casing jackets of rotational equipment should first be disconnected from the system. The inlet lines should be flushed one at a time using flow from the cooling water supply header by operating the upstream isolation valves. Flushing is next done by letting the cooling water flow through the users. After this, the cooling water is made to flow through the users and through the outlet line from the users. Finally, the return line is flushed.

Potable water lines to eyewash and safety showers and drinking water fountains should be flushed until all dirt and foreign matter are removed. Analyses of the water should be made at points of human consumption to ascertain whether the water is suitable for human consumption.

In the fire water system, each fire hydrant and turret should be flushed after removing all nozzles. Before flushing the sprinkler systems, all sprinkler heads should be removed. Finally, before reinstallation, the heads should be inspected to ensure that they are clean.

Condensate lines should be cleaned with water from battery limit with a strong water flow.

Blowing with air: Blowing with air is carried out using air from utility air or a temporary compressor. There are three commonly used methods for air blowing:

- In the direct blowing method, the pipe to be air blown is connected to the air supply source using a temporary pipe or hose. The method often is applied for lines which are small in diameter.

- The air accumulation method is used for lines of larger diameter. Here, a column/drum/vessel associated with the pipe is first pressurized with air. Then, the pipe at the downstream is blown out by rapid opening action of the valve. The method is commonly used for instrument air, plant air, nitrogen fuel gas, hydrocarbon, and cold service lines.

- In large diameter pipes, viz., 8" or more, where it is not possible to get adequate velocities for air blowing, the cardboard blasting method is applied. In this method, a thin film (usually a plastic sheet or a carboard piece covering the fill diameter of the pipe) is placed at one end of the pipe. The pressure is slowly allowed to rise by connecting air through a hose/temporary station. The film finally bursts and the resultant shock flow of the pressurized air carries the dirt along with it. The thickness of the film is slowly increased so that it can sustain higher pressures. The blowing/blasting procedure is repeated until foreign particles are no longer visible in the discharge.

The plant air lines should be blown thoroughly with plant air. The instrument air lines need to remain dry; hence, these lines are blown with instrument air. Thereafter, a service test should be carried out on these lines to check for leaks. After that, these systems are taken into service.

Steam Blowing: Once steam is available at the battery limit, the steam system can be warmed up and blown free of all debris. Steam blowing could be classified into the following three categories:

- Once steaming method is applied to steam tracing lines and hose stations.
- In process steam lines, a warming up operation is first carried out with the valve opened by a few turns before every blowing cycle. Steam blowing is done next for about 10 minutes in general. Thereafter, a cooling operation is carried out for 1 hour or more by closing the valve. The heating up, blowing, and cooling down of the pipe are carried out a number of times resulting in rust, weld slag or spatter to peel off the pipe surface due to repeated expansion and contraction of the pipe. This causes a more effective blowing out.
- For steam lines that feed to steam turbines, the procedure is more stringent. Here, the target plate method is applied. In this method, in addition to the heating/blowing/cooling cycle, a target plate is installed at the blowoff end after three cycles of heating/blowing/cooling. A final blowing cycle is done for 10 minutes after warming up. The target plate is taken out and inspected. Depending upon the number of particles deposited, the blowing procedure is declared successful; otherwise, it is repeated a few more times.

Irrespective of the type of blowing, the steam headers must be blown first. Refer to Figure 16.3 which illustrates a steam header being blown. The blind at the end of the header is removed and a temporary valve and a pre-fabricated pipe piece are installed for these blowing operations.

Before putting steam into the system, all stream traps, control valves, various equipment, instruments, vacuum ejectors and strainers should be removed or blinded.

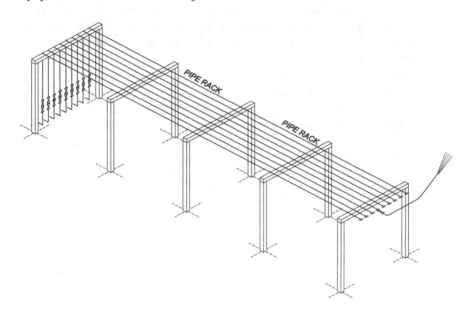

FIGURE 16.3 Flushing and blowing of headers.

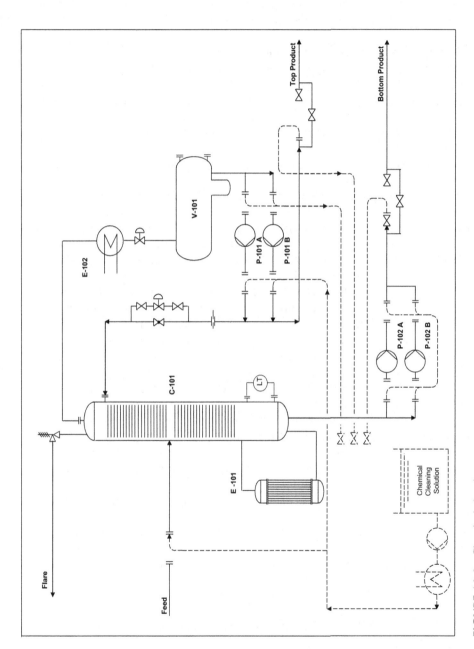

FIGURE 16.4 Chemical cleaning.

As the condensate forms, it must be drained in a controlled manner manually to prevent steam hammering. When blowdown of the steam system is complete, the traps, instruments, strainers and other equipment that were removed prior to the blowoff are reconnected. The system is then reheated and placed in service.

Chemical Cleaning: Certain process units call for further cleaning beyond what is achieved by flushing or blowing. A typical case is chemical cleaning of suction lines of compressors. Pickling of pipes is carried out to remove piping scale, rust, weld slag, spatter and other foreign matter, which might, if unstuck, damage compressors. The operation is usually carried out by circulation of appropriate solutions through the piping and equipment and is usually offloaded to specialized contractors. In certain process units, the entire system needs to be chemically cleaned.

Before commencing the operation, it should be ensured that all control valves, orifice plates, venturi meters and other instrumentation are removed and replaced by spool pieces. The procedure begins with an initial flush with cold water. During this process, the system is filled with water and circulated. The degreasing cycle comes next in which, oil, grease and lubricants that are insoluble in acid are removed by circulating measured quantities of alkali and detergent solution.

Before commencing the acid wash cycle, an intermediate wash with DM water is carried out. The system is next refilled with DM water. To the circulating water, citric acid and corrosion inhibitor are added. The temperature is raised to 60°C and pH maintained at 3.0–3.5. The circulation is continued until the iron concentration stabilizes. At this point, the system is drained under nitrogen blanket.

The next cycle is the acid rinse carried out with 0.2% citric acid at 60°C. When the iron concentration has stabilized, ammonia is added to the system to achieve an alkaline pH. Finally, sodium nitrite (approximately 0.5–1.0%) is added to the circulating solution. The circulation is continued for 4 to 6 hours, and thereafter the system is drained under nitrogen blanket and dried.

A visual inspection of the chemically cleaned surface is carried out. It should be ensured that the internal surfaces of pipes and equipment are uniform and free of deposits and that a shining steel gray magnetite coating is observed on the surface. The chemically cleaned system is next pressurized with nitrogen and maintained under nitrogen blanket until start-up [3].

The chemical cleaning process requires temporary equipment, viz., a storage tank for preparation of the solution, circulating pump, heater as well as several temporary pipe pieces and fittings. Figure 16.4 illustrates a chemical cleaning circuit for a distillation column system. The figure is self-explanatory. It is important to ensure that there is adequate flow in all the process lines, which is achieved by throttling the various manual valves in the system.

16.6   FINAL INSPECTION OF EQUIPMENT

There are two categories of equipment which are particularly inspected by the process engineer (for open arts units) and by the licensor representative for licensed processes, namely, the columns and reactors. These two categories of equipment contribute to the performance of the plant in terms of product purity and throughput. This inspection takes place at a fairly advanced stage of construction and when pre-commissioning activities have already achieved a certain degree of progress.

The construction contractor carefully coordinates with the owner on the dates to have the licensor's personnel at the site, since several critical activities are linked at this stage. The columns and rectors are mechanically cleaned and are offered to the process engineer (or the licensor's representative) one-by-one for carrying out the checks of the internals in order to ensure compliance with specifications. The process engineer (or licensor's representative) thereafter prepares the so-called "punch list," which is basically a list of errors that need to be rectified. Experience shows that normally at this stage no major issues come up. Nevertheless, the small or minor issues pointed by the inspecting engineer need to be taken care of. After rectification, the equipment are once again offered for inspection. After getting the green signal from the inspecting engineer, all manways are boxed up and no further entry of personnel into the equipment is allowed.

16.7   TRAILS FOR ROTATING EQUIPMENT

Before motors are coupled to their respective drives, it is usual practice to make them run for 3 to 4 hours under the supervision of an electrical engineer to check and verify they are operating smoothly. This is called the "no-load test." The following checks are carried out during the mechanical run-in:

- Direction of rotation
- Temperature of bearing
- Vibration
- Other necessary readings.

Before start-up of a unit, all centrifugal pumps are thoroughly checked and, if suitable, run-in properly. Often, pumps are used for water flushing of the discharge lines after carrying out flushing of the suction line. During this flushing operation, the pump is made to run for several minutes, and this is often deemed to be sufficient as function test of the pump. However, one must take care that in many cases high-head pumps, which are selected for hydrocarbons, may not be designed to pump water. To do so can result in damage to the pump/motor. Vendor's specifications should be checked before attempting to run-in pumps with water. Test runs of agitators are also carried out at this stage.

Calibrations are carried out for the metering pumps. At various stroke lengths, the pumping rates are measured by the amount of liquid collected in a measuring vessel in a given time, using a stop watch. Calibration curves are generated by plotting the pumping rate against the stroke length.

16.8   DRYING OF PROCESS LOOPS

Even after completing all the above activities, equipment or piping may still hold water from hydrotest or moisture if blowing was carried out with air that was not moisture free. While residual water needs to be removed anyway, the moisture needs to be removed from those systems where their presence could cause problems during normal operation. Typical examples include refrigeration systems, and carbon steel pipes handling dry chlorine. Beyond a certain moisture level, chlorine is aggressive to carbon steel, and corrosion becomes rapid.

It is normal to carry out drying using instrument air or nitrogen. The plant is divided into sections and dried with instrument air. Dead corners and line ends are dried with special care. In some cases where a fired heater comes in the loop, the drying out becomes very effective if a mild firing is maintained so that the hot air helps to remove the moisture more effectively.

16.9   DRYING OF FIRED HEATERS

Refractory material in fired heaters is not thermally cured in the manufacturing facility. Hence, moisture is normally absorbed by the refractory material during fabrication process and also during transit to the site. The refractory therefore needs to be dried; otherwise, during heating up the moisture will evaporate and may cause cracking and even in some cases may separate from the metal. The dry-out is carried out by slowly expelling the excess moisture from the refractory by gradually raising its temperature before any applicable load is put on the heater. To be assured of a long heater life with minimum maintenance, this work must be done with care.

Before commissioning the fuel gas lines, the furnace firebox must be thoroughly purged with steam till such time a steady plume of steam is seen rising out of the fired heater stack. This is followed by carrying out a gas-free test before lighting the pilot burners. Refer to Figure 11.14 as a typical case. In such systems, the reactor also needs a dry-out run to prevent any possible damage to the catalyst. The refractory is dried using the heat from the burners of the fired heater. In the tubes of the heater, nitrogen is circulated in a closed loop using the recycle gas compressor. The purpose of the nitrogen is to remove any over-heating of the reactor tubes by removing the additional heat, if any, through convective heat transfer. The hot nitrogen dries the inside of the reactor after which it gets cooled by the effluent cooler at the outlet of the reactor and the cooled nitrogen is recycled.

The dry-out cycle is carried out based on instructions from the refractory manufacturer. In summary, after the pilot burners are lit, the arch temperature is raised to 115–125°C at a specified rate by lighting a few main burners and held typically for about 24 hours. The temperature is next increased to 180–220°C at the rate specified by the refractory manufacturer and held for about 12 hours. At this temperature, steam, air or nitrogen is circulated through all the process tubes in order to prevent over-heating. Thereafter, the temperature is increased to 380–420°C and held for 24 hours, and finally to 480–520°C and held for 12 hours (the exact temperatures are determined based on instructions from the refractory manufacturer).

It is normal practice during the dry-out period to rotate the burner operation to ensure an even distribution of heat. During this procedure, instruments and controllers are normally kept on automatic mode to ensure that all alarms, warnings, etc. are functioning properly.

At the end of the dry-out period, the temperature is reduced at the rate specified by the refractory manufacturer. Flow through the tubes is also gradually reduced. At temperatures of about 200°C, the firing may be cut off. The fuel gas lines are blinded. When the firebox is cool enough, the refractory is inspected for any signs of failure.

16.10   CATALYST LOADING

Loading of catalysts is one of the last activities in commissioning. It is carried out in the presence of the process licensor's representative who carries out the supervision. Two methods of catalyst loading are followed in the industry: conventional and high-density.

Conventional Loading Method: This is the simpler of the two methods which does not require either specialized operators or equipment. A stationary hopper is installed on top of the reactor manhole and is equipped with a slide valve. A flexible canvas sleeve is fitted to the hopper and extends to the bottom of the reactor. In parallel, a mobile hopper that can contain 3–4 catalyst drums is kept at ground level. This hopper is filled with catalyst and is lifted by a crane to the top of the stationary hopper, and the contents are emptied. First, a bed of the inert ceramic balls is emptied to the bottom of the reactor through the canvas sleeve and levelling is done. This is followed by emptying the catalyst drums into the reactor followed by levelling. One of the team members is stationed inside the reactor to ensure that that the correct levels are maintained and that the levels of the bed are made uniform. As the height of the catalyst increases inside the reactor, the length of the canvas sleeve is progressively decreased by cutting off the extra portion. In this way, the lower bed is completed and the distributors and other internals are installed. The procedure is repeated for the remaining beds.

High-Density Loading Method: The conventional method described above is the technique generally followed in most industries. However, catalysts can be loaded using a high-density loading technique, in which a higher quantity of catalyst can be loaded in a given volume. A loading kit is used, which consists of a rotating distributor introduced in the reactor from the top flange. The function of the distributor is to sprinkle the catalyst evenly over the bed, allowing it to repose uniformly on the surface. This uniform distribution ensures that the risk of channeling is reduced [3].

16.11   PLANT CHECKOUT AND HANDOVER

Nowadays, with large projects being handled on a lump sum turnkey basis, this concept has become popular. Before handing over a plant to the owner, the contractor offers various sections of the plant to the owner for inspection. During the audit, the owner's engineers carry out a thorough check of the unit to ensure compliance with all relevant specifications with respect to piping, instrumentation, electrical, and structural requirements. The plant is checked thoroughly with respect to the P&IDs. Following the audit, the owner generates a "punch list" of the defects/non-compliance with respect to specifications. At this stage, non-conformities, if any, are normally of a minor nature, e.g., check valve fitted in the reverse direction, incomplete insulation, occasional undersized studs/bolts, etc. The contractor thereafter rectifies the errors, after which the owner carries out a final check. Once the owner is satisfied, the check-out is considered complete.

16.12   TIGHTNESS TEST

The unit is finally tested for tightness. This is important since many process units handle combustibles, as well as toxic substances, the leakage of which could result in disasters, damage or economic loss. It is therefore mandatory to confirm that the plant complies with the required tightness standard before start-up. The following are the three methods which are normally used for carrying out this test:

Pressure Tightness Test: In this method, air or nitrogen is used to pressurize the system. The system is isolated and pressurized to the test pressure (which is normally the operating pressure). Leaks, if any, through flanges, screw connections, valves, etc. are checked using a soap solution. Leakage points found during the test are retightened until the foam formed with the soap solution disappears. In some cases, to stop the leaks, new gaskets need to be installed. Such tests are deemed accepted when the leakage rates fall below the agreed specified value.

Vacuum Hold Test: All vacuum systems need to clear the vacuum hold test. First, the air inside the system is removed to attain the required vacuum. The system is then held under the vacuum while being checked for leakage by the rate of vacuum loss in mmHg/hr. When the vacuum loss falls below the agreed specified value, the test is normally deemed accepted. A vacuum leak test is one of the more difficult tests and may sometimes take several trials to clear.

Service Test: Such tests are performed for those systems where the so-called "leak-age rate" is not a mandatory criterion. Systems such as plant water, compressed air and instrument air fall under this category. In this test, the system is first pressurized with respective operating fluids and then checked for possible leakages. For water lines, leakages are detected visually. For plant air and instrument air, the same is done using a soap solution. Leakage points found during the test are retightened and, if required, certain gaskets may need to be replaced. The test is deemed suc-cessful if no water leakage is observed visually, or if no foam is observed from the soap solution.

Tightness tests are usually performed section-by-section rather than the whole plant at one time. The reason for this is that, after detecting a few leaks, the system needs to be depressurized to attend to them and also change gaskets, if needed. If the system is too large, it takes a proportionally longer time to depressurize making the exercise inefficient. Therefore, it is best to work with small systems. During leak check, it is important to be systematic and document the leaks found. The best way is to start at one end of the section and progressively move to the other end, checking flanges, valves, fittings, instruments, etc. Each leak should be tagged. This makes it easy for the maintenance team and personnel of the next shift to work effectively. The maintenance crew should work together with the leak check team to make it an effective exercise.

It is always advisable to carry out the leak check shortly before start-up. This reduces the possibility of new leaks developing through additional activities.

16.13   INERTIZATION

Inertization is an activity that is particularly required in hydrocarbon industries. Once a system is mechanically complete and all preliminary checks have been carried out, the last activity, i.e., inertization, may need to be done to remove air (oxygen) from the system before the introduction of hazardous process materials, viz., hydrocarbons, in order to prevent the likely formation of explosive mixtures. The following methods are normally used for inertization:

Nitrogen Sweeping Method: In this method, nitrogen is introduced continuously from one end of the system and air is vented from the end opposite to that of nitrogen entry. The air is thus displaced by nitrogen. The vent gas is composed of a mixture of nitrogen-air mixture of gradually decreasing content of oxygen. The process is stopped when oxygen concentration in the system meets the specifications. This method, however, requires a large consumption of nitrogen and is normally used as the last resort, particularly if there is limited availability of nitrogen.

Steam-out Method: In this method, the vent valves are kept open and steam is introduced into the system. The air is displaced as a result. This procedure of steam-out is continued for 4 to 6 hours (depending on the system volume). Gradually, the steam flow is reduced, the inlet and outlet valves are closed and nitrogen is introduced into the system until the desired pressure is reached. Care should be taken to ensure that condensation of the steam does not generate vacuum. The flow rate of nitrogen should be adjusted accordingly. This may not be a cause for concern if the equipment in the system is designed for full vacuum. Periodically, the condensate produced due to condensation is drained.

Evacuation and Replacing with Filling Method: In this method, air from the system is first sucked by means of a steam ejector and then nitrogen is introduced up to slightly positive pressure. This sucking and nitrogen filling is repeated until the oxygen content in the system is reduced to acceptable limits, typically 0.5% O_2.

Pressurization/Depressurization Method: In this method, nitrogen is introduced into the system up to a pressure of 1 to 2 bar (g). The system is allowed to stand for about 30 minutes to allow the gases to mix together. Then the system is depressurized rapidly down to a pressure of 0.1 to 0.2 bar (g). The system is pressurized again. In this way, the oxygen content in the system keeps getting reduced. The procedure is repeated until the oxygen content meets the specifications. The advantage of this method is that if the system volume is known, and the exact pressure built up of nitrogen is decided, the number of cycles required to achieve the inertization can be calculated.

16.14   START-UP ACTIVITIES

There are a number of activities that fall in this category. A discussion on all of these is beyond the scope of this book. The following sections describe the start-up of a typical distillation column illustrated in Figure 2.5. The system consists of a purification column which takes in a crude product and separates it into an overhead product and an impure bottoms product.

Preparation of the System for Hydrocarbon Circulation: Before starting up a column, a circulation of hydrocarbons is started around the column. The circulation is internal, whereby the column bottoms is boiled in the reboiler, the vapors rise up the trays, get condensed in the overhead condenser and are sent back to the column on total reflux. The following activities need to be ensured before starting hydrocarbon circulation:

- All instruments are in line
- All control valves are placed in service, and all bypass valves are closed
- All safety valves are in line
- All valves are lined up in accordance with hydrocarbon circulation
- Cooling water is commissioned in the overhead condensers.

Inventory of the System with Hydrocarbon: The system is first filled with the hydrocarbon to be circulated. The hydrocarbon should be somewhat compatible with the feed and products being handled in the system. The hydrocarbon is lined up from the battery limit, and the column bottom is filled to an 80% level.

Commission the System on Total Reflux: Introduce steam into the reboiler of the purification column so that vapors are generated which move up the trays of the column. Once a level has built up in the reflux drum, commission some reflux back to the column to hold a top temperature at the desired level. Commission the overhead pressure control system to the desired pressure. Continue to increase steam to the reboiler to keep the column bottoms at the desired temperature.

Feed-In: After the column is stable, put controllers in automatic. Take feed into the column and commission the overhead and bottom products to the desired destinations.

This was just an indicative way how a column could be commissioned. The exact operating instructions should come from the process engineer who has worked on the system.

For further reading, the reader can refer to [3, 4].

REFERENCES

1. Mosberger, E. et al., *Ullmann's Encyclopedia of Industrial Chemistry*, Volume B4, VCH Publishers Inc., 1992.
2. Lal, D. P., "Control Commissioning Costs", *Chemical Engineering Progress*, August 2004.
3. Mukherjee, S., "Preparations for Initial Startup of a Process Unit", *Chemical Engineering*, January 2005.
4. Bush, K., Duarte, G., Pohlmann, A., and Zaparoli, A. D., "Improve Startup of an Olefins Complex", *Hydrocarbon Processing*, June 2000.

17 Execution of Large Process Plants

17.1 INTRODUCTION

In Chapter 1, the execution of a process plant was described in a nutshell. The preparation of design basis, appointment of the project management consultant, selection of the process licensor, choice of types of execution strategies, and execution of the project were discussed briefly.

In recent years, with growing emphasis on executing projects on a fast track, the concept of licensing, engineering, procurement and construction (L-EPC) mode of execution is gaining importance. As explained in Chapter 1, the client, instead of selecting the process licensor along with his consultant, delegates the entire project, including licensor selection to the engineering, procurement and construction (EPC) bidder who collaborates with one or more process licensors and offers the entire package.

17.2 EXECUTION STRATEGIES

In connection with the execution of large plants or complexes, we will discuss here two strategies: EPC mode and L-EPC mode. Let us first discuss the EPC mode for a typical refinery expansion project.

17.2.1 EPC MODE OF EXECUTION

Consider a refinery modernization project consisting of a naphtha hydrotreater unit, reforming unit, hydrogenation generation unit and auxiliary units, viz., nitrogen generation unit, cooling tower, bullets and tankage, and effluent treatment plant (refer to Figure 17.1). We will assume that the decision to go ahead has already been made, and we will go directly to the details of execution.

In the refinery sector, the L-EPC mode of execution has not yet been picked up and clients normally wish to keep the selection of the process licensor to themselves. In such cases, however, a strong project management consultant (PMC) is a must. The PMC participates with the client to formulate the design basis of each of these units. It is important to note that there are three process units which are related from the process point of view. The feed to the naphtha reformer comes from the naphtha hydrotreater, and thus the design basis of the former is dependent on the latter. Similarly, the capacity of the hydrogen generation unit is dependent on the quantity of hydrogen required by the naphtha hydrotreater.

While carrying out the licensor selection exercise, the flowrates and compositions of the various interconnecting streams are firmed up. Based on the firmed up details,

DOI: 10.1201/9780429284656-17

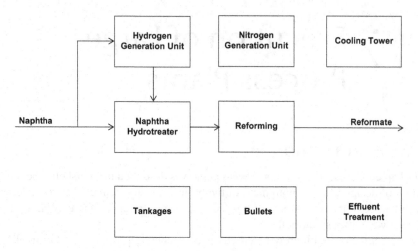

FIGURE 17.1   Block diagram of a refinery expansion.

the design basis of each of the units is prepared. To begin with, the complex is divided into a number of EPC packages. The idea here is not to award the entire project to one EPC contractor. For example, we split the various units in Figure 17.1 as follows:

EPC-1: Naphtha Hydrotreater Unit
EPC-2: Reforming Unit
EPC-3: Hydrogen Generation Unit
EPC-4: Nitrogen Generation Unit
EPC-5: Cooling Tower
EPC-6: Effluent Treatment Plant
EPC-7: Bullets and Tankage

Based on the above, Invitation to Bid (ITB) documents for the above seven packages are prepared and floated by the PMC.

The execution of the entire project is carried out in the following stages:

- Execution of the Basic Engineering Package (BEP) by the licensor
- Development of Front-End Engineering and Design (FEED) documents based on the BEP documents
- Compilation of the ITB documents and floating to EPC bidders
- Preparation of bids by prospective EPC bidders and submission of the bids
- Evaluation of the bids by the PMC along with the owner
- Award of the job to the successful EPC contractor
- Execution of the job by the EPC contractor
- Pre-commissioning and commissioning
- Performance guarantee test run

17.2.2   L-EPC Mode of Execution

As covered in Chapter 1, in this type of contract, the entire range of activities is executed by the EPC contractor(s), including the process license and basic engineering. In this case, instead of the client selecting the process licensor, it is the EPC bidder who collaborates with a process licensor and offers the entire range of services. At times, for small plants, the process licensor takes the responsibility of the EPC bidder as well. However, such cases are not common. The only difference between the EPC and the L-EPC modes of execution is that the interfaces between the basic engineering, FEED and detail engineering are eliminated, and only one party takes the entire responsibility so that the project execution is carried out in a seamless fashion.

17.3   PROJECT SCHEDULE

Figure 17.2 illustrates a typical project schedule for the EPC mode of execution as described in Section 17.1 above. The schedule, in principle, consists of the portion with the licensed packages (EPC 1–3) and the portion with the so-called "open art packages" or utility packages (EPC 4–7). The basic engineering of any licensed package takes anywhere between 4 to 6 months. So, basic engineering is awarded first. Once the basic engineering is completed, the various utility requirements for these units are known. Thereafter, the utility units could be configured and engineered accordingly.

17.4   RESPONSIBILITIES OF THE PMC

Consider the EPC mode of execution case described in Section 17.2.1. Of the seven EPC packages mentioned above, packages EPC 4–7 are normally bid for by EPC contractors who specialize in the design engineering and supply of such packages, and therefore the bidding is straightforward. There is no process licensor involved in these packages. However, packages EPC 1–3 are licensed packages for which a basic engineering document is prepared by the respective process licensor.

- *Responsibilities during Basic Engineering:* The job of the PMC starts right at the beginning of the preparation of the basic engineering stage. In addition to the preparation of the design basis explained earlier in this section, the PMC needs to coordinate with all licensors to ensure that the symbols and guidelines for various documents are uniformly followed. There have been projects where ordering of basic engineering packages to different licensors was carried out by the owners themselves. In the absence of any guidelines, the licensors had prepared their P&IDs and other documents with different symbols and number systems. Later on, while coordinating and integrating the different units, it was a very difficult task to ensure uniformity. Therefore, if the client does not have a big taskforce, it is very important to involve a PMC right at the beginning of the project.

No.	Activity	Year 1												Year 2												Year 3												Year 4											
		1	2	3	4	5	6	7	8	9	10	11	12	13	14	15	16	17	18	19	20	21	22	23	24	25	26	27	28	29	30	31	32	33	34	35	36	37	38	39	40	41	42	43	44	45	46	47	48
	EPC Packages 1 - 3																																																
	Award of Basic Engineering Packages																																																
	Execution of Basic Engineering Packages																																																
	Development of FEED Documents by PMC																																																
	Compilation of ITB for EPC Bids																																																
	Peparation Bids by EPC Bidders																																																
	Bid Evaluation																																																
	Award of Work to EPC Bidder																																																
	Execution by EPC Contractor																																																
	Precommissioning																																																
	Commissioning and Guarantee Run																																																
	EPC Packages 4 - 6																																																
	Development of FEED Documents by PMC																																																
	Compilation of ITB for EPC Bids																																																
	Peparation Bids by EPC Bidders																																																
	Bid Evaluation																																																
	Award of Work to EPC Bidder																																																
	Execution by EPC Contractor																																																
	Precommissioning																																																
	Commissioning and Guarantee Run																																																

FIGURE 17.2 EPC mode project schedule.

- **_Responsibilities during Preparation of FEED Package:_** The contents of the basic engineering package are predominantly basic in the sense that they are driven by process requirements. From the basic level P&IDs, it is not possible to establish a firm EPC price. The PMC needs to update the basic engineering documents to a level that the EPC contractor can provide a firm price in the bid. For this, the PMC needs to update the P&IDs to a reasonable level including all missing information. In addition, the PMC also needs to generate certain engineering drawings. The following is a typical list of value additions that a PMC does on the basic engineering documents:
  - In P&IDs, the following are typically carried out:
    - Inclusion of missing information (viz., line sizes, isolation valves).
    - Inclusion of utility lines (viz., cooling water, nitrogen, drains, closed blowdown, relief outlet, etc.).
    - Inclusion of reducers at appropriate places (for control valves and relief valves, as necessary).
    - Inclusion of vent, drain and relief valve outlet lines with their sizes.
  - Preparation of Utility P&IDs (refer to Chapter 2) so that the EPC bidder can estimate the costs for piping and valves.
  - Preparation of certain additional documents to facilitate EPC bidders to arrive at a cost. These include:
    - Carrying out thermal design of heat exchangers (in some cases, process licensers do not include this in their scope). After doing their thermal design, they need to generate a mechanical engineering drawing of the exchangers.
    - Carrying out mechanical engineering drawing of columns and vessels.
    - Preparing an Electrical Single Line Diagram.
- **_Compilation of ITB:_** Having updated the basic engineering documents, the PMC now has to compile the ITB document. Typically, an ITB consists of two parts: technical and commercial. The following are the typical contents of the technical part of the ITB and are self-explanatory:
  - Description of the project
  - Bidder's scope of work
  - Specifications of raw materials, products and utilities
  - Design philosophies of various engineering disciplines, viz., piping, civil, mechanical, electrical and instrument engineering
  - Project execution philosophy, typically including the following:
    - Project management
    - Engineering
    - Procurement
    - Construction
    - Commissioning
    - Quality assurance
    - Statutory approvals
    - Environmental management
    - Safety

- *Evaluation of Bids:* Once the bids are submitted by the bidders, an evaluation process starts. The technical bid and the documents are evaluated thoroughly by the PMC with respect to the requirements of the ITB. The commercial portion is also evaluated. Queries on technical and commercial issues are also raised to the bidders as required. Bid clarification meetings are held. There are different methods for price evaluation, viz., Net Present Value Method, Internal Rate of Returns Method, etc. A discussion on these is beyond the scope of this book.

17.5   EXECUTION OF BASIC ENGINEERING

Once the bids are evaluated and the job awarded, it is now the turn of the process licensors to start work on the basic engineering packages. As explained earlier, large process plants involve multiple process units. Let us assume the most challenging case that all process units, i.e., EPC 1–3, are awarded to one process licensor. For those process licensors with only one center, they have no other option but to execute all three packages from the same center. However, some licensors have multiple centers across the globe. In such cases, execution may need to be carried out from multiple centers.

Execution from Multiple Centers of a Single Licensing Company: This method of execution is not very uncommon nowadays. All the engineering centers of a process licensing company may not have the requisite expertise to handle every technology in their portfolio. The work is thus shared across centers.

Suppose the naphtha hydrotreater and the reforming units are executed from the center in Europe and the hydrogen generation unit is executed from the center in Asia (refer to Figure 17.1). Each unit will have a dedicated project manager responsible for the unit(s) being executed from the center. In addition, there will be a senior project manager who would be responsible for the execution of all the three units. The latter would be the contact person to the client or the PMC, as the case may be. Periodic meetings are held between the client and the PMC with the process licensor. Normally, a minimum of three meetings are held, namely, the kick-off meeting (refer to Chapter 1 for details), the P&ID review meeting, and the HAZOP (refer to Chapter 13 for details) meeting. The meetings are normally held in the respective centers where the basic engineering is being executed.

Execution from Different Licensing Companies: There could also be a more complicated scenario where the execution of basic engineering could be awarded to different licensing companies. Say, for example, the naphtha hydrotreater and the reforming units are awarded to one licensor and the hydrogen generation unit to another. While the execution sequence and methodology would remain unchanged, it makes things more difficult for the PMC since he has to monitor two different companies.

17.6   EPC EXECUTION

Once the basic engineering of the various licensed units is completed, FEED packages are developed by the PMC, the bidding process is done and the EPC contractor selected, it is time to commence the most challenging part of the project, i.e., the EPC execution. An EPC execution for large chemical complexes could take anywhere between 30 and 36 months. This includes the engineering, procurement and construction phases and finally the commissioning activities.

The EPC contractor is the "single point responsible" party for the whole execution. There are two things which are "fixed" in such an execution, i.e., time schedule and price. With respect to time, the contractor may need to pay the penalty as described in the contract in case the time of execution is not met. Similarly, the EPC price is free of any cost variation and the contract is signed at a fixed price, free of any kind of market variations. Therefore, the price agreed by the EPC contractor is quoted after a detailed deliberation of the risks involved. It is therefore clear from the above discussion that EPC contracts involve a lot of risk for the contractor while, to some extent, reducing the risk of the owner [1].

In an EPC execution, the coordination needed by the EPC contractor and the PMC assumes enormous proportions. Each phase has its own challenges. The engineering phase for large complexes may last 16–24 months and could easily involve 140–150 engineers. The project manager of the EPC contractor has to coordinate with various discipline engineers, including the sub-contractor, if any. Weekly progress review meetings are held. The items which fall in the critical path, viz., compressors, reactors, etc. need to be ordered immediately, and there are related pressures in connection with this activity. The data sheets of such items are expected to be released as early as possible, preferably within one week of receiving the order. The HAZOP is held at an appropriate time and the relevant P&IDs need to be made ready for this activity.

Next comes the procurement. The various vendors who would supply the equipment, procurement engineers, logistics and expediting personnel all are involved in this phase. The bidding process for each set of equipment, viz., columns, heat exchangers, pumps, etc. could be very time consuming and frequently involve multiple bidders. Receiving their bids, evaluating them, and then carrying out negotiations, all form a part of this process. Once orders are placed, frequent follow-up needs to be done. It is very important to ensure that the right item is ordered at the right time. An item which has a short delivery period should not be ordered at an early date. If it delivered to the site very early, there could be major problems regarding its storage and preservation. The review and approval of vendor documents is also a tedious and time-consuming activity.

Logistics is another very important activity in the procurement cycle. Imagine a distillation column 2.5 m in diameter and 50 m in height. The problem here is not the diameter but the height. Proper roadways are needed to assist the transport of such a long item. The problem comes at urning. Adequate bending radius is required. In the absence of proper facilities, the column may have to be fabricated and transported in two pieces and later welded at site. Now imagine a vessel 4.5 m in diameter and 10 m in height. The problem here is the diameter. Taking the height of the trailer on which the column is mounted, the top of the column could well be 5.0 m high. Over-bridges, which may come in the way of road transportation, could cause obstructions to its movement. In such cases, the case may need to be referred to the owner asking for two parallel vessels with suitably reduced diameters.

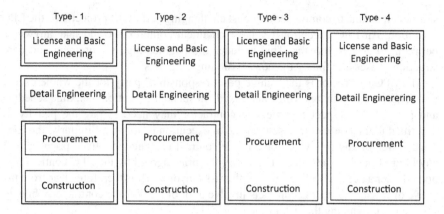

FIGURE 17.3   Different types of execution strategies.

Construction and commissioning activities complete the list. The pre-construction planning is critical during which needs are established and possible roadblocks are analyzed. This phase should have the participation of all parties, viz., the project manager, construction manager, sub-contracting personnel, scheduler, and finally the safety manager. The following are important steps in a construction operation [2]:

- Selection of the construction contractor
- Preparation of the schedule
- Selection of equipment
- Setting up the site office
- Team mobilization
- Commencement of construction activities

Needless to say, proper construction planning goes a long way in achieving a successful and on-time completion.

Figure 17.3 summarizes the various execution strategies that could be adopted while executing large projects.

REFERENCES

1. https://ekainfra.com/blog/engineering-procurement-construction-epc-contracts/
2. https://pmtips.net/article/the-6-phases-of-a-construction-project-life-cycle

18 Cost Estimation

18.1 INTRODUCTION

While a process engineer is exposed to a lot of core subjects during his chemical engineering course in the university, what is somehow missed is a proper exposure to cost estimation. How much do process equipment cost? What is the cost of piping components? How much do the engineering efforts cost? Finally, what would be the total installation cost of a process plant? These are the questions that a process engineer may have, and it is important to provide the answers. A feel for costs of plant and machinery is therefore essential for a process engineer.

This chapter attempts to provide an overview of cost estimation. It is expected that after reading this chapter, the engineer would get a fairly good idea on this essential area of process engineering.

18.2 CAPITAL COST ESTIMATES

Capital cost estimates used in the industry can be classified under five different headings. Various types of estimates, in increased levels of accuracy, are [1, 2]:

- Order-of-Magnitude
- Study (factored)
- Preliminary Design
- Definitive
- Detailed

It should, however, be kept in mind that irrespective of the type of estimate, all the requirements of the particular classification are rarely met until one arrives at the final detailed estimate. In other words, estimates require combinations of methods for attaining any desired level of accuracy [1].

Before we delve into cost estimate for a complete plant, let us discuss how to estimate the costs of individual equipment.

18.3 COST ESTIMATES OF MAJOR EQUIPMENT

Once the process engineer has sized the major equipment, there are several ways to arrive at the cost of each of them. The most accurate method is, of course, to get priced offers from vendors. At the initial stages of a project, this may be difficult to do because many equipment sizes may not be frozen. Vendors are normally very supportive in providing verbal prices or priced offers and, in many cases, are the only reliable source of cost information [3].

DOI: 10.1201/9780429284656-18

The other estimating method is to collect equipment cost information from similar projects and to correlate this data by size, weight, design conditions and materials of construction. Costs of equipment arrived by this method, however, need to be adjusted for inflation to bring all costs information to one base time. Adjusting costs for inflation is discussed later under Section 18.3.6.

18.3.1 ESTIMATING VESSELS

One of the methods to estimate cost of vessels and columns is to correlate historical data on costs of similar equipment vs. weight. Many methods can be found in the literature for estimating the weight and costs of vessels and columns [3].

Over a period of time, companies have developed in-house correlations based on data from projects executed in the past. The usual process is as follows:

-   Knowing the dimensions of the vessel and the design temperature and pressure, the weight of the shell and heads is calculated.
-   The cost of the bare bone vessel is calculated by multiplying the weight by the cost of the material (e.g., carbon steel, stainless steel, etc.).
-   The labor cost is added based on standard rates from the databank of the company.
-   The cost of skirts, ladders and platforms, special internals, nozzles, and manholes are taken from the databank and added to the get the total cost of the vessel.

There are other methods in which the cost of columns can be correlated as a function of the number of trays with diameter as the parameter at various design pressures. Column costs can also be based on the volume of the vessel with the operating pressure as a parameter. Such methods, however, require a great deal of data and do not give as good a correlation as the weight method [3].

There are some other methods which are less accurate. The technique here is to take the cost values of similar equipment from past projects after making suitable corrections for the equipment capacity and the time gap for which the cost data is available. These two types of corrections are described in Sections 18.3.5 and 18.3.6, respectively.

Although the above methods of estimating the cost of vessels and columns can be used, the weight method is usually the best.

18.3.2 HEAT EXCHANGERS

Ideally, the price for air coolers should be best obtained from vendors. The second option would be to use in-house correlations developed based on data on a basis of cost per square meters ($/m^2$) of surface area. While using such correlations, care should be taken to ensure that correction factors for materials of construction have been used. The third option is to use literature data on air coolers, but it should be used when other options are exhausted. In any event, at least one or two air coolers in each project should be priced based on a vendor-priced offer. This serves as a check for costing arrived at using correlations [3].

The cost of a shell-and-tube exchanger depends upon the type of exchanger, i.e., fixed tube, U-tube or floating head. The tubeside pressure, shellside pressure and materials of construction also affect the price. In cases where quotes cannot be obtained from vendors, correlating in-house data by plotting $/m² with correction factors for the variables that affect price will yield costs with a fair degree of accuracy. Similar to the case of air coolers, in the absence of enough in-house data, use of literature may be resorted to [3].

18.3.3 PUMPS AND DRIVERS

Cost for pumps and their drivers (normally the motor) are sometimes obtained from the respective vendors, i.e., pump vendor and motor vendor. For estimation purposes, this is the easier route. However, there is a distinct advantage in procuring from a single source. This fixes the responsibility on one vendor for matching and coupling the motor and the pump. Thus, once a contract is awarded, many companies prefer to follow this route since it is more beneficial from the point of view of reducing the number of interfaces.

Many companies have developed in-house correlations using flow rate vs. cost with head as a parameter. Such correlations are developed for different types of pumps since, for example, a vertical multi-stage pump may be quite different from a horizontal multi-stage pump [3].

The costs of electric motors are normally correlated as a function of kilowatt with voltage, speed, and type of construction as parameters. Correction factors are also developed for different types of protection, viz., flameproof, intrinsically safe, etc. (refer to Chapter 10, Section 10.6 for different types of protection).

The other type of drivers is the steam turbines. Here, the costs are normally correlated as a function of kilowatt with steam inlet and outlet pressure as variables.

While using correlations, it is advisable to cross-check the costs arrived at by getting vendor offers of a few of the larger, more expensive pumps [3].

18.3.4 STORAGE TANKS

Cost of fixed roof and floating roof tanks is usually correlated against volume, for different materials of construction. However, the cost of internal heating coils, insulation, agitators, etc. is considered separately and does not form a part of these correlations [3].

18.3.5 EFFECT OF CAPACITY ON EQUIPMENT COST

While carrying out cost estimates of plant or equipment, it may so happen that the estimator has access to the cost of the equipment of a different size. In such cases, he can apply the Six-Tenths Rule to arrive at the cost of equipment of the desired size.

The equation is as follows:

$$C_2 = C_1 \left(\frac{E_2}{E_1} \right)^n$$

(18.1)

The limitation of the Six-Tenths Rule is that the exponent changes with the size of the equipment; therefore, it is recommended to restrict the use of the equation to a capacity ratio of 2:1 and as a worst case never above 5:1 [1]. In addition, the estimator should ensure that the two equipments of different sizes are similar with respect to type, design conditions and material of construction [5].

18.3.6 EFFECT OF TIME ON EQUIPMENT COST

While carrying out cost estimates, the estimator may have the cost of a plant that was constructed a few years back. The updated cost of such a plant or equipment can be arrived at using the equation:

$$C_y = C_x \left(\frac{I_y}{I_x} \right)$$

(18.2)

Such indices give fairly accurate results as long as the time span is within 4–5 years. For time spans over 5 years, such indices may be used only for carrying out order-of-magnitude estimates where accuracy levels are low. In any case, such indices should never be used for time spans over 10 years [1]. However, these indices do not take into account factors such as technological advancements over the period in consideration, or local conditions.

Many different types of cost indexes are published regularly. Some can be used for estimating equipment costs; others apply specifically to labor, construction, materials, or other specialized fields. The most common of these indexes are the Marshall and Swift all-industry and process-industry equipment indexes, the Nelson-Farrar Refinery Construction Index and the Chemical Engineering Plant Cost Index. The Nelson-Farrar Refinery Construction Index is published the first week of each month in the *Oil and Gas Journal*. The Chemical Engineering Plant Cost Index is published each month in *Chemical Engineering* magazine.

The Marshall and Swift Equipment Cost Index and the Chemical Engineering Plant Cost index are recommended for process equipment and chemical-plant investment estimates. These two cost indexes give very similar results. Similarly, the Nelson-Farrar Refinery Construction Index gives good results in refinery applications. Table 18.1 provides a compilation of the Nelson-Farrar cost indexes for the last several years.

18.3.7 EFFECT OF LOCATION ON EQUIPMENT COST

It has been a standard practice to provide plant cost data on a U.S. Gulf Coast or Northern Europe basis. The reason for this is that, in the past, these have been the

TABLE 18.1
Nelson-Farrar Cost Indexes

Year	Nelson-Farrar Refinery Construction Index Index Value (1946) - 100
1990	1225.7
1995	1392.1
2000	1542.7
2001	1579.7
2002	1642.2
2003	1710.4
2004	1833.6
2005	1918.8
2006	2008.1
2007	2106.7
2008	2251.5
2009	2217.7
2010	2337.6
2011	2435.6
2012	2465.2
2013	2489.5
2014	2555.2
2015	2550.2
2016	2598.7

Source: Reproduced with permission from *Oil and Gas Journal.*

main centers of chemical process industry. The cost of a plant at a location other than the U.S. Gulf Coast can be estimated by:

$$C_z = C_{USGC} \times LF_z \qquad (18.3)$$

Location factors for plants at various locations can be obtained from *Richardson International Cost Factors* manual. Table 18.2 provides location factors at several locations across the world. As a result of globalization, over the past couple of decades, the location factors for various locations are converging towards unity [4, 6].

18.3.8 EFFECT OF MATERIAL OF CONSTRUCTION ON EQUIPMENT COST

At times, the estimator may have access to the cost of equipment of identical size and design conditions, say, for carbon steel, while his requirement is that for a higher grade of material for similar size and design conditions. In that case, he needs to apply suitable correction factors. Table 18.3 lists the cost factors C_{ex} for various exotic materials relative to carbon steel as the base material. The reader is, however, advised that the table provides only indicative numbers. For example, the relative difference between the cost of equipment made of carbon steel vis-à-vis stainless

TABLE 18.2
Plant Location Factors

Country	Region	Location Factor
United States	Richardson	1.00
	Houston	0.88
	Kansas City	0.98
	San Francisco	1.18
	Atlanta	0.88
Canada	Montreal	1.00
	Toronto	1.02
Mexico	Mexico City	1.01
Brazil	Rio de Janeiro	1.15
The Netherlands	Amsterdam	1.22
France	Paris	1.18
Germany	Frankfurt	1.10
Italy	Milan	1.19
United Kingdom	London	1.26
Russia	Moscow	1.42
China	Guangzhou	0.83
Japan	Kobe	1.25
India	Mumbai	1.01
UAE	Abu Dhabi	1.10
Singapore	Singapore	1.14
Australia	Sydney	1.32

Source: Adapted with permission from *Richardson International Cost Factors Manual*, 2021© Edition, by Cost Data on Line, Inc.

TABLE 18.3
Cost Factors for Exotic Materials with Carbon Steel as Base Material

Material	C_{ex}
Carbon Steel	1.0
Aluminum and Bronze	1.07
Cast Steel	1.1
304 Stainless Steel	1.3
316 Stainless Steel	1.3
321 Stainless Steel	1.5
Hastelloy C	1.55
Monel	1.65
Nickel and Inconel	1.7

Source: Reproduced from *Chemical Engineering Design Principles, Practice, and Economics of Plant and Process Design*, by G. Towler and R. Sinnott, Copyright Elsevier 2008.

steel is shown as 1.3–1.5. This may hold well in some countries. In other countries, the difference may be larger.

$$C_{ex} = \left(\frac{purchased\ cost\ of\ item\ in\ exotic\ material}{purchased\ cost\ of\ item\ in\ carbon\ steel\ material} \right) \quad (18.4)$$

18.4   UTILITIES AND OFFSITES

Once the mail process units have been configured, it is easy to estimate the quantities of the utilities required for each unit and also the total thereof. Having established utility summary for the complex, the offsite facilities need to be estimated. Utility requirements can vary greatly from one plant to the other. Hence, careful study needs to be done to assess the loads of cooling water, steam, electric power, process water, etc. for the various process units in a complex and also in the offsite units to make sure that the cost estimate for the utilities is accurate enough.

There is a lot of variation in the consumption of utilities from one process complex to the other and also between "grassroots plants" and "revamp projects" at existing sites, the use of factors for estimating the utilities is not recommended [3].

Site preparation and soil analysis are important factors for estimating the cost for grassroots plants. If it is an early stage of the project and no site has been selected, sufficient allowance for site preparation needs to be included. Later on, once the site has been selected, this part of the estimate should be revisited. If the soil is soft, an estimate of the added cost for piling, compacting, or any additional requirement must be included.

The following items are usually included in the offsite estimates:

Land

1. Site Preparation
2. Grading
3. Roads
4. Sewers
5. Fencing

Utilities

1. Power Generation
2. Cooling Towers
3. Storage and Tankages
4. Water Treatment
5. Sub-station
6. Instrument Air
7. Plant Air
8. Effluent Treatment
9. Flare System

Interconnecting Piping: Interconnecting piping may amount to a substantial portion of the total cost of the project. However, it depends upon the number of units in the complex. While there may not be ready-made factors to estimate such costs, this issue cannot be taken lightly and may have to be estimated in a rigorous way.

Total Offsite Costs: Of the total cost of a project, the offsite costs could range from 20 to 50%. If a preliminary build-up estimate of the offsites is less than 30% of the total cost, it may need to be reevaluated. Unless the offsites are very well defined, it would be better to use a factor of 50 to 75% of battery limits estimate as the cost of offsites.

18.5 CAPITAL COST ESTIMATES

As listed in Section 18.2, there are essentially five different types of capital cost estimates. Let us discuss them one by one.

18.5.1 ORDER-OF-MAGNITUDE ESTIMATES

Also known as preliminary estimates, such estimates are used in arriving at a decision on whether to go ahead with a particular project or consider alternate options. In other words, such estimates are used for obtaining financial approvals from the management. If at a later stage, it is found that a project would cost more, the management is requested to approve the additional amount [5].

In this type of estimate, cost estimation for a proposed process plant is derived from the cost of similar plants built in the past. In such estimates, however, the accuracies are limited from −20% to +40% for relatively simple plants and sometimes even worse for large complicated plants having multiple units with back and forth process integration. In this method, if the cost data of a similar plant of a different capacity is available, then the cost of the proposed plant can be estimated using upscaling and downscaling methods available, such as the Six-Tenths Rule (Equation 18.1) explained in Section 18.3.5 [2].

The exponent n in Equation 18.1 is typically in the range 0.8–0.9 for solids-handling plants. For refinery and petrochemical units, n is usually about 0.7. For small-scale units but with a high degree of instrumentation n is in the range 0.4–0.5. Thus, on average, across the chemical process industry, an average for n of 0.6 is used [6].

In view of the above, the Six-Tenths Rule is more accurate when the cost estimation is carried out for large complexes rather than for individual units. Certain units may have an n greater than 0.6, while some less than 0.6. Thus, for a full complex that covers various types of equipment, these differences tend to even out [2].

The Six-Tenths Rule has another limitation. The value of n is normally not constant but changes with equipment size. Therefore, Equation 18.1 should never be used at capacity ratios above 5:1. For best accuracy, a maximum ratio of 2:1 is recommended [1].

One more thing that needs to be emphasized in this method is that the cost data of the reference plant must be adjusted to make it consistent with that of the proposed plant. Cost indexes are available to update such data up to present date [1]. This has already been covered in Section 18.3.6.

In summary, when cost data is limited, but a quick estimate is needed for screening of capital project proposals, the order-of-magnitude estimate is a preferred procedure. In addition, the method is inexpensive; however, such a method should not be used for arriving at major economic decisions [1].

***

EXAMPLE 18.1

The capital cost of a refinery unit of capacity 6.00 MMTPA in 2012 is \$235 million in the U.S. Gulf Coast. Estimate the capital cost of a similar plant of capacity 8.00 MMTPA in 2016 in the same location.

Nelson-Farrar Index for 2012: 2465.2

Nelson-Farrar Index for 2016: 2598.7

Capital cost for an 8.00 MMTPA plant in 2016 is:

$$C_x = 235 \left(\frac{8.00}{6.00} \right)^{0.6} \left(\frac{2598.7}{2465.2} \right)$$

= \$294 million.

***

EXAMPLE 18.2

The capital cost of a refinery unit of capacity 6.00 MMTPA in 2012 is \$235 million in the U.S. Gulf Coast. Estimate the capital cost of an identical plant in 2016 in France.

Location factor for U.S. Gulf Coast: 1.00

Location factor for France: 1.18

Nelson-Farrar Index for 2012: 2465.2

Nelson-Farrar Index for 2016: 2598.7

The capital cost of an identical plant in 2016 in France is given by:

$$C_x = 235 \left(\frac{1.18}{1.00} \right) \left(\frac{2598.7}{2465.2} \right)$$

= \$292 million.

***

18.5.2 STUDY (FACTORED) ESTIMATES

As the name implies, this method is based on the cost of all equipment items in the plant. The cost of the process plant is arrived at by multiplying the total cost of equipment by established factors.

Apart from equipment, the other elements in a process plant typically include the following:

- Bulks (viz., piping, instrumentation, electrical, spare parts, etc.)
- Construction and commissioning
- Engineering fees
- Transport, insurance, etc.
- Financing, bank charges, etc.
- Contingencies

Table 18.4 illustrates the various components of a fixed-investment as well as the ranges of the various cost components. A range is provided because the cost components would vary with the technology. For example, industries such as steel and cement are heavy on machinery components. Refineries and petrochemicals are heavy on static equipment and piping components, and so on. In case of licensed technologies, there may be a license fee as well, and this would get added as another cost component.

Engineering companies which have executed multiple projects related to a particular technology have developed in-house cost factors of the various cost components for that technology. In such cases, once the equipment costs are established, the additional cost components are estimated based on the percentages of equipment cost. The accuracies of such an estimating method are in the range of –20% to +30% [2].

For such an estimating method, a firm equipment list with specifications is mandatory. An equipment list could either be generated from past projects or could be established from first principles. In the latter case, the process description and the material and energy balance are to be established, based on which the process engineer will size all equipment. In this kind of estimate, it is essential that the bare

TABLE 18.4
Factors for Estimating Capital Cost

Sl. No.	Item	Percentage of Total Cost
1	**Equipment**	**22–32**
2	**Bulks**	**18–26**
	- Piping	10–14
	- Instrumentation	5–10
	- Electrical	3–5
	- Spare parts	0.2–0.5
3	**Construction and Commissioning**	**24–30**
4	**Engineering Fees** (process, mechanical, instrumentation, electrical, civil and structural)	**10–12**
5	**Indirect Costs** (packaging, ocean freight, inland transportation, taxes and duties, insurance, etc.)	**8–12**
6	**Contingencies**	**2–3**

equipment costs are determined accurately since the accuracy of the entire estimating system is based on the accuracy of the bare equipment costs.

In case the estimator does not have access to the various cost components of the plant, there is a shortcut method available. The technique which is available for many years is the use of Lang factors [7]. These factors take into account the type of operation involved in a plant:

3.10 × delivered cost of equipment for solid processing plant
3.63 × delivered cost of equipment for solid-fluid processing plant
4.74 × delivered cost of equipment for fluid processing plant

For example:

In a coal washery plant (solid processing plant), where equipment costs are estimated at $20,000,000, the total cost of the plant would be about $93,000,000.

In a coal-to-fuel plant (solid-fluid processing plant), where equipment costs are estimated at $300,000,000, the total cost of the plant would be about $1,090,000,000.

In a petrochemical complex (fluid processing plant), where equipment costs are estimated at $600,000,000, the total cost of the plant would be about $2,844,000,000.

Finally, those estimators who do not have access to reliable cost data or estimating software, Towler and Sinnott [6] have provided a correlation to estimate the same:

$$C_e = a + bS^n \qquad (18.4)$$

These constants are furnished in the form of a table. However, the reader is cautioned that such estimates are valid for a particular point in time. Suitable escalation needs to be taken while using them.

The method of factored estimates can be summarized as follows:

- Prepare material and energy balances and draw a preliminary process flow diagram (PFD)
- Size all equipment based on process requirements and the PFD
- Select the materials of construction
- Prepare the equipment list with specifications
- Estimate the purchased cost of the all equipment items after applying all necessary correction factors
- Calculate the installed fixed capital cost of the plant

Once the process data sheets of equipment are available, the estimator may or may not choose to get some detail engineering done (at least in terms of minimal information like calculating the shell thickness). In the latter case, the same has to be left to be carried out by the vendor. The vendor, in turn, may regret to carry out elementary detail engineering since it involves both time and cost. Once this is sorted out, the vendor is requested to give a priced offer. The offer may be firm or preliminary but should at least be to a level of accuracy of ±5%, so that the overall accuracy of the plant cost estimate remains at −20 to +30% by this method.

There are several publications on factors used to "build up" a plant cost estimate; however, the most reliable is to develop such factors based on data from several similar plants by experienced cost estimation engineers.

18.5.3 PRELIMINARY DESIGN ESTIMATES

This type of estimate involves further detailed engineering work. Process engineering should be nearly complete. Equipment should be more accurately sized that those used in study estimates. Vendor quotes must be available at least for major equipment.

In addition to PFDs and materials and energy balances, at least the preliminary piping and instrumentation diagrams (P&IDs) should be completed. The P&IDs should as a minimum include line sizes and control loops, control valves and pressure relief valves. A preliminary equipment layout showing plan and elevation views should be made along with estimates of bulks for piping, as well as instrumentation items. In addition, a preliminary electrical single-line diagram should also be available showing the sub-stations, transformers, circuit breakers, distribution lines and motor control centers [1].

In view of the depth of engineering work referred to above, the utilities and offsite facilities should also be developed in greater detail. While this estimation could be done either using in-house or using vendor quotes, accuracy of the estimates should be greater than those for purely factored estimates.

Accuracies of such preliminary estimates which have been well executed could be in the range of −15 to +25% [2].

18.5.4 DEFINITIVE ESTIMATES

In a definitive cost estimate, we talk of accuracies in the range of −7 and +15%. Therefore, such estimates should be carried out when detail engineering is at least 40% complete and/or vendor quotes have been received for at least 90% of the major equipment [2]. The extent of engineering should cover preliminary specifications of all equipment including sketches, final P&IDs, equipment layout drawings with elevations, structural drawings, offsite definition and the final utility balance summary.

The definitive estimate also requires a project schedule covering the major milestones until mechanical completion and their expected date of completion. The construction costs should also be covered. This covers the cost of the construction contractors including the setup at site, tools and machineries required at site, salaries of construction staff, travel costs, etc. In addition, cost of the supervising engineers of the engineering contractor deputed to site is also to be included.

18.5.5 DETAILED ESTIMATES

As the description suggests, a detailed estimate can only be carried out once the detail engineering is complete. In such an estimate, there is no room for any approximations, projections or factoring as is done for some of the estimates covered above. It can only be carried out based on actual equipment cost, piping, instrumentation and electrical of the plant piping model, and civil and structural costs. The accuracy of such an estimate is expected between −4 and +6% [2].

During the normal evolution of a project, some of the above categories of estimates could also be covered. The estimation of a process plant is actually a dynamic process and is continually updated and corrected as the project progresses.

Symbols

a	Constant for particular type of equipment
b	Constant for particular type of equipment
C	Cost of equipment
C_1	Cost of plant or equipment of size E_1
C_2	Cost of plant or equipment of size E_2
C_{ex}	Cost factors for various exotic materials with carbon steel
C_x	Cost of plant or equipment at time when cost is known
C_y	Cost of plant or equipment at time when cost is unknown
C_z	Cost of plant or equipment at location z
C_{USGC}	Cost of plant or equipment at U.S. Gulf Coast
I_x	Cost index at time when cost is known Cost
I_y	index at time when cost is unknown Location
LF_z	factor at location z
n	Cost exponent (a value of 0.6 is normally taken)

REFERENCES

1. Humphreys, K. K. and Wellman, P., *Basic Cost Engineering*, Marcel Dekker Inc., 1996.
2. Turton, R., Bailie, R. C., Whitting, W. B. and Shaeiwitz, J. A., *Analysis, Synthesis, and Design of Chemical Processes'*, Prentice Hall International Series, 1998.
3. Branan, C. R., *Rules of Thumb for Chemical Engineers*, 4th Edition, Elsevier Inc., 2005.
4. https://en.wikipedia.org/wiki/Chemical_plant_cost_indexes
5. Peters, M. S., Timmerhaus, K. D. and West, R. E., *Plant Design and Economics for Chemical Engineers*, 5th edition, McGraw Hill, 2003.
6. Towler, G. and Sinnott, R., *Chemical Engineering Design Principles, Practice and Economics of Plant and Process Design*, Butterworth-Heinemann, 2008.
7. Lang, H. J., "Simplified Approach to Preliminary Cost Estimates," *Chemical Engineering*, June 1948.

Appendix 1
Process Specification

A1.1 GENERAL PROJECT DATA

XYZ Company Ltd. intends to install a new hydrogen generation unit (HGU) consisting of a steam reformer and pressure swing adsorption (PSA) packages under the Fuel Quality Upgradation Project.

The objective of the HGU is to produce hydrogen from natural gas through steam reforming followed by shift conversion and PSA purification.

A1.2 BLOCK FLOW DIAGRAM

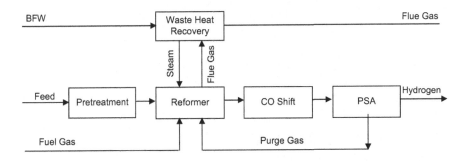

A1.3 PLANT CAPACITY AND TURNDOWN

The HGU, including the PSA, shall be designed for a nominal capacity of 100,000 Nm^3/hr of purity 99.99%.

The unit shall be capable of operating at a turndown ratio of 30% of the nominal throughput capacity whilst maintaining the product specification and recovery efficiency.

A1.4 ON-STREAM FACTORS

Number of on-stream hours per year: 8000
Number of on-stream hours per day: 24

A1.5   FEED SPECIFICATIONS

The HGU shall be designed to process feed gases as specified below.

Natural Gas	Unit	Characteristics
Temperature (min/nor/max)	°C	–/ambient/–
Pressure (min/nor/max)	kg/cm²g	36/38/40
Composition		
- hydrogen	vol%	–
- carbon dioxide	vol%	5.90
- nitrogen	vol%	0.21
- moisture	vol%	–
- oxygen	vol%	–
- argon	vol%	–
- methane	vol%	87.60
- ethane	vol%	5.45
- propane	vol%	0.80
- n-butane	vol%	0.04
Trace Components		
- total sulphur (S)	ppmv	12
- hydrogen sulphide	ppmv	9
Metals		
- mercury (Hg)	ppmv	Nil
Average molecular weight	kg/kmol	18.62
Low Heating Value	MJ/kg	42.93
High Heating Value	MJ/kg	47.46
Dew Point	°C	–63

A1.6   PRODUCT SPECIFICATIONS

The HGU shall be designed for the following product specifications:

Hydrogen	Unit	Characteristics
Temperature (min/nor/max)	°C	–/ambient/–
Pressure (min/nor/max)	kg/cm²g	–/22/–
Product Purity	vol%	99.96%
Impurities		
- nitrogen	ppmv	balance
- moisture	ppmv	20 max
- $CO + CO_2$	ppmv	10 max
- total sulphur	ppmv	1.5 max
- chlorine + chlorides	ppmv	1 max
- CH_4	ppmv	30

Note: The above specifications are typical and have no relation
        with any operating plant or technology.

Appendix 2
Basic Design Data

A2.1 GENERAL PROJECT DATA

Project Name	Fuel Quality Upgradation Project
Client	XYZ Company Ltd.
Location	
Plant Type	Hydrogen Generation Unit
Capacity	100,000 Nm^3/hr hydrogen

Design Life of Plant Equipment:

Description	Unit	
Heavy walled reactors and separators	Years	25
Columns, vessels, heat exchanger shells and similar services	Years	20
Piping, furnace tubes, high alloy heat exchanger tube bundles	Years	10
Furnace tubes including reformers	Hours	80,000
Carbon steel/low alloy heat exchanger tube bundles	Years	5
Rotating equipment	Years	20

A2.2 METEOROLOGICAL AND GEOLOGICAL DATA

Atmospheric Data:

Specification	Unit	Min.	Max.	Avg.
Elevation above mean sea level	m			125
Atmospheric pressure	mmHg			740
Ambient temperature	°C	7	45	–
Relative humidity	%		89 at t_{max}	35 at 45°C
Rainfall data				
- for 1-hr period	mm		136	82.3
- for 24-hr period	mm		266	
Wind velocity	km/hr		23	
Design dry bulb temperature	°C		46	
Design wet bulb temperature			29	
Low ambient temperature for MDMT	°C		5	
Coincident temperature and relative humidity for air blower/air compressor design	°C		36% at 45 °C	

Wind, Rainfall and Snow Data:

Wind velocity	km/hr	20 max
Maximum rainfall over 24-hr period	mm	260
Maximum rainfall over 1-hr period	mm	82.3
Lightening to be considered		Yes
Snow load	kN/m²	Nil
Snow depth	mm	Nil
Seismic design		
- For civil engineering		Yes
- For metallic structures		Yes
- Pipes		Yes
Seismic zone		IV
Soil report available		Yes

Analysis of Ambient Air:

	Unit	Composition
Nitrogen	vol%	78.084
Oxygen	vol%	20.942
Argon	vol%	0.934
Carbon dioxide	vol%	0.040
Total suspended particulates	µg/m³	45.83
- Particulate matter PM_{10}	µg/m³	–
- Particulate matter $PM_{2.5}$	µg/m³	–
Corrosive atmosphere		Yes
Dusty atmosphere		Yes
Marine environment		No

Note: MDMT: Minimum design metal temperature.

A2.3   BATTERY LIMIT CONDITIONS

In the following section, the battery limit conditions of the process and utility streams are provided.

The battery limit for an incoming process or utility stream shall mean 1 m before the first isolation valve. The battery limit for an outgoing process or utility stream shall mean 1 m after the first isolation valve.

The battery limit for an outgoing condensate stream shall mean 1 m after the last steam trap.

The battery limit for flare connections shall mean the outlet flange of the respective safety or control valve.

Battery Limit Conditions for Process Streams:

Service	Normal Operating		Maximum Operating		Mechanical Design	
	kg/cm²g	°C	kg/cm²g	°C	kg/cm²g	°C
Feed natural gas	4	40	4	42		
Hydrogen for start-up	20	40	22	40		

(Continued)

Service	Normal Operating		Maximum Operating		Mechanical Design	
	kg/cm²g	°C	kg/cm²g	°C	kg/cm²g	°C
Hydrogen export	22	ambient	24	ambient		
Treated-off gas export	6	38	6	40		
HP steam export	60	420	62	430	70/(FV)	440
Condensate export	1	60	1	60		

Note: FV: full vacuum.

Battery Limit Conditions for Utility Streams:

Service	Normal Operating		Maximum Operating		Mechanical Design	
	kg/cm²g	°C	kg/cm²g	°C	kg/cm²g	°C
Superheated HP steam	58	400	62	430	70/(FV)	470
Superheated MP steam	10	240	12	250	17/(FV)	300
Saturated LP steam	3	150	4	170	5/(FV)	210
Condensate	5	100	6	100	14	160
BFW import	17	120	18	130	35	150
BFW export	17	110	17	120		
Demineralized water	5	40	6	40	13	65
Plant water	3	ambient	4	ambient	6	65
Cooling water supply	4.5	33	5	35	7	65
Cooling water return	3.5	43	4	45	7	65
Instrument air	4.5	ambient	5.5	ambient	10	65
Plant air	4	ambient	5	ambient	10	65
Nitrogen	5	ambient	7	ambient	10	65
Fuel gas	3	40	4.5	45	7	80

Note: BFW: boiler feed water; FV: full vacuum.

A2.4 UTILITY SPECIFICATIONS

Water:

Parameter	Unit	Raw Water	Demin Water	Boiler Feed Water	Cooling Water
pH		7.2–8.2	6.5–7.3	8–9	7.5–8.6
Turbidity	NTU	9–10			12–25
Total dissolved solids	mg/l	750–1400	<3		<200
M-Alkalinity	mg/l	150–430			150–500
Calcium hardness, CaCO₃	mg/l	105–320			
Magnesium hardness, CaCO₃	mg/l				
Total hardness, CaCO₃	mg/l	210–425	<0.1		300–550
Silica, SiO₂	mg/l	33	<0.03	<0.05	80–100
Chlorides, Cl	mg/l	240–480		<1	350–1050
Sulphates, SO₄	mg/l	10–15			800
Iron, Fe	mg/l		<0.03	<0.3	
Conductivity at 20°C	µS/cm	1050–1800	<0.5	<2	

Air and Nitrogen:

Parameter	Unit	Plant Air	Instrument Air	Nitrogen
Nitrogen	vol%	78.084	78.084	99.5
Oxygen	vol%	20.942	20.942	0.01
Argon	vol%	0.934	0.934	
Carbon dioxide	vol%	0.04	0.04	0.1
Dew point	°C		(–) 40	

Note: For natural gas, refer to Appendix 1, Section A1.5.

A2.5 UNITS OF MEASUREMENT

Measurement	Unit	Abbreviation
Mass	kilogram	kg
	tonne (metric)	t
Length	meter	m
	millimeter	mm
Volume	cubic meter	m³
	liter	l
Volume, vapour	normal cubic meter	Nm³
Temperature	degree celsius	°C
Pressure	kilogram per square centimeter, gauge	kg/cm²g
Pressure difference	kilogram per square centimeter, gauge	kg/cm²
Time	hour	hr
	minute	min
	second	s
Mass flow	kilogram per hour	kg/h
Volumetric flow liquid	cubic meter per hour	m³/h
Volumetric flow vapour	normal cubic meter per hour	Nm³/h
Heat energy	kilocalorie	kCal
Heat flow rate	kilowatt	kW
	megawatt	mW
Enthalpy	kilojoule per kilogram	kJ/kg

A2.6 TRANSPORT LIMITATIONS

Equipment sizes shall follow the following limits of the following weight and dimensions constraints and without any requirement of site welding:

Length (overall including head and skirt), m	30
Diameter/width (including all nozzle projections), m	5.0
Weight limitation consideration multi-model transport, MT	700

Appendix 3
Frequently Used Conversion Factors

Length

    1 m   = 100 cm
    1 m   = 3.3208 ft
    1 m   = 1.0936 yard
    1 m   = 0.001 km
    1 km = 0.621371 mile

Area

    1 m^2 = 10.7639 ft^2
    1 m^2 = 1.19599 $yard^2$
    1 acre = 4046.86 m^2
    1 km^2 = 0.3861 $mile^2$

Volume

    1 m^3 = 35.3147 ft^3
    1 m^3 = 1000 liters
    1 m^3 = 264.172 US gallons
    1 m^3 = 219.969 UK gallons
    1 m^3 = 6.28981 barrels

Temperature

    $1°C = [(1 \times 1.8) + 32]°F$
    $1°F = [(1-32)/1.8]°C$
    $1°C = (1+273.15) K$
    $1°F = (1+460)°R$

Calorific Value

    1 kCal/kg = 4.1868 kJ/kg
    1 kCal/kg = 1.7988 BTU/lb
    1 $kCal/m^3$ = 4.1868 kJ/m^3
    1 $kCal/m^3$ = 0.1123 BTU/ft^3

Heat Transfer Coefficient

1 kCal/hr.m^2.^{o}C = 1.163 W/m^2.^{o}C

1 kCal/hr.m^2.^{o}C = 0.024816 BTU/hr.ft^2.^{o}F

1 W/m^2.^{o}C   = 0.859845 kCal/hr.m^2.^{o}C

1 W/m^2.^{o}C   = 0.17611 BTU/hr.ft^2.^{o}F

Energy

	kCal	kJ	BTU	kWh
kCal	1	4.1868	3.96567	0.001163
kJ	0.239	1	0.94782	0.000278
BTU	0.2522	1.05506	1	0.000293
kWh	859.846	3600	3412.14	1

Power

1 kW = 1.341 HP

1 HP = 0.7457 kW

Viscosity

1 Poise   = 1 gm/cm.sec

1 Poise   = 1 dyne.sec/cm^2

1 Centipoise = 0.01 gm/cm.sec

1 Centipoise = 0.001 kg/m.sec

1 Centipoise = 3.6 kg/m.hr

1 Centipoise = 0.000672 lb/ft.sec

1 Centipoise = 2.4191 lb/ft.hr

1 Centipoise = 3.6 kg/m.hr

Pressure

	atm	kg/cm^2	bar	psi	kPa
atm	1	1.0332	1.0133	14.69	101.325
kg/cm^2	0.9678	1	0.9807	14.223	98.0665
bar	0.9869	1.0197	1	14.5	100.000
psi	0.06804	0.07031	0.0689	1	6.8948
kPa	9.869×10^{-3}	10.197×10^{-3}	1×10^{-2}	145×10^{-3}	1

Appendix 4
List of YouTube Links

Chapter 4: Piping and Mechanical Considerations
Climbing up a Distillation Column
   SDENG - Oil Refinery Tower Ascending - YouTube

Chapter 6: Vessels and Tanks
Pressure/Vacuum Vents
   PRESSURE VACUUM RELIEF VALVE (TANKS SAFETY EQUIPMENT)
   Finekay® - YouTube

Chapter 7: Heat Exchangers
Heat Exchanger Automated Tubeside Cleaning
   Sentinel™ Automated Cleaning Technology by StoneAge - YouTube

Assembling a Floating Head Heat Exchanger
   FLOATING HEAD SHELL & TUBE HEAT EXCHANGER - YouTube

Assembling a U-Tube Head Heat Exchanger
   "U TUBE TYPE" SHELL & TUBES HEAT EXCHANGER - YouTube

Chapter 8: Air Coolers and Fired Heaters
Assembling an Air-Cooled Heat Exchanger
   Kelvion Air Fin Cooler - Petrochemical Application - YouTube

Assembling an Air-Cooled Heat Exchanger
   AIR COOLER HEAT EXCHANGER - ANIMATED ASSEMBLY - YouTube

Chapter 9: Mass Transfer Equipment
Columns – General
Distillation column working guide details of packing and tray columns - YouTube

Chapter 11: Instrumentation and Controls
Orifice Type and Venturi Type Flowmeters
   The Differential Pressure Flow Measuring Principle (Orifice-Nozzle-Venturi)
   - YouTube

Electromagnetic Flowmeter
   The Electromagnetic Flow Measuring Principle - YouTube

Ultrasonic Flowmeter
   The Ultrasonic Flow Measuring Principle - YouTube

Coriolis Flowmeter
    The Coriolis Flow Measuring Principle - YouTube

Vortex Flowmeter
    The Vortex Flow Measuring Principle - YouTube

Time-of-Flight Method (ultrasonic and radar type)
    Time-of-Flight measuring principle animation - YouTube

Chapter 12: Safety and Relief System Design in Process Plants
Pilot Operated Pressure Relief Valve
    Pilot Operated Relief Valve Animation - YouTube

Chapter 16: Pre-commissioning, Commissioning and Start-up
Erection of World's Largest Distillation Column
    World's largest Crude distillation column, Dangote Refinery, Nigeria - YouTube

Index

Page numbers in *Italics* refer to figures; page numbers **bold** refer to tables

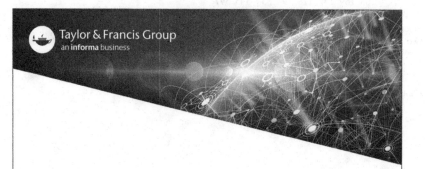

Printed in the United States
by Baker & Taylor Publisher Services